VOLUME FORTY FIVE

ADVANCES IN CHEMICAL ENGINEERING

Modeling and Simulation of Heterogeneous Catalytic Processes

ADVANCES IN CHEMICAL ENGINEERING

VOLUME FORTY FIVE

ADVANCES IN CHEMICAL ENGINEERING

Modeling and Simulation of Heterogeneous Catalytic Processes

Edited by

ANTHONY G. DIXON
Department of Chemical Engineering,
Worcester Polytechnic Institute,
Worcester, MA

AMSTERDAM • BOSTON • HEIDELBERG • LONDON
NEW YORK • OXFORD • PARIS • SAN DIEGO
SAN FRANCISCO • SINGAPORE • SYDNEY • TOKYO
Academic Press is an imprint of Elsevier

Academic Press is an imprint of Elsevier
225 Wyman Street, Waltham, MA 02451, USA
525 B Street, Suite 1800, San Diego, CA 92101-4495, USA
32 Jamestown Road, London NW1 7BY, UK
The Boulevard, Langford Lane, Kidlington, Oxford OX5 1GB, UK

First edition 2014

ISBN: 978-0-12-800422-7
ISSN: 0065-2377

For information on all Academic Press publications
visit our website at store.elsevier.com

CONTENTS

CONTRIBUTORS

Olaf Deutschmann
Institute of Catalysis Research and Technology, and Institute for Chemical Technology and Polymer Chemistry, Karlsruhe Institute of Technology (KIT), Karlsruhe, Germany

Claudia Diehm
Institute of Catalysis Research and Technology, Karlsruhe Institute of Technology (KIT), Karlsruhe, Germany

Milorad P. Dudukovic
Chemical Reaction Engineering Laboratory (CREL), Department of Energy, Environmental & Chemical Engineering (EECE), Washington University in St. Louis (WUStL), St. Louis, Missouri, USA

Fausto Gallucci
Chemical Process Intensification, Eindhoven University of Technology, Eindhoven, The Netherlands

Matthias Hettel
Institute for Chemical Technology and Polymer Chemistry, Karlsruhe Institute of Technology (KIT), Karlsruhe, Germany

Hüsyein Karadeniz
Institute for Chemical Technology and Polymer Chemistry, Karlsruhe Institute of Technology (KIT), Karlsruhe, Germany

Canan Karakaya
Institute for Chemical Technology and Polymer Chemistry, Karlsruhe Institute of Technology (KIT), Karlsruhe, Germany

John Mantzaras
Paul Scherrer Institute, Combustion Research, Villigen, Switzerland

Patrick L. Mills
Department of Chemical and Natural Gas Engineering, Texas A&M University-Kingsville (TAMUK), Kingsville, Texas, USA

Ivo Roghair
Chemical Process Intensification, Eindhoven University of Technology, Eindhoven, The Netherlands

Martin van Sint Annaland
Chemical Process Intensification, Eindhoven University of Technology, Eindhoven, The Netherlands

PREFACE

The Topsøe Catalysis Forum is an annual 2-day meeting hosted by the Haldor Topsøe Company on various subjects of interest in the field of catalysis. The Forum is by invitation, but it is nonconfidential, and the meeting program and presentations are openly available on the company Web site.

The 11th Topsøe Catalysis Forum was held at Munkerupgaard, Denmark, on August 29–30, 2013 on the topic "Modeling and simulation of heterogeneous catalytic processes" with the aim of reviewing state-of-the-art modeling of catalytic reactors and processes. The first day of the meeting focused on overview lectures on such topics as CFD model development, kinetics modeling and analysis, simulation of microreaction technology, fixed-bed reactor optimization and design, fixed-bed heat transfer modeling, and modeling catalytic combustion. On the second day, three parallel sessions were formed on the areas of multiphase reactors, CFD for chemical reactor engineering, and fixed-bed and microreactors, each session having three talks. The closing session featured an overview of multiscale process engineering concepts in reaction engineering.

As this very successful and enjoyable conference concluded, the *Advances in Chemical Engineering Series* Editor Guy Marin suggested that a volume on the topic of the conference that included chapters based on selected talks would be both current and of interest to the broader catalysis and reaction engineering community. With the agreement of Haldor Topsøe, this volume is the result, containing four chapters reflecting a sample of the contributions to the Forum.

Although the nominal titles of both the Forum and this volume emphasize modeling and simulation, the authors of all four contributions (as well as several of the talks not able to be included) found it was essential to include discussions of experimental and measurement techniques. This illustrates both the need for experimental validation of the modeling efforts and the use of the modeling to interpret and understand the experimental results. As the power of computers has grown, enabling more detailed and realistic numerical models by CFD and other techniques, so too have the demands for more accurate experimental measurements by such methods as NMR/MRI, CARPT, and others. This is especially true in multiphase processes. The interplay between modeling and experiment is explicitly addressed throughout the chapters of this volume.

Chapter 1 presents a vigorous and thought-provoking argument by Dudukovic and Mills for the development and implementation of more advanced experimental and modeling methods in industrial practice. The authors illustrate their points with case studies from industrial processes, and add to those their own points of view from the CREL at Washington University and their extensive industrial experience. The chapter comments on and exemplifies the lack of real progress in the adoption of advanced modeling and design tools in reaction engineering, from the perspectives of two highly qualified and respected practitioners in this field.

The second chapter, from the group of Deutschmann at KIT, expands on the interconnection between modeling and experiment. The limitations and interpretation of data from two experimental techniques, stagnation flow on a catalytically coated disk and flow through a catalytically coated honeycomb monolith, are elucidated using CFD modeling of test reactions. A feature of the CFD approach used is the incorporation of detailed surface-reaction mechanisms into the CFD codes.

Continuing the theme of coupling extensive kinetics models to CFD simulations, the third chapter by Mantzaras of the Paul Scherrer Institute provides a look at recent developments in hydrogen hetero-/homogeneous combustion. Dr. Mantzaras also considers both multidimensional numerical modeling and *in situ* spatially resolved measurements, tied in to hetero-/homogeneous chemistry. The author's experience and standing in this field have resulted in an authoritative survey and some insights into future directions in hydrogen combustion.

The fourth contribution, from van Sint Annaland and his group at Eindhoven, presents recent advances in the integration of membrane and fluidization technologies. The authors discuss hydrodynamics and scalar transport in fluidized beds, with an emphasis on the modeling and experimental challenges posed by the multiscale nature of the phenomena involved. There is a strong focus on novel experimental techniques and the application of multiscale modeling of both the fluidized bed and the associated membranes.

My thanks are due to the authors and reviewers for all their hard work, to the editorial team at Elsevier, and to Series Editor Guy B. Marin for the invitation to act as Guest Editor for this volume of *Advances in Chemical Engineering*.

Anthony G. Dixon
Department of Chemical Engineering,
Worcester Polytechnic Institute

CHAPTER ONE

Challenges in Reaction Engineering Practice of Heterogeneous Catalytic Systems

Milorad P. Dudukovic*, Patrick L. Mills†

*Chemical Reaction Engineering Laboratory (CREL), Department of Energy, Environmental & Chemical Engineering (EECE), Washington University in St. Louis (WUStL), St. Louis, Missouri, USA

†Department of Chemical and Natural Gas Engineering, Texas A&M University-Kingsville (TAMUK), Kingsville, Texas, USA

Contents

Abstract

The Topsøe Catalysis Forum was created as a framework for an open exchange of views on catalysis in fields of interest to Haldor Topsøe. The forum scope included a discussion of new catalytic reactions and new principles of catalysis in an attempt to jointly look beyond the horizon (Topsoe catalysis forum, 2013). The 2013 meeting was dedicated to *Modeling and Simulation of Heterogeneous Catalytic Processes* and provided an opportunity to review and discuss the current state of the art in the engineering practice of heterogeneous catalytic systems (Topsoe catalysis forum, 2013). The primary objective of this chapter is to capture key elements of our conference presentation (Dudukovic, 2013) that were focused on multiscale reaction engineering concepts and to what extent these have been applied in the commercial implementation of multiphase heterogeneous catalytic reacting systems. Of particular interest is to identify common approaches and tools used in practice, and to examine their effectiveness in the scale-up and development of more efficient, environmentally friendly catalytic

Advances in Chemical Engineering, Volume 45
ISSN 0065-2377
http://dx.doi.org/10.1016/B978-0-12-800422-7.00001-7

processes. Current practice is limited by the availability of experimental tools to increase the reliability of scale-up, and by the lack of more robust models for analysis and optimization of reactor systems for existing processes or the design of new reactor systems for implementation of new catalytic chemistries. From an economic perspective, the pursuit of short-term financial objectives favors the use of existing reactors with minimal modifications with performance analysis based upon simplified approaches. A longer-term perspective on the development and implementation of more advanced experimental techniques and modeling approaches for reactor analysis that are applicable to commercial reactor conditions would accelerate the development of new process technologies and result in reduced risk with associated lower costs.

1. INTRODUCTION

Multiscale process engineering (MPE) attempts to describe various physiochemical phenomena in process systems over a large range of time and length scales using various modeling approaches to provide robust predictions of process system behavior. MPE is gaining increased momentum as the preferred approach for developing robust process models that can be utilized for efficient development of sustainable solutions for emerging technologies. This approach has been advocated by experts in the reaction engineering field (de Lasa et al., 1992; Dudukovic et al., 2002; Krishna and Sie, 1994; Lerou and Ng, 1996; Schouten, 2008; Tunca et al., 2006) and has been the subject of recent conferences dedicated to multiscale multiphase process engineering (MMPE) (MMPE, 2011, 2014) as well as monographs on multiscale process modeling (Li et al., 2013). However, implementing this approach for existing and new process technologies, whose profitability is always impacted by variables such as feedstock price and composition, competition from other companies, and other business platform dynamics, could potentially benefit by synergistic collaborations between universities and industry than commonly practiced today (Huesemann, 2003; Schouten, 2008). These observations are reinforced by more than 40 years of experience of interactions between academia and industry at Chemical Reaction Engineering Laboratory (CREL) at the Washington University in St. Louis (Dudukovic, 2009; Mills and Duduković, 2005a, 2005b).

Since 1974, the objectives of the CREL have been to advance the state of the art of multiphase reaction engineering via education and research involving students and to transfer these advances to industrial practice. Recognizing that reaction engineering, as an academic discipline, can only develop new advances if it is related to industrial practice, cooperation and financial support was sought and received from numerous companies located in five continents as well as from various national government funding agencies (e.g., NSF, DOE, USDA, and DARPA). The financial support for research in CREL originated from various industrial sectors, including petrochemical processing, bulk and specialty chemicals, pharmaceuticals, semiconductor grade silicon production, biotechnology, and specialty materials. Many research programs involved heterogeneous-catalyzed kinetics, catalysis, and the analysis of catalytic reactors (Chemical Reaction Engineering Laboratory, 2014).

It is noteworthy that the precompetitive research generated by CREL in-house initiatives using consortium funding from various companies over the past four decades produced many graduates that collectively represented a strong reaction engineering workforce that accomplished notable technical advances in a host of diverse technologies (Chemical Reaction Engineering Laboratory, 2014). Instead of embracing this model of pooling resources, sharing the knowledge acquired in a broad consortium like CREL, utilizing that knowledge for in-house projects, and creating new initiatives for reaction engineering research, industrial trends for the last 10–15 years or so have largely placed decreased emphasis on precompetitive research and the development of core competency groups in reaction engineering and the process sciences. Various reasons for reducing emphasis can be identified, but it can often be traced back to an assessment of costs for supporting these groups within a company's organization. For example, when the DuPont Company sold the fibers and textiles unit in 2003 (Brubaker, 2003), various core competency groups located in the corporate science and engineering research laboratory were eventually downsized, reorganized, or disbanded owing to reduced needs from the remaining business platforms to provide improved technology or plant operational support. Other business examples can be identified in the open literature for similar or other related reasons. In addition, it has been observed by one of us (P. L. M.) over the years as an industrial practitioner that changes in business direction has sometimes lead to the loss of researchers with strong skills in heterogeneous catalysis and reaction engineering, or decreased support for a core group with reaction engineering expertise. Nevertheless, the need still exists in companies to have a skilled technology workforce that has the knowledge and experience to either drive the development of next-generation processes, or to provide expert guidance and critique of work performed by external engineering technology businesses who may be contracted to perform it as part of a larger project on process development (Ericsson et al., 2007).

Another key aspect that has impacted support and development of new reaction engineering principles for next-generation heterogeneous processes is connected to increased emphasis on the ownership of intellectual property by both companies and universities. The literature on this topic is extensive, but a recent article highlights some key issues from a university perspective (Hallet, 2014). Generally, development of an acceptable legal agreement can consume notable time and resources. However, examples can be cited on successful agreements between universities and industrial partners (Glicksman, 2003). The above-cited developments and others that are not set forth here for brevity have generally had a negative effect on the reliability of scale-up of new processes. Thus, it is appropriate to first examine the difference in current prevailing approaches and what is needed in order to improve the technologies for heterogeneous catalytic processes.

In the distant past, profit maximization was the primary guiding principle in process development and operation. However, the key future challenge for our profession includes meeting the global energy, environmental, and material needs of the world using more efficient processes that can eliminate or dramatically reduce wastes while maintaining profitable (Dudukovic, 2010; Schouten, 2008). This is necessitated by the realization that global damage to the environment is the product of three factors: overall process inefficiency, consumption per capita (which is proportional to Gross Domestic Product (GDP)), and total population (Dudukovic, 2009, 2010). As political and economic pressures prevent any foreseeable actions in curtailing the last two factors, the only strategy for the future that can make our processes sustainable is to work relentlessly to increase all types of process efficiency. Use of multiphase catalytic systems is prevalent in most processes that chemically convert various raw materials to final products for the market (Fig. 1.1; Dudukovic, 2009). The keys to an economically, environmentally friendly, and energy-efficient process are to identify the preferred chemical transformation, reactor type, and most efficient separations, and to successfully scale-up these transformations so they can be safely operated on a commercial scale to achieve or exceed both company business metrics and customer needs. This can only be accomplished by increased use of science in reducing risk of innovation and in changing the political priorities and societal climate regarding acceptable practices. For existing processes, sufficient in-depth understanding of the interplay among reaction kinetics, transport effects, fluid mixing, and equipment scale is needed to choose the optimal operating conditions.

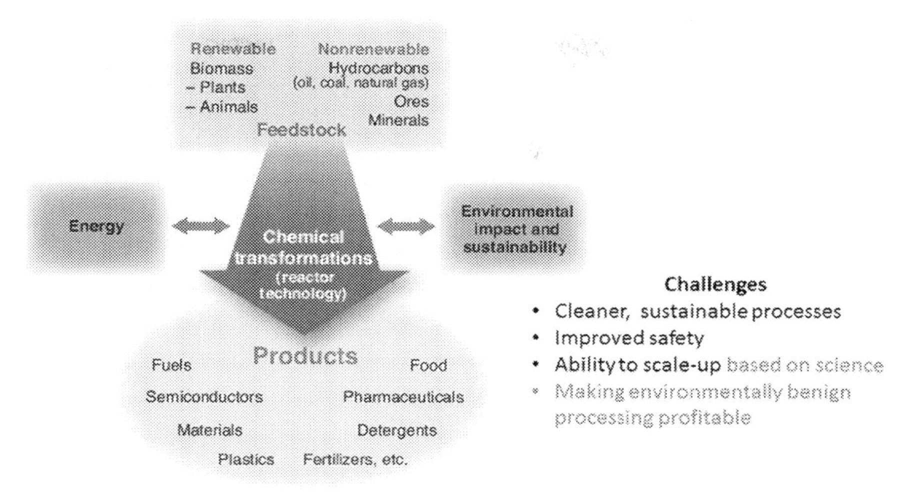

Figure 1.1 Use of multiphase systems is prevalent in all processes dealing with transformation of raw materials and impacts of the environment. *Figure originally published in Dudukovic (2009).*

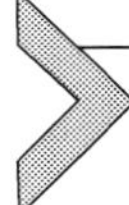

2. MULTISCALE CHARACTER OF HETEROGENEOUS CATALYTIC PROCESSES

Fig. 1.2 provides evidence that multiphase catalytic reactors are present across a spectrum of process technologies. They constitute over 98% of reactors employed in practical processes and are employed in numerous industrial sectors, such as processes for manufacture of advanced materials (e.g., composites, nanomaterials, optical fibers, plastics, and semiconductors), bio-based materials, bulk chemicals, catalysts, environmental remediation, fine and specialty chemicals, pharmaceuticals, plastics, polymers, natural gas derivatives, petroleum-refined products, and transportation fuels. These products represent a large contribution to the GDP in both the United States and elsewhere in the world (Tunca et al., 2006).

Many researchers pointed out the multiscale character of catalytic processes (Krishna and Sie, 1994; Tunca et al., 2006). For example, the temporal and length scales involved in functioning heterogeneous catalysts and the reactors in which they are used span several orders of magnitude as described by Centi and Perathoner (2003). Hence, the difference in length scale between an active catalyst site and the reactor in which it is expected to operate efficiently is approximately 10 orders of magnitude. Accordingly,

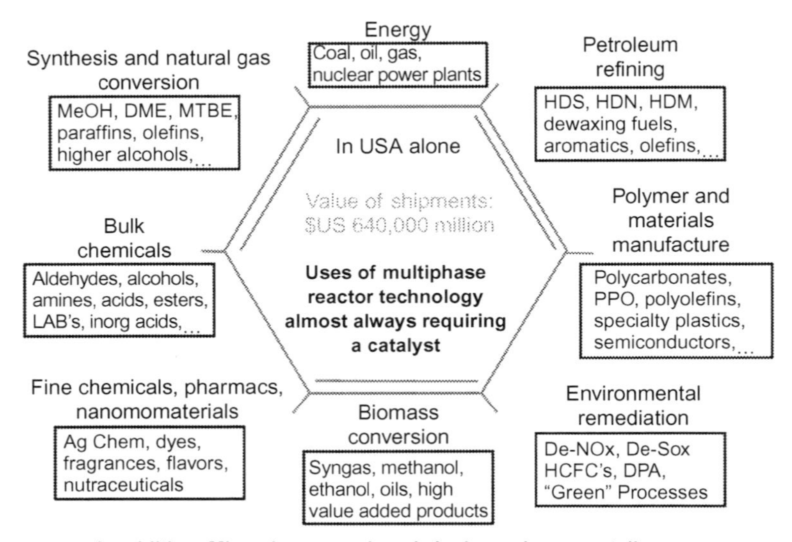

Figure 1.2 Industrial sectors that use multiphase reactors.

characteristic times for events on these scales vary from less than 1 ms to hours. The depth of understanding of molecular-scale events varies depending on the reaction and catalyst type. Yet, the reaction engineering methodology calls for a quantitative and detailed description of reaction kinetics to support development reactor design, scale-up, and safe operation.

The above example provides the multiscale basis for understanding and modeling chemical reactors, which is depicted in Fig. 1.3. This concept became the starting basis for nearly all research done in CREL (Dudukovic, 2010). It offers a rational way of quantifying reactor performance based on mass, energy, and momentum balances by relating reactor scale phenomena to the relevant multiscale transport effects and kinetic phenomena. Quantitative understanding of these multiscale transport–kinetic fluid interactions is the key to the selection of the best catalyst, best reactor type for a given chemistry, and is required for successful scale-up. The ability to capture the influence of local transport effects and reactor flow pattern on the reactor-averaged apparent reaction rate is necessary to accurately determine the volumetric productivity and selectivity of the reactor. These, in turn, are the key figures of merit for designing a sustainable economic process since they can be readily connected to business metrics.

Even though the reactor typically represents between 5% and 15% of capital and operating costs of typical process plant, its choice determines both

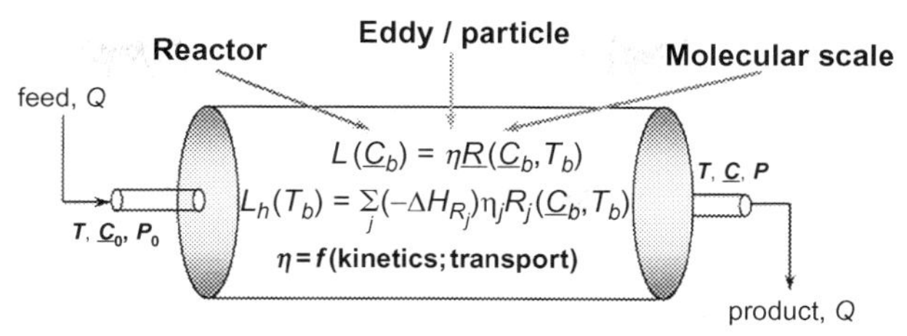

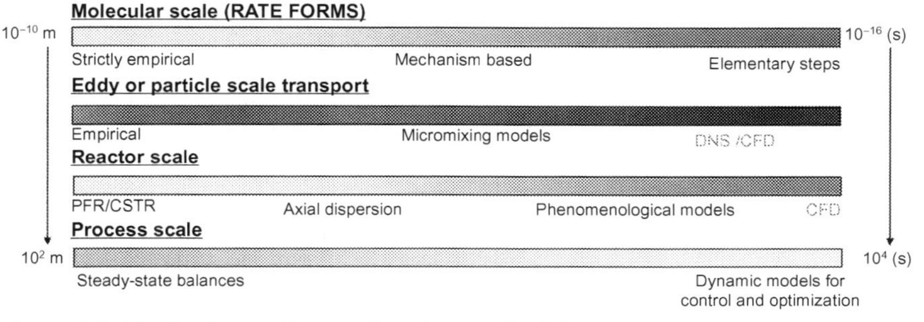

Figure 1.3 Multiscale reaction engineering methodology.

the number of and load on prereactor and postreactor separation units and dictates the cost of the whole process (Tunca et al., 2006). This is why the choice of the proper reactor type is essential and it should involve a rational approach based on a validated reactor model. Such a model must capture the events on a multitude of scales at the right level of scrutiny, and provide the ability to scale-up lab-scale discoveries to commercial processes. The complexity arises from the fact that the interactions of physical events on various scales are often dependent on the scale of the equipment. This makes it increasingly necessary, due to the increased development of more active and selective catalysts, to quantitatively understand the change of the flow pattern with reactor scale and the interaction of it with mesoscale transport. Hence, moving the level of understanding and of quantitative description at all scales affecting reactor performance from the left to right on Fig. 1.3 can lead to models having greater predictability, which are needed for safer scale-up and design of next-generation reactor systems.

The key to successful heterogeneous catalytic reactor design and operation is to have a quantitative reactor model capable of predicting the effect of reactor scale and operating conditions on volumetric productivity and selectivity of the reactor. This does not imply that an *ab initio* model for each scale has to be merged into a detailed, complex model for the whole reactor. It means that a working phenomenological model, developed based on first principles and properly validated by experiments, should be developed and tested on various scales. This model will contain sufficient information across various subreactor scales to meet the objectives set for it. Such models are all too often missing in the current practice.

Table 1.1 illustrates the quantities that affect reactor volumetric productivity and selectivity as presented in some classical reaction engineering textbooks (Kramers and Westerterp, 1963; Levenspiel, 1999). By developing multiscale models to account for these subreactor scale phenomena, CREL has proved their value for processes involving manufacture of diverse materials, such as the desulfurization of fuels, hydrogenations of specialty chemicals, polysilicon production, curing of reinforced resins, gas-to-liquid conversions, and synthesis of pharmaceuticals, to name a few (Chemical Reaction Engineering Laboratory, 2014). The same approach can be invaluable in developing science-based multiscale models for commonly used reactor types, which then can be used for any specific chemistry and catalyst (Tunca et al., 2006).

Based on the available open literature, industrial technologists have not generally reported the development of state-of-the-art reactor models that combine intrinsic reaction kinetics, transport effects, and fluid hydrodynamics across the required range of temporal and length scales. If available, they have been kept as company confidential for business reasons. However, Shin and coworkers (2007) describe a hybrid model for a multitubular reactor used in terephthaldehyde manufacture that couples a commercial CFD

Table 1.1 Phenomena Affecting Reactor Volumetric Productivity, Average Reactor Process Rate, and Reactor Selectivity

Reactor volumetric productivity is defined as	Moles of product produced per unit reactor volume and unit time = average process rate for the reactor
Average reactor process rate is a function of	*Kinetic rate*, which depends on catalyst turnover number, local temperature and composition and their distributions; *Local mass and heat transfer effects* on the kinetic rate; *Reactor flow and fluid contacting patterns*, which determine the composition–temperature–pressure field in the reactor
Reactor selectivity is a function of	Kinetics, local transport effects, and global contacting patterns

package to describe the fluid dynamics on the shell side with an advanced process modeling tool to model the catalytic chemical reactions and related phenomena on the tube side. These models were executed simultaneously in which each model calculated key input information for the other code. The authors provide data that suggest that this approach allows scale-up using data from a single tube to a reactor containing thousands of tubes.

It has been the collective experience of the authors that within the past decade or so, factors such as changes in company business direction, selling of various technology-based business units, increased outsourcing of engineering projects, changes in company support for core competency groups, and retirements of various internal company consultants, to name a few, has generally reduced the technical knowledge experience in most companies and reduced the number of experts in reaction engineering of heterogeneous catalytic systems. In addition, an increased number of new technology managers do not have formal training in the field, but nevertheless, they must make critical decisions on evolution of process development. An analogy can be made here between this situation and management of public-funded research as explained by Heller (2014). A key suggestion that emerges is that well-experienced and seasoned managers with notable track records in the development of products, processes, and/or services should be the ones that manage applied research.

Advances in computer technology have led to a proliferation of powerful process flow-sheeting programs based on mass and energy conservation laws and other supporting relationships (Seider et al., 2009). For example, programs such as those offered by AspenTech® (AspenTech, 2014) are often used not only to determine the basis for process design but also to evaluate the environmental impact analysis of new processes or process modifications. These tools and others like it are very useful in taking the drudgery out of mass and energy balances and provide powerful methodologies for both simulation of various process system components and various engineering activities associated with process design and analysis. However, the available reactor models in these codes are mainly limited to those having ideal flow patterns with a single phase, or multiphase systems that can be treated as a pseudohomogeneous system. Hence, they do not account for transport–kinetic interactions and complex fluid hydrodynamics that often occur in pilot and commercial scale reactors with multiphase gas–liquid flow in the presence of stationary or moving catalysts.

To perform more realistic process simulations, robust reactor models are needed along with a systematic multiscale approach to reactor selection and optimal operation (Krishna and Sie, 1994; Tunca et al., 2006). Current algorithms provide little or no guidance on selecting the preferred reactor and its operation. To accomplish this, improved reactor models for describing transport–kinetic interactions and flow patterns in reactors must be developed and then used to predict reactor productivity and selectivity as function of process operating conditions and on various reactor scales. These models are notably missing from current industrial practice since many practical reactors are still analyzed in process simulations by assuming perfect fluid backmixing or plug-flow for description of the reactor scale flow pattern, or utilize the outdated axial dispersion model under conditions when it has no predictive ability.

As recently pointed out by Stitt and coworkers (Mills and Duduković, 2005a), use of computers for producing realistic commercial reactor design is still in the Stone Age. Figs. 1.4A and B (Stitt et al., 2013) summarizes key points that the use of heuristics is so prevalent that the compounded addition of the "safety" factor of 20% leads to 75% reactor oversizing and to a 50% increase in capital requirement. A change in current approaches is strongly encouraged with an increased reliance on science-based models (Stitt et al., 2013). New modeling approaches have been advocated by academics for decades (Derksen and Van den Akker, 1999; Hoekstra et al., 1999; Mudde and Van Den Akker, 2001; Venneker et al., 2002), but it remains to be seen if these will be implemented by practitioners by the end of the current decade. Competition between various companies will help drive the development and implementation of new tools. Those companies that

A

Mr. 20%—the scourge of capital productivity

- Uncertainty over transport coefficients
  - Designer includes 20% "design margin"
- Uncertainty over reaction rate equation
  - Designer adds 20% "design margin"
- Uncertainty over calculated reactor size
  - Designer adds 20% "design margin"
- Total effect on reactor volume: $(1.2)^3 = 1.75$
  - Reactor is 75% oversized due to "Mr. 20%"
  - This inflates capital cost by 50%!
- 20%? ...... This is not a *"Design Factor"*
  - It is a Factor of Ignorance
  - AND IT COSTS MONEY!

B

The industrialists' problem

- We've been using and building a tool box of "traditional" methods for over 100 years
  - We know they're not great—but we've learnt to use them
  - And we have a very large file of case studies
    - And known "design margins"
  - We trust them!!!
- So what is needed to drive change?
  - New modelling approaches need to provide us with a similar amount of confidence
    - Validation, cross referencing

Figure 1.4 (A) Penalty to pay for lack of models. (B) Call for development of models with validation. *Panel (A) is adapted from Stitt et al. (2013) and panel (B) is adapted from Stitt et al. (2013).*

utilize them effectively will gain process advantages over those that continue to use outdated empirical or overly simplified approaches. By returning to our assertion that commercial process technology should be judged using measures of process efficiency, environmental footprint, safety, and profitability, next-generation process models should be developed that can incorporate realistic rate information and the required multiphysics versus simplistic ideal models with no or minimal predictability.

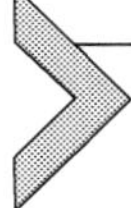

3. NEEDED INTEGRATED APPROACH TO CATALYTIC PROCESS DEVELOPMENT

The existing approach to scale-up is compared to the ideal approach in Fig. 1.5. Current practice relies heavily on existing reactor designs along with performance data obtained from operating process plants using existing catalysts. This approach is viewed as cost effective in the short term, but it is subject to failure when a catalyst supplier loads a new formulation that has both higher activity and selectivity since no model often exists to predict higher rates of heat generation and its effect on mass transport and hydrodynamics. Other scenarios can be defined that lead to greater risk and the potential failure to meet target reactor design production, but the final recommendations being preached here will be the same so these are omitted for brevity.

It can be argued that past and current industrial practices for catalytic process development and operation have significantly contributed to a reduced rate of new advances in reaction engineering that are applicable to commercial reactor systems and their associated processing conditions. Greater emphasis is often placed on developing new or improved catalyst compositions for a given process chemistry without investing the time required to develop robust kinetic models for incorporating into advanced reactor models. Catalyst vendors or catalyst development groups within a company often rely upon pilot or commercial scale data from their customers or end-users to obtain global performance data without performing robust kinetic assessments. When pilot-plant data are available, it is usually collected under nonideal conditions, such as nonisothermal behavior in the axial and/or radial directions. These data are unique to that particular set of operating conditions and reactor configuration, and cannot be readily scaled-up to a reactor having a notably larger diameter or geometry. Scaling-up to a unit having a notably larger diameter and length, which often operates adiabatically, is now subject to higher temperatures, and

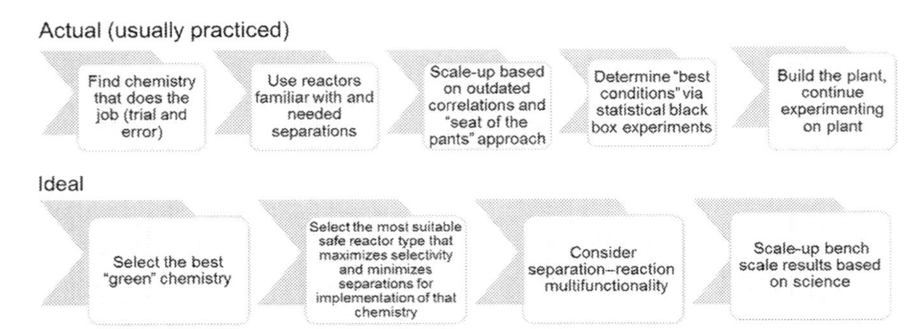

Figure 1.5 Actual (upper) and ideal (lower) process R&D scale-up approaches.

hence catalyst performance metrics will generally be different. The preferred approach is to operate the pilot unit under adiabatic operation, but few studies have been reported where this has been achieved (de Klerk, 2005). For this and other related reasons, working models and quantitative understanding of reactors in most commercial processes often do not exist, sometimes even after decades of operation. Process improvement are often achieved very slowly owing to the need to use small incremental changes in process settings and observing the plant response before deciding the next steps. Model guidance is lacking, so resortment to empirical approaches is required.

It is now useful to discuss how the proper science-based approach to scale-up can be made more attractive since it minimizes risk and opens the door to new technologies. A brief perspective on catalyst design and development is provided first since catalytic kinetics is the primary engine that ultimately drives reactor performance.

For many years, catalyst discovery and development was largely based upon a trial-and-error approach with guidance provided using various empirical approaches (Bartholomew, 2006). However, numerous efforts have resulted in advancing the scientific basis for catalyst design and characterization during the past few decades (Borges and Díaz, 2012; Jahangiri et al., 2014; Poncelet et al., 1995; Wang and Barteau, 2001). With new instrumental methods and instruments, it is now possible to monitor catalytic surface processes at the nanoscale (Zhang et al., 2013). In addition, atomic- and molecular-level modeling that was previously limited by available computer hardware became more prevalent with new advances in computational power and the associated software. Density functional theory

(DFT) (Hohenberg, 1964; Kohn and Sham, 1965) and other models have provided elegant examples on both why and how some known catalytic test reactions work. Advances in the DFT theory now allow the description of catalytic reactions at surfaces with the detail and accuracy required for computational results to compare favorably with experiments. In addition, theoretical methods can be used to describe surface chemical reactions in detail and to understand variations in catalytic activity from one catalyst to another. Advances in DFT and computational methods have been the subject of a recent review by Nørskov et al. (2009) to which the reader is directed for details. It is evident from the above references and those cited therein, as well as other extensive open literature, that notable advances are occurring in terms of catalyst design principles, catalyst characterization, and modeling of surface processes. Discoveries in these areas are expected to accelerate as additional understanding of surface processes occurs from both theoretical models and experimental data. Reaction engineering can benefit from these developments since it will allow the development of robust microkinetic models that provide more in-depth description on the formation of desired and undesired reaction products.

In the middle to late 1990s, combinatorial catalyst screening methods and detailed catalyst testing gained notable attention from the perspective of both developers of the technology as well as end-users. Development of the FASTCAT reactor system for rapid screening (Mills and Nicole, 2004) and Multiple Automated Reactor System (MARS) for detailed screening (Mills and Nicole, 2005a, 2005b) are examples of the technology from the mid 1990s for gas-phase catalyzed reactions. Advances also occurred in environmental catalytic processes, zeolite catalysis, metallocene catalysts for production of polyolefins and for other systems (Dudukovic et al., 1994). Before combinatorial methods were popularized, catalyst screening was largely performed using experimental systems containing a single laboratory reactor or a small number of reactors that also required significant manual labor. Once a preferred catalyst recipe was identified, scale-up to meet demonstration-scale or commercial reactor requirements was typically performed in batch equipment and largely based on past previous experience using empirical approaches.

Recent advances in high-speed screening technologies for heterogeneous catalysts have been recently summarized by Zheng and Zhou (2011). Laboratory reactors and various aspects associated with catalyst testing from the perspective of obtaining intrinsic reaction kinetics has been the subject a recent review by Kapteijn and Moulijn (Kapteijn and Moulijn, 2008). The ability to perform both rapid catalyst screening and detailed catalyst performance testing has provided a more robust approach for obtaining both new catalyst materials more reliable reaction kinetics needed for reactor design and analysis than previously possible.

The path to follow for systematic catalyst and environmentally clean catalytic process development is also sketched in Fig. 1.5. Development of the understanding of the selected green chemistry should be followed by a combined effort in molecular modeling and small-scale experimentation to identify the best catalyst. The dynamics of the reaction should then be studied using the preferred commercial catalyst in an appropriate bench-scale reactor. The previously cited review by Kapteijn and Moulijn (2008) provides an excellent summary of various laboratory reactor systems, such as the steady-state isotopic transient kinetic analysis approach and the temporal analysis of products (TAP). For gas–solid catalyzed reactions, the TAP is a powerful system to identify possible reaction pathways and to connect the catalyst state with the observed kinetic rates (Gleaves et al., 1988, 1997; Shekhtman et al., 1999). The various TAP operating modes enable a comparison between experimental and model-predicted pathways as a means of discarding all mechanisms that are not feasible and to identify a region of best operating regimes for the catalyst (Redekop et al., 2011; Yablonsky et al., 2007). A suitable continuous-flow reactor should then be used to discriminate between various kinetic rate equations based upon various proposed reaction mechanisms. Such a reactor should be gradient-less in composition and temperature, if possible, or the transport resistances should be well-quantified using the techniques such as those summarized by Kapteijn and Moulijn (2008). In addition, microkinetic dynamic studies using temperature-programmed adsorption, desorption, and reaction can add value in developing a quantitative description of the reaction rate (Campbell, 1994; Dumesic, 1993; Losey et al., 2001). Combined separation and reaction should also be considered as well as the potential for dynamic operation versus steady-state continuous processing. The final result should be a robust quantitative description of the rate form and guidance for the best mode of operation and reactor type.

Classical heterogeneous catalytic reactor types used in various process technologies include packed beds, wall-catalyzed reactors, bubble columns, stirred tanks, risers, and fluidized beds. Monoliths and micro reactors have also made inroads in the last couple of decades. Novel designs attempt to combine reaction and separation via reactive distillation, catalytic distillation, *in situ* adsorption, use of membrane reactors, and they offer the promise of process intensification via improved volumetric productivity and selectivity as indicated in Table 1.2 (Jin et al., 2000; Smith, 1980; Taylor and Krishna, 2000). The pioneering work of Eastman Chemical in which a multicolumn process for methyl acetate synthesis was replaced by a single reaction–distillation column provides a clear indication of the achievable benefits (Agreda and Partin, 1984). In another example, the AIChE Reaction Engineering Practice award was awarded to Larry Smith in 2007 for implementing catalytic distillation in numerous refinery operations, thus improving efficiencies. His proposed process modifications (Smith, 1980, 1984) consisted of placing bags of catalysts in existing reactor shells, which paid for themselves within one or two quarters and resulted in long-term increased profitability. Yet, in many existing large-scale processes where process intensification could dramatically improve efficiencies and reduce wastes, very little if anything has been done since depreciated plants still operate profitably but continue to waste energy and materials. Thus, economic barriers are still prevalent to implementation of process

Table 1.2 Methods of Process Intensification for Improved and More Efficient Reactor Performance

Combination of reaction and separation	Reactive/catalytic distillation (Harmsen, 2007) *In situ* adsorption (Koppatz et al., 2009) Membrane reactors (Marcano and Tsotsis, 2002)
Miniaturization and transport enhancement	Micro reactors (Jensen, 2001; Mason et al., 2007; Mills et al., 2007) Rotating packed bed (Li et al., 2008; Munjal et al., 1989) Structured packing (Nigam and Larachi, 2005)
Dynamic forced periodic reactor operation	Gas–solid exothermic reactions (Kulkarni and Dudukovic, 1997) Gas–solid exothermic and endothermic reaction coupling (Ramaswamy et al., 2006) Gas–liquid–solid systems (trickle beds) (Silveston and Hanika, 2002)
Use of solvents and supercritical media	Phase transfer catalysis (Liotta, 1978) Enhancement of selectivity and ease of catalyst separation (Hamley et al., 2002; Subramaniam and McHugh, 1986)

intensification. This paradox is the result from the imperfect market system that places higher values only short-term goals.

4. SCALE-UP STRATEGIES

Ultimately, the primary task is replicating results obtained in a bench-scale unit in a commercial scale system. Unless replication or expansion of an existing process is being attempted, it is highly advisable to avoid so-called rules-of-thumb in scale-up. Instead, development and implementation of modern scale-up methodology based on multiscale consideration of the system is advocated. This means that robust models for the kinetic rate, transport effects on the particle or single eddy scale, and reactor flow and phase contacting patterns should be developed for the targeted reaction system. These models should be tested on bench-scale and/or pilot-plant units. The key scale-up issue can be stated as follows: once the reaction system was successfully run in the laboratory to produce the desired ranges for conversion, yield, and selectivity, how does one reproduce the results at a commercial scale? So-called *horizontal scale-up* (scale-up in parallel, scale-up by multiplication, or scale-out) offers one alternative, while *vertical scale-up* provides another alternative. In the former, units that were studied in the laboratory are connected in parallel and uniform loading of the catalyst in each tube and uniform flow distribution into each tube is performed. In the latter, one must account for the effect of equipment scale on the interplay of transport and kinetics. The former approach maintains the same geometry, flow pattern, fluid contacting pattern and flow regime, but has to consider the logistics of system integration consisting of multiple units and flow distribution, which is a nontrivial matter for multiphase flows. In the latter, relying on rules-of-thumb can lead to major problems as without proper quantitative understanding of the system, relying solely on heuristics or statistical approaches has a high likelihood of failure.

The *first scale-up requirement* for the same feed inlet temperature and composition is that the mean residence time (or mean contact time in heterogeneous systems) is the same in the bench-scale and large-scale unit. This ensures that the Damköhler number, which is ratio of characteristic flow time to characteristic time for intrinsic reaction, is about the same when the transport effects are either negligible or the same in the two systems. Otherwise, one must account for the change in transport rates with reactor scale. The *second scale-up requirement* is that the dimensionless variance of the residence time distribution must be the same in the two units. For example, this parameter is twice the reciprocal of the axial Péclet number in the case of the axial dispersion model. More precisely, for multiphase systems, the covariance of dimensionless sojourn times in the phases must be the same in the large and small reactor. Finally, the heat transfer area per unit volume should be the same. These criteria can be matched while working with a single reactor scale for scale-up in parallel. This explains the popularity of using multitubular packed beds that share the same coolant in a common shell versus a fluidized-bed reactor where gas and solids flow patterns can be quite complex. Scale-up using empty or packed microreactors and monoliths follow this scale-up strategy. Still, pitfalls exists which will not be elaborated on here.

Vertical scale-up requires knowledge of how the flow and contacting patterns change with scale and is difficult to predict *a priori*. Neither theory nor computational tools for modeling multiscale, multiphase reactors are fully available at the moment to accomplish this task. Thus, insight must be gained into the physics of multiphase flows to develop phenomenological models for coupling with kinetics that can be used for scale-up with validation. Examples of challenges faced in the scale-up of three different processes for three different reactors, namely, a trickle-bed, a solid–liquid riser, and a gas–solid riser, are described below.

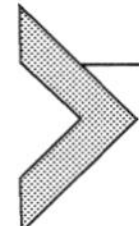

5. EXAMPLE 1. HYDROGENATION IN A TRICKLE-BED REACTOR—SCALE-UP SURPRISES TO AVOID

A robust hydrogenation catalyst in the form of extruded pellets with long on-stream life was selected and tested for hydrogenation of an aldehyde. After successfully running a laboratory trickle-bed (diameter = 0.0341 m and height = 0.235 m) that produced the desired aldehyde conversion of 90% using a targetted hydrogen-to-liquid feed ratio with a liquid hourly space velocity (LHSV) of 1.3 h^{-1}, it was decided to forego any further testing in large-scale equipment. Scale-up was implemented using the same LHSV, which is a well-established heuristic. This approach assumes that the larger diameter reactor will exhibit the same or improved performance as the smaller laboratory-scale reactor. As evident from Table 1.3(B), the larger diameter reactor ($d = 0.455$ m), when operated at steady-state using the same LHSV of 1.3 h^{-1}, consistently produced a much lower conversion of 40%. Yet, Table 1.3(A) indicates that even much larger scale-up factors of 248,000 (as opposed to the aldehyde hydrogenation scale-up of merely 16,000) are readily achieved in petroleum hydrotreating when the same heuristic of constant LHSV is used. This proves that scale-up should never be based on heuristics, as is prevalent industrial practice, without understanding their origin and their region of validity.

It is well known that trickle-bed performance depends on the effectiveness at which gas and liquid reactants contact the solid catalyst (Dudukovic et al., 2002). At low liquid irrigation velocities, the external catalyst area is partially dry since it is intermittently exposed directly to the flowing gas. These interactions result in so-called *partial wetting*, which is characterized by incomplete liquid–solid external contacting. At sufficiently high liquid velocities, all external catalyst area is effectively wetted by flowing liquid so the liquid–solid wetting efficiency is unity (Beaudry et al., 1987; Dudukovic, 1977; Mills and Dudukovic, 1981). At high enough pressures, sufficient hydrogen solubility ensures that the limiting reactant is in the

Table 1.3 Success and Failure in Scale-Up of Trickle Bed Reactors with LHSV = Const

	Lab/Bench	**Commercial**
(A) Hydrotreating at high pressure (liquid limited reaction), S = 248,000		
Catalyst volume (m^3)	0.001	248
Bed diameter (m)	0.025	3.25
Bed length (m)	1.3	25.0
LHSV (h^{-1})	1.4	1.4
Contacting efficiency	0.65	1.0
Resulting conversion	$x_{A,Lab} < x_{A,Com}$	
(B) Hydrogenation of a chemical (gas-limited reaction), S = 16,000		
Catalyst volume (m^3)	0.0002	3.15
Bed diameter (m)	0.0341	0.455
Bed length (m)	0.235	19.4
LHSV (h^{-1})	1.3	1.3
Contacting efficiency	0.51	1.0
Conversion achieved	0.90	0.40

liquid phase. Typically, this is the case in all hydrodesulfurization of heavy fractions and in many other hydrotreating operations encountered in petroleum processing. Under such conditions, the rate of supply of the liquid reactant to the individual catalyst particles is enhanced at higher liquid velocities, which result in higher liquid–solid contacting efficiencies (Al-Dahhan and Dudukovic, 1995). At constant LHSV, the ratio of liquid velocity in the larger reactor to that in the smaller one is equal to the ratio of their heights.

Table 1.3 also reveals that the larger reactor is always taller and has much larger contacting liquid–solid efficiency. This improves the effectiveness factor of the particles in the larger reactor, and it overcomes the possible drop in conversion due to increased bypassing in the large reactor as long as the rules for periodic liquid redistribution are followed. This is the reason why scale-up at constant LHSV in petroleum hydrotreating is such a success story. In contrast, the limiting reactant for aldehyde hydrogenation is gaseous hydrogen since the total pressure is notably lower so the concentration of dissolved hydrogen in the liquid is less than that encountered in petroleum hydrotreating. At a constant LHSV, the contacting efficiency approaches unity in the much taller and larger hydrogenation reactor. In addition, the liquid film created around the catalyst particles provides an additional resistance to transport of gas-phase hydrogen to the surface of the catalyst particle. This results in a dramatically reduced performance of the larger reactor, as evident from Table 1.3(B). It should also be noted that both liquid-phase and gas-phase limited reactions have been successfully modeled and these phenomenological models are sufficient for proper scale-up (Khadilkar et al., 1999, 2005).

The current state of the art for trickle-bed technology was recently described in the newest edition of Ullman's Encyclopedia (Dudukovic et al., 2014). Interested readers will find the key references for most important topics. Among these are the safety concerns regarding prevention of runaways in trickle beds since all reactions conducted in this reactor are exothermic and commercial reactors typically operate adiabatically. Either excess hydrogen or a volatile solvent are used to either control or limit the increase in temperature, which often works well with conventional catalysts. However, more active catalysts can result in temperatures that greatly exceed the adiabatic temperature rise based solely on the desired reaction due to subsequent over hydrogenation of other compounds. The increased temperatures can also result in loss in selectivity but not in activity. In the absence of radial mixing, this can cause excessive hot spots down the reactor.

Prevention of liquid misdistribution and hot spots has not been resolved at the moment (Hanika, 1999; Jaffe, 1976; McManus et al., 1993; Sie and Krishna, 1998), but much work is needed before rational design can be implemented. No notable research on these issues has been performed for the last several decades in the belief that existing reactors can be operated with impunity using much more active catalysts. The authors are aware of failures using this approach in companies, but these are never revealed to external sources. Research has been initiated on modeling gas and liquid flow in trickle beds using CFD with appropriate closure principles (Jiang et al., 1999). This has been further advanced by others (Boyer et al., 2005; Kuipers and van Swaaij, 1997; Ranade, 2002; Ranade et al., 2011). However, in the absence of an accurate 3D geometrical model for the packed-bed structure, all these models lack the predictability required for scale-up and design.

The teaching associated with this example is that scale-up in the absence of an appropriate phenomenological model should not be attempted. The reaction kinetics should be described by an appropriate model. Also, highly active catalysts should not be used in reactors where flow maldistribution is possible and where over hydrogenation can create a serious loss of yield with extremely high temperatures. At much higher temperatures, catalyst effectiveness factors, if the catalyst pore volume is filled with reaction vapor, can be many orders of magnitude higher leading to dangerous hot spots (Hanika, 1999; Jaffe, 1976; McManus et al., 1993; Sie and Krishna, 1998). In addition, trickle beds are also prone to hot spots and thermal runaway, and the common practice of overdesign only promotes this behavior.

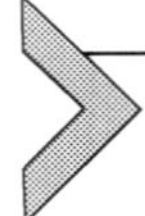

6. EXAMPLE 2. SOLID ACID ALKYLATION—SOLIDS FLOW PATTERN IN A RISER

Production of high octane fuels and detergents via alkylation is traditionally realized using strong mineral acids such as HF and H_2SO_4 as homogeneous catalysts in a liquid system. The complex catalytic cycle involving a number of proton exchanges has been reasonably well understood and modeled (Albright and Kranz, 1992; Brooks et al., 1955). For the purposes of engineering analysis, the main alkylation reaction can be represented by a reaction between an olefin and a paraffin, which is of first-order in each reactant or of second-order overall. The undesired set of yield-reducing reactions resulting from the reaction of olefins is represented by a second-order reaction in the concentration of olefin. Catalyst deactivation is proportional to this undesired reaction (Nayak, 2009). A large feed ratio of paraffin to olefin that necessitates large recycle of paraffin has been traditionally used in stirred tanks and required recycle pumps for the acid phase recycle. The process has a large environmental footprint and safety issues because of the large burden of spent acid and due to possible acid leaks around the reactor and pump shafts. The newest technology implemented at the turn of the millennium has an improved flow pattern involving reactors-in-series where olefin is fed into each to keep its concentration low, while products and paraffin flow continuously through the system (Nayak, 2009). Safety has been improved, and this process reflects the best available technology when using liquid acids as catalysts.

Solid acid catalysts offer a potential advantage since they produce both a smaller environmental footprint and improved process safety. Their successful utilization depends on finding the solid acid catalyst that can be readily regenerated and also identifying a reactor configuration that can be readily scaled-up. One of the options that attracted attention is the use of the liquid–solid riser. A possible effective means for conducting this process involves contacting the fresh solid catalyst with the proper paraffin–olefin mixture in a riser, transforming the olefins to the desired high conversion, and then recycling the catalyst upon regeneration. The volumetric productivity, selectivity, and the extent of catalyst deactivation in the riser depend on the flow pattern of the two phases. The available literature, including reactor vendor manuals, suggests that at the high fluid and particle Reynolds numbers, both liquid and solid catalysts can be safely considered in plug-flow with the slip velocity between the two related to the mean solids holdup. Whatever reaction and deactivation kinetic model is used, the productivity and selectivity in the riser, as well as the extent of catalyst deactivation, depend on the extent of solids back mixing in the riser. A thorough experimental and modeling investigation of the liquid–solid riser constituted the thesis of Roy (2000).

To properly test the flow pattern in the cold flow model of the riser (Fig. 1.6; Roy, 2000), a conventional tracer study was first conducted that showed the liquid was essentially in plug-flow with a RTD equivalent to greater than about 20 tanks-in-series for the given operating conditions. The Computer Automated Radioactive Particle Tracking (CARPT) facility in CREL was used to determine the solids flow pattern in the riser, and the

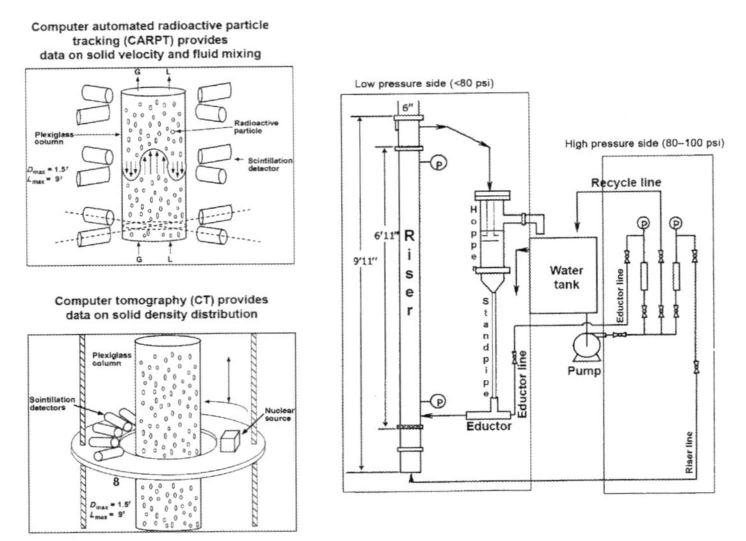

Figure 1.6 Schematic of CARPT and CT (left); schematic of liquid–solid riser (right) (Roy, 2000).

γ-ray computed tomography (CT) technique was used to establish the solids velocities and distribution in the riser. These techniques are sketched in Fig. 1.6 while their detailed description is provided elsewhere (Roy and Dudukovic, 2001). The experimental results and comparison with an Euler–Euler 3D simulation with proper closures are shown in Fig. 1.7 (Roy and Dudukovic, 2001; Roy et al., 2000, 2005a, 2005b). Projections of the paths of three radioactive particles, which are traced one-at-a-time in the mixture of solids that they were equal to in size and density, is shown in the upper left corner. Clearly, individual particles are not in plug-flow and meandering during their upward trajectory spans the entire column diameter. A single radioactive particle, monitored by CARPT, exhibits a tortuous path through the riser during a single visit. Multiplying the difference in subsequent positions with the sampling frequency yields instantaneous velocities. Averaging 2000 or more of such pathways yields a symmetric solids ensemble-averaged velocity profile with particles rising in the middle and falling by the wall. It is evident that solids preferentially flow up in the center of the riser and slow down or, at some conditions, flow downward at the wall. The examination of the pdf (probability density function) of the particle velocities (not shown) establishes the increased number of negative velocities close to the wall. In addition, particle eddy

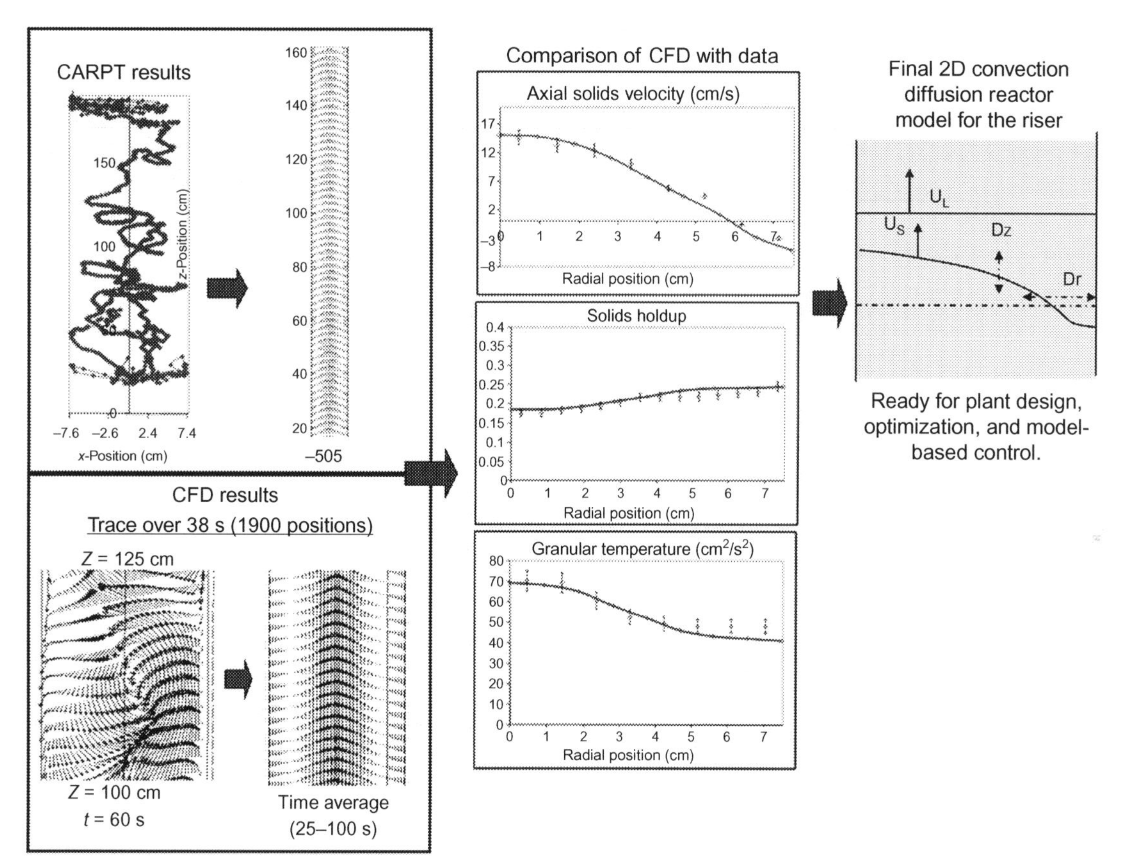

Figure 1.7 Quantification of solids flow and distribution via CARPT–CT and comparison with 3D CFD computations (left); comparison of axial solids velocity, solids holdup, and granular temperature (middle); convection–diffusion reactor model (right). *Figure originally published in Roy and Dudukovic (2001) and Roy et al. (2005b).*

diffusivities can be obtained from the particle trajectories directly. Results for an instantaneous solid velocity field are displayed in the lower left corner. CFD computations reveal a highly complex 3D instantaneous swirling flow structure that explains the single particle trajectory, but 75 s of averaging produces the symmetric flow pattern of rising solids in the middle and falling by the wall.

With the help of CT information, one can now quantitatively compare the experimental data with model predictions as shown in the three insets in the middle of Fig. 1.7. Both the axial solids velocity profile and the radial solids holdup profile are accurately predicted. The agreement between simulations and data is excellent for solids velocity, solids holdup distribution and for the granular temperature, i.e., solids kinetic energy. While other techniques may provide data for average quantities, only CARPT provides the information essential for the dynamic comparison. The experimental information obtained provides the proper understanding of the flow pattern in the system and allows tuning of the appropriate CFD code. Since further increases in riser diameter for the commercial scale are modest, it is safe to assume that the needed dynamics has been captured in this study. Hence, CFD computations can now be used to generate additional data for development of proper phenomenological models for coupling with kinetic information in scale-up.

For solid acid alkylation, the phenomenological model for the riser consists of plug-flow of liquid, fully developed solids velocity and solids holdup profile with superimposed axial and radial solids diffusivities (Ramaswamy et al., 2005). These eddy diffusivities can be computed from the 3D CFD model. This model can now be coupled with appropriate reaction and deactivation kinetics. CFD can now provide all the required model parameters. In summary, a direct measurement of the solids RTD is also obtained from CARPT data. As evident from Fig. 1.8 (Roy, 2000), the flow pattern is not ideal plug-flow as assumed by many, and the RTD of solids depends on the conditions used and varies from two to six tanks-in-series.

This example illustrates research that should be done prior to scale-up to avoid process complications. Relying on the available data base would not have been sufficient. A detailed understanding of the two-phase flow is needed and can be used to assess whether or not the available catalyst is suitable for the riser. In the absence of an accurate phenomenological model, wrong conclusions can be reached. Swinging periodic operation in packed beds could be considered (Exelus, Inc., 2014) as an alternative arrangement for solid acid alkylation.

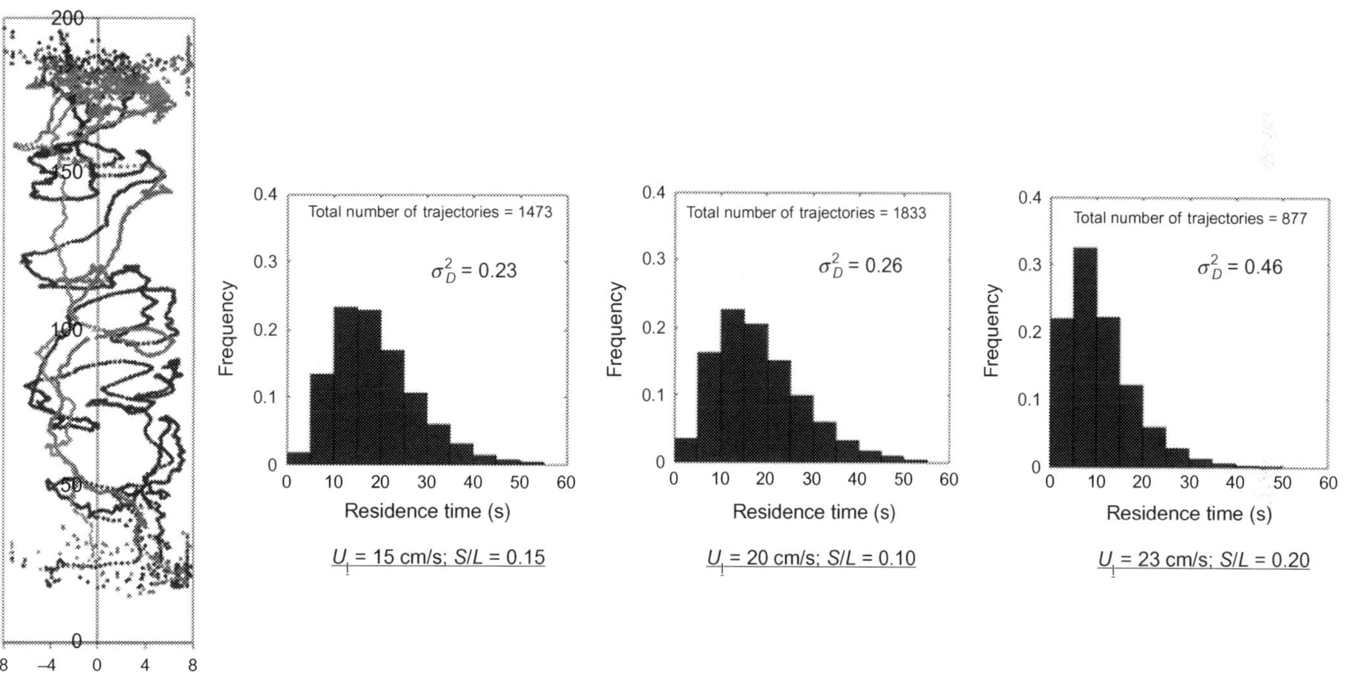

Figure 1.8 Solids particle trajectories in the riser (left); solids residence time under different operating conditions (right). (*Roy, 2000*).

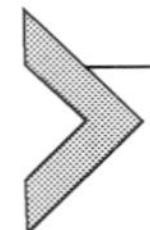

7. EXAMPLE 3. PARTIAL OXIDATION—NEED FOR THE PROPER MODEL OF THE CIRCULATING FLUIDIZED-BED SYSTEM

A third and final example of problematic scale-up occurs in the production of maleic anhydride by the partial oxidation of butane with air. This was selected to illustrate the pitfalls of scale-up without appropriate reactor-level models. All of the excellent work in catalyst development and on proper selection of the best reactor type was nullified by expecting the scale-up to be based on established rules-of-thumb and using empirical models to fit pilot-plant performance data.

The advantages of this process over the traditional benzene route are illustrated in Table 1.4. Much improved mass and carbon efficiencies are possible and less toxic reactants are involved in the *n*-butane route. Hence, this process was described in several green engineering books as environmentally superior and was advertised, based on assumed selectivity, as a non-carbon dioxide producer (Allen and Shonnard, 2002). However, whenever hydrocarbons are contacted with air at elevated temperatures, the potential exists for either partial or total combustion to CO and CO_2. Table 1.5 illustrates a more realistic stoichiometry for this system. How much the two yield lowering reactions will affect the desired reaction depends upon the catalyst capacity for oxygen, the ratio of catalyst to butane, the intimacy of mixing between fresh catalyst and butane, and the inlet temperature. All reactions are highly exothermic, and the excess thermal heat is removed by cooling of the reactor wall in the case of a tubular reactor or via heat exchange pipes if a fluidized-bed reactor is utilized. The amount of butane in the feed (1.8% for packed beds and 4% fluidized beds) is controlled

Table 1.4 Potential for Improved Process Efficiency of Maleic Anhydride Synthesis from Butane

	Benzene route	***n*-Butane route**
Reaction	$2C_6H_6 + 9O_2 \xrightarrow{V_2O_5\ MoO_3} 2C_4H_2O_3 + H_2O + 4CO_2$	$C_4H_{10} + 3.5O_2 \xrightarrow{(VO)_5P_2O_5} C_4H_2O_3 + 4H_2O$
Atom economy	$18/42 = 0.43$	$9/21 = 0.43$
Mass efficiency	$\frac{2(4)(12)+2(3)(16)+2(2)(1)}{2(6)(12)+9(2)(16)+2(6)(1)} \times 100\% = 44.4\%$	$\frac{(4)(12)+(3)(16)+(2)(1)}{(4)(12)+3.5(2)(16)+(10)(1)} \times 100 = 57.6\%$

Table 1.5 Reaction Pathways and Reactors Tried for Partial Oxidation of *n*-Butane

Reaction Pathways on Vanadium Pentoxide Catalyst	Industrial Reactors (Felthouse et al., 2001)
$C_4H_{10} + \frac{7}{2}O_2 \rightarrow C_4H_2O_3 + 4H_2O$	1. Packed beds
$C_4H_2O_3 + O_2 \rightarrow 4CO + H_2O$	2. Fluidized beds
$C_4H_{10} + \frac{11}{2}O_2 \rightarrow 2CO + 2CO_2 + 5H_2O$	3. CFB reactor

to prevent formation of an explosive mixture. Hence, a low concentration of butane results in low concentrations of maleic anhydride (1% in the product gas), which requires a costly separation of the product mixture. Much effort was spent on developing the vanadyl pyrophosphate (VPO) catalyst which has a better capacity for storing active oxygen and improved selectivity then previously ones that were used (Centi et al., 1988).

In terms of which reactor type to use, existing designs for the benzene process based on multitubular, wall-cooled reactors were selected, which were plagued by hot spots and low productivity. Better productivity can be achieved by improving the wall heat transfer, and this topic received considerable attention in the last decade or so. The work of Dixon et al. (2005, 2006) should be singled out as providing a systematic study of the phenomena involved using first principles and also introduced new computational algorithms, thus laying the foundation for future improved designs. However, these wall-cooled packed tubular reactors operate at steady-state, and the catalyst activity and selectivity performance metrics are greatly reduced compared to the fresh catalyst. Moreover, catalyst deactivation demands regeneration or repacking of the tubes. To provide for periodic catalyst replenishment, fluidized beds were tried and found to be inefficient since they had large inventory of dead nonselective catalyst. This led to the conclusion that circulating fluidized beds (CFBs) clearly are the best choice. In principle, they should be able to maintain the high catalyst activity and selectivity in the riser where the freshly reoxidized catalyst is contacted by butane and some air, if needed. The reduced VPO catalyst can be readily regenerated in a bubbling fluidized bed.

The schematic for a CFB process for partial oxidation of *n*-butane is shown in Fig. 1.9. Very high selectivity and conversion (>90%) was shown experimentally in both the bench-scale (1/8 in diameter × 16 ft tall tube) and pilot-plant riser units (ca. 4-in diameter) (Contractor, 1999; Patience

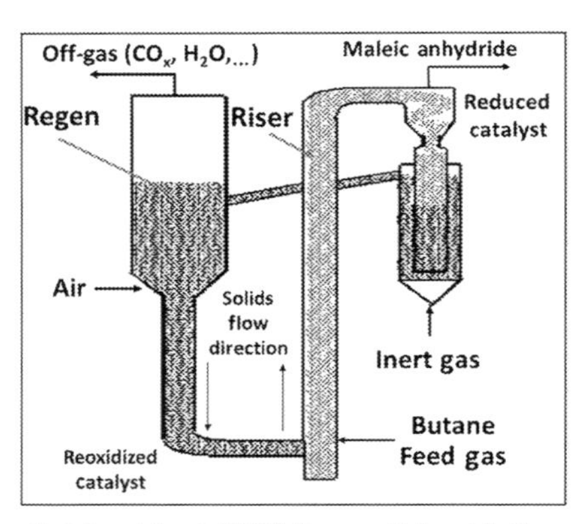

Figure 1.9 Circulating fluidized bed (CFB) for partial oxidation of *n*-butane to maleic anhydride.

and Bockrath, 2010). An extensive study related to scale-up was in order since the needed diameter increase of the riser for the commercial plant was substantial (several feet in diameter) and no reliable model for partial oxidation in such a system existed. Despite notable efforts, the resulting commercial riser never met the design capacity and selectivity goals for economical operation. Due to an oversupply of maleic anhydride that occurred during process operation, it was more economical to purchase maleic anhydride from external suppliers and use this as the feedstock to the second-step maleic acid hydrogenation process for manufacture of tetrahydrofuran (THF). The riser was eventually shut down and dismantled for business reasons.

One can only speculate what prevented the commercial riser from meeting the design specifications. CFBs have been used for decades in catalytic cracking, biomass conversion by gasification, power generation by coal combustion, and are now popular for possible applications in chemical looping combustion. Unfortunately, experience gained with such systems does not provide all the rules needed for successful application in partial oxidations. It is possible that mixing of fresh catalyst with *n*-butane and the control of local hot spots in the presence of added air may have contributed to operational issues. If the VPO catalyst had sufficient carrying capacity for oxygen, perhaps this problem could have been avoided. In addition, catalyst inventory in large plants is often minimized due to cost and that does not scale-up proportionally to other subsystems. Mixing of fresh catalyst with butane at the inlet would be very different in a large diameter riser than in a bench-scale unit and pilot-plant. The catalyst-to-feed gas ratio and mixing rates should have been varied in the pilot-plant to examine conditions that lead to deteriorating performance rather than be focused exclusively on targeted conversion and yield. Thus, not meeting the design specifications would appear to be a detractor for using CFBs as a reactor system for partial oxidation. This is a paradox because a properly designed CFB should be a good system for the desired series reactions since gas approaches plug-flow and solids are not completely backmixed. At the start, it was also mentioned that in all scale-ups, one must match the contact time distribution in the small and large unit. This is challenging to accomplish because to the present day, reliable models for predicting solids RTD in risers and reliable techniques are lacking for its experimental determination (Bhusarapu, 2005). Considering that risers have been in industrial practice for over half a century, a more disciplined approach to scaling would have likely led to a successful process for the manufacture of THF.

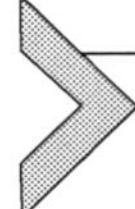

8. CHALLENGES FOR IMPROVED SCALE-UP OF MULTIPHASE REACTORS

The above discussion and examples illustrate that reactor scale models have not advanced much during the past decades and this hinders our ability to reduce risk in implementation of new more efficient catalytic technologies. To make progress, it is necessary to develop improved and more accurate descriptions of flow and mixing in typical multiphase reactors. Multiphase reactors frequently encountered in practice (Fig. 1.10) such as risers, bubble columns, fluidized beds, packed beds, and stirred tanks are opaque so that not all flow visualization tools (Table 1.6) are suitable.

Computer Automated Radioactive Particle Tracking (CARPT) and γ-ray-based CT techniques provide unique data for quantifying the hydrodynamics of opaque multiphase systems (Chaouki et al., 1997). These techniques are capable of providing experimental data over a wide range of operating conditions in different multiphase reactors as shown in Fig. 1.10 over the entire domain of the flow. Data collected from CARPT–CT were successfully used for the validation of multiphase CFD codes for flow pattern and mixing evaluation (Devanathan et al., 1990; Khopkar et al., 2005; Kumar et al., 1997; Rammohan et al., 2001a, 2001b).

Recently, a novel hybrid radioactive particle tracking (RPT) facility has been developed where the calibration step for the radioactive particle was eliminated (Khane and Al-Dahhan, 2013). The new technique is claimed to minimize the dynamic bias associated with the original CARPT

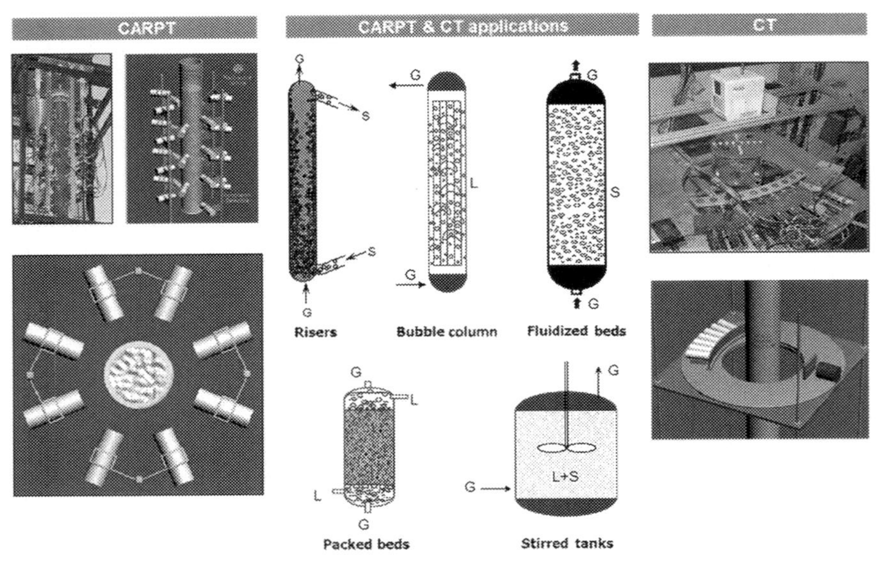

Figure 1.10 Opaque multiphase reactors on which CARPT–CT are used to obtain mixing and phase distribution information.

Table 1.6 Multiphase Reactor Visualization Tools

	Invasive	Noninvasive
Local measurements	Needle (optical and impedance) probes, heat transfer probe, pilot tube	Visualization techniques (photographic, radiographic, particle image velocimetry, NMR imaging), laser Doppler anemometry, polarographic technique, radioactive particle tracking, and tomographic techniques
Global measurements	–	Pressure drop/fluctuation, dynamic gas disengagement technique, tracing technique, conductimetry, and radiation attenuation techniques

Summarized from Boyer et al. (2002).

technique. However, no details have been provided yet in the open literature for subsequent review and analysis.

A shared facility for training of students and development of improved reactor models would provide an opportunity for development of future advances. The power of the RPT techniques can be best appreciated from Fig. 1.11 (Degaleesan et al., 1996). The animation based on this figure

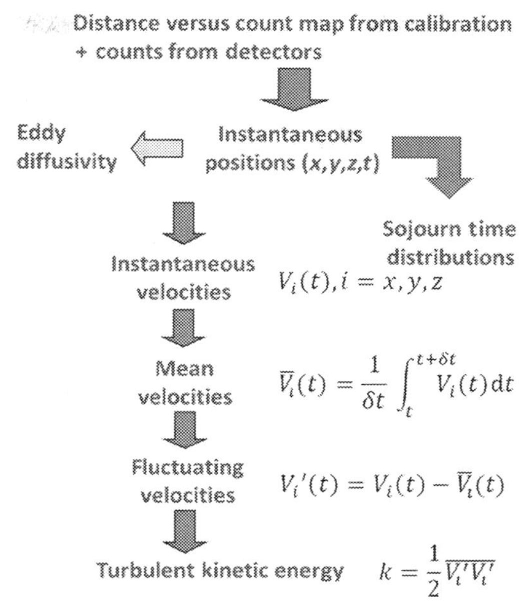

Figure 1.11 Quantification of the flow field by CARPT.

available at (http://crelonweb.eece.wustl.edu) views a slurry bubble column with 20% solids loading at high superficial gas velocity of 45 cm/s. As the radioactive particle moves about in the reactor, the instantaneous position of the tracer particle is determined from the counts received by the detectors and by utilizing the distance versus count map obtained from calibration. The time differentiation of two successive particle positions yields the instantaneous Lagrangian velocity of the particle. The mean velocity is then computed by ensemble averaging the instantaneous velocities obtained during the CARPT run. Once the mean velocity at each location is calculated, the fluctuating components as well as the turbulent kinetic energy are obtained from the mean and instantaneous velocities. Apart from that, the instantaneous position data for the tracer particle can also be processed to obtain the sojourn time distributions in different regions in the reactor, which can provide insight regarding dead zones. The Lagrangian tracer particle trajectories also yield directly the eddy diffusivities in the system.

The CARPT–CT techniques have been introduced at CREL to provide cold flow modeling information on slurry bubble column flows operated in churn turbulent regime needed for gas-to-liquid conversion processes (Degaleesan, 1997; Degaleesan and Dudukovic, 1998; Degaleesan et al.,

1996, 2001). The data produced were instrumental in establishing a proper phenomenological model for the bubble column. This model was able to predict both liquid and gas tracer curves obtained in an AFDU pilot-plant column in LaPorte, Texas, for three different reaction systems: methanol synthesis, dimethyl ether synthesis, and Fischer–Tropsch synthesis (Chen et al., 1998; Devanathan et al., 1990; Gupta et al., 2001a, 2001b). Unfortunately, the desire to use improved science in scale-up gas-to-liquid fuels processes to large diameter bubble columns disappeared when liquefaction of natural gas became more economically attractive.

It should be noted that CARPT experiments in the gas–solid riser produced for the first time the definitive solids residence time distribution in the riser itself (Fig. 1.12). Precise monitoring of the time when the tracer particle enters the system across the inlet plane, and the time when it exits across either inlet or exit plane, provides its actual residence time in the riser. Ensemble averaging for several thousand particle visits yields the solids RTD. The first passage time distribution is also readily be obtained. This information cannot be obtained by measuring the response at the top of the riser to an impulse injection of tracer at the bottom. By using CARPT, true descriptions of solids residence time distributions can be obtained in the riser (Bhusarapu et al., 2004, 2006). One task of CFD modelers is to develop codes that can predict the experimental observations of CARPT. Here, it

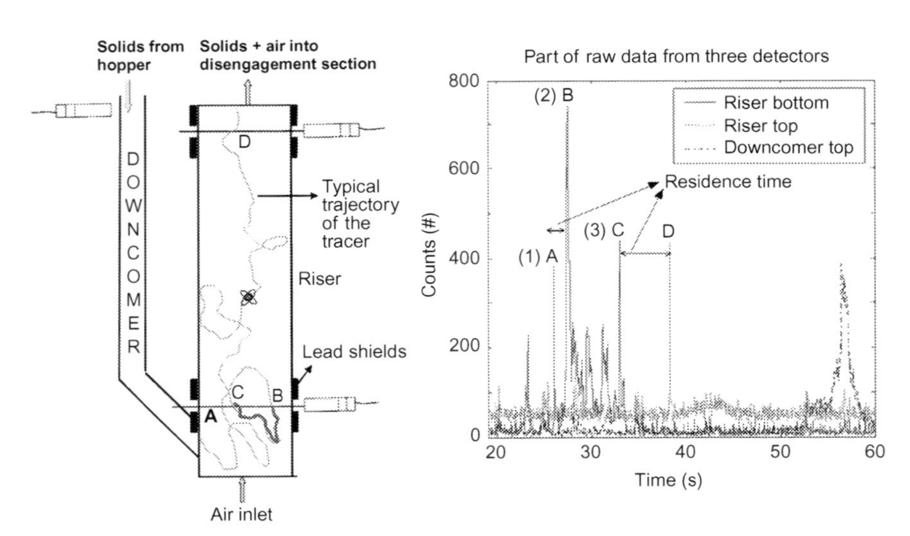

Figure 1.12 Determination of solids residence time distribution in the riser: schematic of the CARPT experiment in the riser (left); part of raw data from three detectors (right).

should be mentioned that obtaining the variance of the solids RTD in the riser is not the best way for describing the magnitude of solids backmixing. It is customary to calculate the axial dispersion coefficient from the variance of the RTD, yet the responses obtained clearly indicate by their shape that the axial dispersion model is not the best model to describe them (Bhusarapu et al., 2004, 2006). If this was performed, then it would be concluded that for the fast fluidization (FF) regime, the Peclet number is ca. 2.3 and axial dispersion coefficient is $0.62\ m^2/s$, thus indicating significantly more backmixing of solids than observed in the discrete particle transport (DPT) regime where Peclet number of 6.2 and an axial dispersion coefficient of only $0.37\ m^2/s$ are reported. A much better measure of solids back mixing in the riser is the macromixing index of Villermaux (M) which compares the mean of the trajectory length experienced by catalyst particles in the riser to the length of the riser. Clearly, M, which is directly calculable from CARPT data, is about twice as large in the DPT as in the FF regime. A properly validated CFD model should be able to predict M. The task remains to develop improved models for the riser for coupling with kinetics that occurs on the catalyst particle.

9. CONCLUSIONS

Scale-up techniques used in heterogeneous catalysis are mainly outdated and ineffective. The knowledge base must be expanded, and a new workforce must be both educated and trained in the multiscale approach to build more fundamental, science-based models.

Most processes in heterogeneous catalysis involve multiphase reactors and require scale-up, which in turn requires improved knowledge base of flow, mixing, kinetics and transport.

Development of fundamentally based phenomenological models for reactors with two (three) moving phases is possible (e.g., bubble columns, riser, stirred tank, etc.). Availability of such models adds value and significantly reduces scale-up risks. It enables rational approach for the improvement of existing processes.

CARPT–CT techniques provide a unique tool for evaluation of holdup and velocity distribution in multiphase systems and for validation of CFD codes. A shared facility should be established.

CFD codes based on Euler–Euler interpenetrating fluid model with appropriate closures, upon validation, provide the means for effective calculation of reactor flow and mixing parameters in industrial scale reactors and could become valuable scale-up tools.

Phenomenological reactor models are capable of predicting tracer impulse responses. Thus, they can predict reactor performance for linear kinetics exactly and offer a good starting point for assessment of performance of nonlinear systems. Much more effort should be spent on their systematic development for commonly used reactor types.

ACKNOWLEDGMENTS

Haldor Topsøe deserves our gratitude for providing the opportunity for exchange of ideas at their Science Forum. The authors are indebted to Yujian Sun, Boung Wook Lee, and Onkar Manjrekar, graduate students at Washington University in St. Louis, for their expert assistance in organizing the references and figures for this manuscript in record time. We also offer special gratitude to the guest editor, Professor Tony Dixon, for dealing with our tardiness and his useful comments while preparing the final manuscript.

REFERENCES

Agreda VH, Partin LR: Acetic acid as reactant and extractive agent, US4435595 A, 1984.

Albright LF, Kranz KE: Alkylation of isobutane with pentenes using sulfuric acid as a catalyst: chemistry and reaction mechanisms, *Ind Eng Chem Res* 31:475–481, 1992. http://dx.doi.org/10.1021/ie00002a004.

Al-Dahhan MH, Dudukovic MP: Catalyst wetting efficiency in trickle-bed reactors at high pressure, *Chem Eng Sci* 50:2377–2389, 1995. http://dx.doi.org/10.1016/0009-2509(95)00092-J.

Allen DT, Shonnard D: *Green engineering: environmentally conscious design of chemical processes,* Upper Saddle River, NJ, 2002, Prentice Hall.

AspenTech: *Optimize chemical processes with Aspen Plus(r)*, Last accessed, August 12, 2014, http://www.aspentech.com/products/aspen-plus.aspx.

Bartholomew CH: *Fundamentals of industrial catalytic processes,* ed 2, Hoboken, NJ, 2006, Wiley.http://www.wiley.com/WileyCDA/WileyTitle/productCd-0471457132.html.

Beaudry EG, Dudukovic MP, Mills PL: Trickle-bed reactors: liquid diffusional effects in a gas-limited reaction, *AIChE J* 33:1435–1447, 1987. http://dx.doi.org/10.1002/aic.690330904.

Bhusarapu R: *Solids flow mapping in gas–solid risers* (Doctoral dissertation), St. Louis, MO, 2005, Washington University.

Bhusarapu S, Al-Dahhan M, Dudukovic MP: Quantification of solids flow in a gas–solid riser: single radioactive particle tracking, *Chem Eng Sci* 59:5381–5386, 2004. http://dx.doi.org/10.1016/j.ces.2004.07.052.

Bhusarapu S, Al-Dahhan MH, Dudukovic MP: Solids flow mapping in a gas–solid riser: mean holdup and velocity fields, *Powder Technol* 163:98–123, 2006. http://dx.doi.org/10.1016/j.powtec.2006.01.013.

Borges ME, Díaz L: Recent developments on heterogeneous catalysts for biodiesel production by oil esterification and transesterification reactions: a review, *Renew Sustain Energy Rev* 16:2839–2849, 2012. http://dx.doi.org/10.1016/j.rser.2012.01.071.

Boyer C, Duquenne A-M, Wild G: Measuring techniques in gas–liquid and gas–liquid–solid reactors, *Chem Eng Sci* 57:3185–3215, 2002. http://dx.doi.org/10.1016/S0009-2509(02)00193-8.

Boyer C, Koudil A, Chen P, Dudukovic MP: Study of liquid spreading from a point source in a trickle bed via gamma-ray tomography and CFD simulation, *Chem Eng Sci* 60:6279–6288, 2005. http://dx.doi.org/10.1016/j.ces.2005.03.049.

Brooks BT, Boord CE, Kurtz SS, Schmerling L, editors: *The chemistry of petroleum hydrocarbons* (vols. 2 and 3). New York, NY, 1955, Reinhold.

Brubaker H: *DuPont to sell its fibers and textiles unit for $4.4 billion DuPont to sell textiles division*, Philadelphia, PA, 2003, Philadephia Inq. http://articles.philly.com/2003-11-18/business/25462661_1_dupont-shares-stainmaster-invista.

Campbell CT: Future directions and industrial perspectives micro- and macro-kinetics: their relationship in heterogeneous catalysis, *Top Catal* 1:353–366, 1994. http://dx.doi.org/10.1007/BF01492288.

Centi G, Perathoner S: Catalysis and sustainable (green) chemistry, *Catal Today* 77:287–297, 2003. http://dx.doi.org/10.1016/S0920-5861(02)00374-7.

Centi G, Trifiro F, Ebner JR, Franchetti VM: Mechanistic aspects of maleic anhydride synthesis from C4 hydrocarbons over phosphorus vanadium oxide, *Chem Rev* 88:55–80, 1988. http://dx.doi.org/10.1021/cr00083a003.

Chaouki J, Larachi F, Dudukovicc MP: *Non-invasive monitoring of multiphase flows*, Amsterdam, 1997, Elsevier. Accessed June 25, 2014. http://www.sciencedirect.com/science/book/9780444825216.

Chemical Reaction Engineering Laboratory, Accessed, August 12, 2014, http://crelonweb.eece.wustl.edu.

Chen JW, Gupta P, Degaleesan S, Al-Dahhan MH, Dudukovic MP, Toseland BA: Gas holdup distributions in large-diameter bubble columns measured by computed tomography, *Flow Meas Instrum* 9:91–101, 1998. http://dx.doi.org/10.1016/S0955-5986(98)00010-7.

Contractor RM: Dupont's CFB technology for maleic anhydride, *Chem Eng Sci* 54:5627–5632, 1999. http://dx.doi.org/10.1016/S0009-2509(99)00295-X.

de Klerk A: Adiabatic laboratory reactor design and verification, *Ind Eng Chem Res* 44:9440–9445, 2005. http://dx.doi.org/10.1021/ie050212a.

de Lasa HI, Doğu G, Ravella A, editors: *Chemical reactor technology for environmentally safe reactors and products*, Dordrecht, Germany, 1992, Kluwer Academic Publishers.

Degaleesan S: *Turbulence and liquid mixing in bubble columns* (Doctoral dissertation), St. Louis, MO, 1997, Washington University.

Degaleesan S, Dudukovic MP: Liquid backmixing in bubble columns and the axial dispersion coefficient, *AIChE J* 44:2369–2378, 1998. http://dx.doi.org/10.1002/aic.690441105.

Degaleesan S, Roy S, Kumar SB, Dudukovic MP: Liquid mixing based on convection and turbulent dispersion in bubble columns, *Chem Eng Sci* 51:1967–1976, 1996. http://dx.doi.org/10.1016/0009-2509(96)00054-1.

Degaleesan S, Dudukovic M, Pan Y: Experimental study of gas-induced liquid-flow structures in bubble columns, *AIChE J* 47:1913–1931, 2001. http://dx.doi.org/10.1002/aic.690470904.

Derksen J, Van den Akker HEA: Large eddy simulations on the flow driven by a Rushton turbine, *AIChE J* 45:209–221, 1999. http://dx.doi.org/10.1002/aic.690450202.

Devanathan N, Moslemian D, Dudukovic MP: Flow mapping in bubble columns using CARPT, *Chem Eng Sci* 45:2285–2291, 1990. http://dx.doi.org/10.1016/0009-2509(90)80107-P.

Dixon AG, Nijemeisland M, Stitt EH: CFD study of heat transfer near and at the wall of a fixed bed reactor tube: effect of wall conduction, *Ind Eng Chem Res* 44:6342–6353, 2005. http://dx.doi.org/10.1021/ie049183e.

Dixon AG, Nijemeisland M, Stitt EH: Packed tubular reactor modeling and catalyst design using computational fluid dynamics. In Marin GB, editor: *Advances in chemical engineering*, 2006, Elsevier, pp 307–389. Accessed June 25, 2014, http://linkinghub.elsevier.com/retrieve/pii/S0065237706310058.

Dudukovic MP: Catalyst effectiveness factor and contacting efficiency in trickle-bed reactors, *AIChE J* 23:940–944, 1977. http://dx.doi.org/10.1002/aic.690230624.

Dudukovic MP: Frontiers in reactor engineering, *Science* 325:698–701, 2009. http://dx.doi.org/10.1126/science.1174274.

Dudukovic MP: Reaction engineering: status and future challenges, *Chem Eng Sci* 65:3–11, 2010. http://dx.doi.org/10.1016/j.ces.2009.09.018.

Dudukovic MP: Multi-scale process engineering (MPE) concepts in reaction engineering practice of catalytic systems. In *Topsoe 2013 catalysis forum: modeling and simulation of heterogeneous catalytic processes*, Munkerupgaard, Denmark. 2013. http://www.topsoe.com/sitecore/shell/applications/%20~/media/PDF%20files/Topsoe_Catalysis_Forum/2013/Milorad_dudukovic_closing.ashx.

Dudukovic M, Mills P, Bell A, Manzer L: Preface—symposium on catalytic reaction engineering for environmentally benign processes and US-Russia workshop on environmental catalysis, *Ind Eng Chem Res* 33:2885–2886, 1994. http://dx.doi.org/10.1021/ie00036a600.

Dudukovic MP, Larachi F, Mills PL: Multiphase catalytic reactors: a perspective on current knowledge and future trends, *Catal Rev Sci Eng* 44:123–246, 2002. http://dx.doi.org/10.1081/CR-120001460.

Dudukovic MP, Kuzeljevic ŽV, Combest DP: Three-phase trickle-bed reactors. In *Ullmanns encyclopedia of industrial chemistry*, Weinheim, Germany, 2014, Wiley-VCH Verlag GmbH & Co. KGaA, pp 1–40.

Dumesic JA, editor: *The microkinetics of heterogeneous catalysis*, Washington, DC, 1993, American Chemical Society.

Ericsson KA, Prietula MJ, Cokely ET: The making of an expert, *Harv Bus Rev* 85(7–8):114–121, 2007.http://hbr.org/2007/07/the-making-of-an-expert/ar/1.

Exelus, Inc., Accessed, August 12, 2014, http://www.exelusinc.com/.

Felthouse TR, Burnett JC, Horrell B, Mummey MJ, Kuo Y-J: Maleic anhydride, maleic acid, and fumaric acid. In Howe-Grant M, editor: *Kirk-Othmer encyclopedia of chemical technology*, 4th edition, New York, NY, 2001, Wiley.

Gleaves JT, Ebner JR, Kuechler TC: Temporal analysis of products (TAP)—a unique catalyst evaluation system with submillisecond time resolution, *Catal Rev Sci Eng* 30:49–116, 1988. http://dx.doi.org/10.1080/01614948808078616.

Gleaves JT, Yablonskii GS, Phanawadee P, Schuurman Y: TAP-2: an interrogative kinetics approach, *Appl Catal, A* 160:55–88, 1997. http://dx.doi.org/10.1016/S0926-860X(97)00124-5.

Glicksman R, Comm. Chair: *MIT's industrial partnerships*, 2003, Massachusett Institute of Technology. http://web.mit.edu/chancellor/IndlPartnershipsRpt.pdf.

Gupta P, Al-Dahhan MH, Dudukovic MP, Toseland BA: Comparison of single- and two-bubble class gas–liquid recirculation models—application to pilot-plant radioactive tracer studies during methanol synthesis, *Chem Eng Sci* 56:1117–1125, 2001a. http://dx.doi.org/10.1016/S0009-2509(00)00329-8.

Gupta P, Ong B, Al-Dahhan MH, Dudukovic MP, Toseland BA: Hydrodynamics of churn turbulent bubble columns: gas–liquid recirculation and mechanistic modeling, *Catal Today* 64:253–269, 2001b. http://dx.doi.org/10.1016/S0920-5861(00)00529-0.

Hallet N: Fighting to keep ideas, *Minn Dly*, 2014. http://www.mndaily.com/news/campus/2014/04/17/fighting-keep-ideas.

Hamley PA, Ilkenhans T, Webster JM, et al: Selective partial oxidation in supercritical water: the continuous generation of terephthalic acid from para-xylene in high yield, *Green Chem* 4:235–238, 2002. http://dx.doi.org/10.1039/b202087b.

Hanika J: Safe operation and control of trickle-bed reactor, *Chem Eng Sci* 54:4653–4659, 1999. http://dx.doi.org/10.1016/S0009-2509(98)00532-6.

Harmsen GJ: Reactive distillation: the front-runner of industrial process intensification, *Chem Eng Process Process Intensif* 46:774–780, 2007. http://dx.doi.org/10.1016/j.cep.2007.06.005.
Heller A: Not all research is equal, *Angew Chem Int Ed Engl* 53:2782–2783, 2014. http://dx.doi.org/10.1002/anie.201310269.
Hoekstra AJ, Derksen JJ, Van Den Akker HEA: An experimental and numerical study of turbulent swirling flow in gas cyclones, *Chem Eng Sci* 54:2055–2065, 1999. http://dx.doi.org/10.1016/S0009-2509(98)00373-X.
Hohenberg P: Inhomogeneous electron gas, *Phys Rev* 136:B864–B871, 1964. http://dx.doi.org/10.1103/PhysRev.136.B864.
Huesemann MH: The limits of technological solutions to sustainable development, *Clean Technol Environ Policy* 5:21–34, 2003.
Jaffe SB: Hot spot simulation in commercial hydrogenation processes, *Ind Eng Chem Process Des Dev* 15:410–416, 1976. http://dx.doi.org/10.1021/i260059a011.
Jahangiri H, Bennett J, Mahjoubi P, Wilson K, Gu S: A review of advanced catalyst development for Fischer–Tropsch synthesis of hydrocarbons from biomass derived syn-gas, *Catal Sci Technol* 4(8):2210–2229, 2014. http://dx.doi.org/10.1039/c4cy00327f.
Jensen KF: Microreaction engineering—is small better? *Chem Eng Sci* 56:293–303, 2001. http://dx.doi.org/10.1016/S0009-2509(00)00230-X.
Jiang Y, Khadilkar MR, Al-Dahhan MH, Dudukovic MP: Two-phase flow distribution in 2D trickle-bed reactors, *Chem Eng Sci* 54:2409–2419, 1999. http://dx.doi.org/10.1016/S0009-2509(98)00360-1.
Jin W, Gu X, Li S, Huang P, Xu N, Shi J: Experimental and simulation study on a catalyst packed tubular dense membrane reactor for partial oxidation of methane to syngas, *Chem Eng Sci* 55:2617–2625, 2000. http://dx.doi.org/10.1016/S0009-2509(99)00542-4.
Kapteijn F, Moulijn JA: 9.1 laboratory catalytic reactors: aspects of catalyst testing. In Ertl G, Knozinger H, Schuth F, Weitkamp J, editors: *Handbook of heterogeneous catalysis*, Weinheim, Germany, 2008, Wiley-VCH Verlag GmbH & Co. KGaA. Accessed June 30, 2014, http://onlinelibrary.wiley.com/doi/10.1002/9783527610044.hetcat0108/abstract.
Khadilkar MR, Mills PL, Dudukovic MP: Trickle-bed reactor models for systems with a volatile liquid phase, *Chem Eng Sci* 54:2421–2431, 1999. http://dx.doi.org/10.1016/S0009-2509(98)00503-X.
Khadilkar MR, Al-Dahhan MH, Dudukovic MP: Multicomponent flow-transport-reaction modeling of trickle bed reactors: application to unsteady state liquid flow modulation, *Ind Eng Chem Res* 44:6354–6370, 2005. http://dx.doi.org/10.1021/ie0402261.
Khane V, Al-Dahhan MH: Experimental and computational investigation of slow and dense granular flow in moving/pebble bed type reactors. In AIChE 2013 annual meeting, November 3–8, San Francisco, CA. 2013.
Khopkar AR, Rammohan AR, Ranade VV, Dudukovic MP: Gas-liquid flow generated by a Rushton turbine in stirred vessel: CARPT/CT measurements and CFD simulations, *Chem Eng Sci* 60:2215–2229, 2005. http://dx.doi.org/10.1016/j.ces.2004.11.044.
Kohn W, Sham LJ: Self-consistent equations including exchange and correlation effects, *Phys Rev* 140:A1133–A1138, 1965. http://dx.doi.org/10.1103/PhysRev.140.A1133.
Koppatz S, Pfeifer C, Rauch R, Hofbauer H, Marquard-Moellenstedt T, Specht M: H_2 rich product gas by steam gasification of biomass with in situ CO_2 absorption in a dual fluidized bed system of 8 MW fuel input, *Fuel Process Technol* 90:914–921, 2009. http://dx.doi.org/10.1016/j.fuproc.2009.03.016.
Korekazu Ueyama, K. (conference chair): *1st International symposium on multiscale multiphase process engineering (MMPE)*, Kanazwawa City, Japan, 2011. http://www.mmpe.jp.
Kramers HA, Westerterp KR: *Elements of chemical reactor design and operation,* Amsterdam, 1963, Netherlands University Press.

Krishna R, Sie ST: Strategies for multiphase reactor selection, *Chem Eng Sci* 49:4029–4065, 1994. http://dx.doi.org/10.1016/S0009-2509(05)80005-3.
Kuipers JAM, van Swaaij WPM: Application of computational fluid dynamics to chemical reaction engineering, *Rev Chem Eng* 13:1–118, 1997. http://dx.doi.org/10.1515/REVCE.1997.13.3.1.
Kulkarni MS, Dudukovic MP: Periodic operation of asymmetric bidirectional fixed-bed reactors: energy efficiency, *Chem Eng Sci* 52:1777–1788, 1997. http://dx.doi.org/10.1016/S0009-2509(97)00023-7.
Kumar SB, Moslemian D, Dudukovic MP: Gas-holdup measurements in bubble columns using computed tomography, *AIChE J* 43:1414–1425, 1997. http://dx.doi.org/10.1002/aic.690430605.
Lerou JJ, Ng KM: Chemical reaction engineering: a multiscale approach to a multiobjective task, *Chem Eng Sci* 51:1595–1614, 1996. http://dx.doi.org/10.1016/0009-2509(96)00022-X.
Levenspiel O: *Chemical reaction engineering,* ed 3, New York, NY, 1999, Wiley.
Li X, Liu Y, Li Z, Wang X: Continuous distillation experiment with rotating packed bed, *Chin J Chem Eng* 16:656–662, 2008. http://dx.doi.org/10.1016/S1004-9541(08)60137-8.
Li J, Ge W, Wang W, et al: *From multiscale modeling to meso-science: a chemical engineering perspective,* New York, NY, 2013, Springer.
Liotta C: *Phase transfer catalysis: principles and techniques,* New York, NY, 1978, Academic Press.
Losey MW, Schmidt MA, Jensen KF: Microfabricated multiphase packed-bed reactors: characterization of mass transfer and reactions, *Ind Eng Chem Res* 40:2555–2562, 2001. http://dx.doi.org/10.1021/ie000523f.
Marcano JGS, Tsotsis TT: *Catalytic membranes and membrane reactors,* Weinheim, Germany, 2002, Wiley-VCH.
Mason BP, Price KE, Steinbacher JL, Bogdan AR, McQuade DT: Greener approaches to organic synthesis using microreactor technology, *Chem Rev* 107:2300–2318, 2007. http://dx.doi.org/10.1021/cr050944c.
McManus RL, Funk GA, Harold MP, Ng KM: Experimental study of reaction in trickle-bed reactors with liquid maldistribution, *Ind Eng Chem Res* 32:570–574, 1993. http://dx.doi.org/10.1021/ie00015a021.
Mills PL, Dudukovic MP: Evaluation of liquid–solid contacting in trickle-bed reactors by tracer methods, *AIChE J* 27:893–904, 1981. http://dx.doi.org/10.1002/aic.690270604.
Mills PL, Duduković MP: A pioneer in multiphase reaction engineering, part 1, *Ind Eng Chem Res* 44:4841–4845, 2005a.
Mills PL, Duduković MP: A pioneer in multiphase reaction engineering, part 2, *Ind Eng Chem Res* 44:5869–5872, 2005b.
Mills PL, Nicole JF: A novel reactor for high-throughput screening of gas–solid catalyzed reactions, *Chem Eng Sci* 59:5345–5354, 2004. http://dx.doi.org/10.1016/j.ces.2004.07.109.
Mills PL, Nicole JF: Multiple automated reactor systems (MARS). 1. A novel reactor system for detailed testing of gas-phase heterogeneous oxidation catalysts, *Ind Eng Chem Res* 44:6435–6452, 2005a.
Mills PL, Nicole JF: Multiple automated reactor systems (MARS). 2. Effect of microreactor configurations on homogeneous gas-phase and wall-catalyzed reactions for 1,3-butadiene oxidation, *Ind Eng Chem Res* 44:6453–6465, 2005b.
Mills PL, Quiram DJ, Ryley JF: Microreactor technology and process miniaturization for catalytic reactions—a perspective on recent developments and emerging technologies, *Chem Eng Sci* 62:6992–7010, 2007. http://dx.doi.org/10.1016/j.ces.2007.09.021.

Michael Schlüter, M. (conference chair): 2nd *International symposium on multiscale multiphase process engineering (MMPE)*, Hamburg, Germany, 2014. http://processnet.org/en/MMPE14.html.
Mudde FR, Van Den Akker HEA: 2D and 3D simulations of an internal airlift loop reactor on the basis of a two-fluid model, *Chem Eng Sci* 56:6351–6358, 2001. http://dx.doi.org/10.1016/S0009-2509(01)00222-6.
Munjal S, Dudukovc MP, Ramachandran P: Mass-transfer in rotating packed beds-I. Development of gas–liquid and liquid–solid mass-transfer correlations, *Chem Eng Sci* 44:2245–2256, 1989.
Nayak SV: *Zeolites for cleaner processes: alkylation of isobutane and n-butene* (Doctoral dissertation), St. Louis, MO, 2009, Washington University.
Nigam KDP, Larachi F: Process intensification in trickle-bed reactors, *Chem Eng Sci* 60:5880–5894, 2005. http://dx.doi.org/10.1016/j.ces.2005.04.061.
Nørskov JK, Bligaard T, Rossmeisl J, Christensen CH: Towards the computational design of solid catalysts, *Nat Chem* 1:37–46, 2009. http://dx.doi.org/10.1038/nchem.121.
Patience GS, Bockrath RE: Butane oxidation process development in a circulating fluidized bed, *Appl Catal, A* 376:4–12, 2010. http://dx.doi.org/10.1016/j.apcata.2009.10.023.
Poncelet G, Martens J, Delmon B: *Preparation of catalysts VI: scientific bases for the preparation of heterogeneous catalysts*, Burlington, CA, 1995, Elsevier. Accessed June 25, 2014, http://public.eblib.com/EBLPublic/PublicView.do?ptiID=318228.
Ramaswamy RC, Ramachandran PA, Dudukovic MP: Modeling of solid acid catalyzed alkylation reactors, *Int J Chem React Eng* 3, 2005. http://dx.doi.org/10.2202/1542-6580.1255.
Ramaswamy RC, Ramachandran PA, Dudukovic MP: Recuperative coupling of exothermic and endothermic reactions, *Chem Eng Sci* 61:459–472, 2006. http://dx.doi.org/10.1016/j.ces.2005.07.019.
Rammohan AR, Kemoun A, Al-Dahhan MH, Dudukovic MP: A Lagrangian description of flows in stirred tanks via computer-automated radioactive particle tracking (CARPT), *Chem Eng Sci* 56:2629–2639, 2001a. http://dx.doi.org/10.1016/S0009-2509(00)00537-6.
Rammohan AR, Kemoun A, Al-Dahhan MH, Dudukovic MP: Characterization of single phase flows in stirred tanks via computer automated radioactive particle tracking (CARPT), *Chem Eng Res Des* 79:831–844, 2001b. http://dx.doi.org/10.1205/02638760152721343.
Ranade VV: *Computational flow modeling for chemical reactor engineering,* San Diego, CA, 2002, Academic Press.
Ranade VV, Chaudhari RV, Gunjal PR: *Trickle bed reactors: reactor engineering and applications*, Oxford, UK, 2011, Elsevier. Accessed June 25, 2014. http://app.knovel.com/web/toc.v/cid:kpTBRREA03.
Redekop EA, Yablonsky GS, Constales D, Ramachandran PA, Pherigo C, Gleaves JT: The Y-procedure methodology for the interpretation of transient kinetic data: analysis of irreversible adsorption, *Chem Eng Sci* 66:6441–6452, 2011. http://dx.doi.org/10.1016/j.ces.2011.08.055.
Roy S: *Quantification of two-phase flow in liquid–solid risers* (Doctoral dissertation), St. Louis, MO, 2000, Washington University.
Roy S, Dudukovic MP: Flow mapping and modeling of liquid–solid risers, *Ind Eng Chem Res* 40:5440–5454, 2001. http://dx.doi.org/10.1021/ie010181t.
Roy S, Dudukovic MP, Mills PL: A two-phase compartments model for the selective oxidation of n-butane in a circulating fluidized bed reactor, *Catal Today* 61:73–85, 2000. http://dx.doi.org/10.1016/S0920-5861(00)00352-7.
Roy S, Kemoun A, Al-Dahhan MH, Dudukovic MP, Skourlis TB, Dautzenberg FM: Countercurrent flow distribution in structured packing via computed tomography, *Chem Eng Process Process Intensif* 44:59–69, 2005a. http://dx.doi.org/10.1016/j.cep.2004.03.010.

Roy S, Kemoun A, Al-Dahhan MH, Dudukovic MP: Experimental investigation of the hydrodynamics in a liquid–solid riser, *AIChE J* 51:802–835, 2005b. http://dx.doi.org/10.1002/aic.10447.

Schouten JC: Chemical reaction engineering: history, recent developments, future scope. In Komiyama H, editor: Green chemical reaction engineering for a sustainable future. Proceedings of the 20th international symposium on chemical reaction engineering (ISCRE-20), Kyoto, 2008.

Seider WD, Seader JD, Lewin DR, Widagdo S: *Product and process design principles: synthesis, analysis, and evaluation,* ed 3, Hoboken, NJ, 2009, Wiley.

Shekhtman SO, Yablonsky GS, Chen S, Gleaves JT: Thin-zone TAP-reactor—theory and application, *Chem Eng Sci* 54:4371–4378, 1999. http://dx.doi.org/10.1016/S0009-2509(98)00534-X.

Shin SB, Han SP, Lee WJ, et al: Optimize terephthaldehyde operations, *Hydrocarbon Processing* 86:83–90, 2007.

Sie ST, Krishna R: Process development and scale up: III. Scale-up and scale-down of trickle bed processes, *Rev Chem Eng* 14:203, 1998. http://dx.doi.org/10.1515/REVCE.1998.14.3.203.

Silveston P, Hanika J: Challenges for the periodic operation of trickle-bed catalytic reactors, *Chem Eng Sci* 57:3373–3385, 2002. http://dx.doi.org/10.1016/S0009-2509(02)00206-3.

Smith LA Jr: Catalytic distillation process, US4232177 A, 1980.

Smith LA Jr: Catalytic distillation structure, US4443559 A, 1984.

Stitt H, Marigo M, Dixon A: Just because the results are in colour, it does not mean they are right. In *9th European congress of chemical engineering*, The Hague, The Netherlands, 2013, http://www.ecce2013.eu/keynotes.php.

Subramaniam B, McHugh MA: Reactions in supercritical fluids—a review, *Ind Eng Chem Process Des Dev* 25:1–12, 1986. http://dx.doi.org/10.1021/i200032a001.

Taylor R, Krishna R: Modelling reactive distillation, *Chem Eng Sci* 55:5183–5229, 2000. http://dx.doi.org/10.1016/S0009-2509(00)00120-2.

Topsoe catalysis forum: modeling and simulation of heterogeneous catalytic processes, Munkerupgaard, Denmark, 2013. http://www.topsoe.com/research/~/media/PDF%20files/Topsoe_Catalysis_Forum/2013/Topsøe_catalysis_forum_%202013_link3.ashx.

Topsoe catalysis forum, Munkerupgaard, Denmark, Last accessed August 12, 2014, http://www.topsoe.com/research/Topsoe_Catalysis_Forum.aspx.

Tunca C, Ramachandran PA, Dudukovic MP: Role of chemical reaction engineering in sustainable process development. In Abraham MA, editor: *Sustainability science and engineering: defining principles*, San Diego, CA, 2006, Elsevier.

Venneker BCH, Derksen JJ, Van den Akker HEA: Population balance modeling of aerated stirred vessels based on CFD, *AIChE J* 48:673–685, 2002. http://dx.doi.org/10.1002/aic.690480404.

Wang D, Barteau MA: Kinetics of butane oxidation by a vanadyl pyrophosphate catalyst, *J Catal* 197:17–25, 2001. http://dx.doi.org/10.1006/jcat.2000.3061.

Yablonsky GS, Constales D, Shekhtman SO, Gleaves JT: The Y-procedure: how to extract the chemical transformation rate from reaction-diffusion data with no assumption on the kinetic model, *Chem Eng Sci* 62:6754–6767, 2007. http://dx.doi.org/10.1016/j.ces.2007.04.050.

Zhang S, Nguyen L, Zhu Y, Zhan S, (Frank) Tsung CK, (Feng) Tao F: In-situ studies of nanocatalysis, *Acc Chem Res* 46:1731–1739, 2013.

Zheng ZQ, Zhou XP: High speed screening technologies in heterogeneous catalysis, *Comb Chem High Throughput Screen* 14:147–159, 2011. http://dx.doi.org/10.2174/138620711794728725.

CHAPTER TWO

Spatial Resolution of Species and Temperature Profiles in Catalytic Reactors: *In Situ* Sampling Techniques and CFD Modeling

Claudia Diehm*, Hüsyein Karadeniz†, Canan Karakaya†, Matthias Hettel†, Olaf Deutschmann*,†

*Institute of Catalysis Research and Technology, Karlsruhe Institute of Technology (KIT), Karlsruhe, Germany
†Institute for Chemical Technology and Polymer Chemistry, Karlsruhe Institute of Technology (KIT), Karlsruhe, Germany

Contents

Advances in Chemical Engineering, Volume 45
ISSN 0065-2377
http://dx.doi.org/10.1016/B978-0-12-800422-7.00002-9

Abstract

Spatial resolution of species and temperature profiles can provide valuable information for understanding, design, and optimization of catalytic reactors. The combination of experimental investigation and CFD modeling does not only improve our knowledge but also helps to discover uncertainties and limitations of novel scientific techniques for an adequate interpretation of the observations. Two lab-scale reactor configurations with *in situ* capillary techniques are investigated experimentally and numerically for the resolution of spatial species and temperature profiles: the stagnation flow on a catalytically coated disc and the flow through a catalytically coated honeycomb monolith, in which CO is totally and CH_4 is partially oxidized, respectively, over Rh/Al_2O_3 catalysts. CFD simulations reveal two significant items for the interpretation of the measured profiles: internal mass transport inside the catalyst in the stagnation flow reactor and the impact of the capillary probe in the honeycomb monolith.

1. INTRODUCTION

The understanding and optimizing of heterogeneous catalytic reactors requires a detailed knowledge of the heterogeneous surface reactions and the interaction of the active surface with the surrounding reactive flow. Consequently, the heterogeneously catalyzed gas-phase reactions must be analyzed together with potential homogeneous gas-phase reactions, mass transport in the gas-phase, as well as heat transport in both the gas-phase and solid structures. In addition, inlet and boundary conditions, sometimes changing in time, need to be taken into account. Figure 2.1 illustrates the complexity

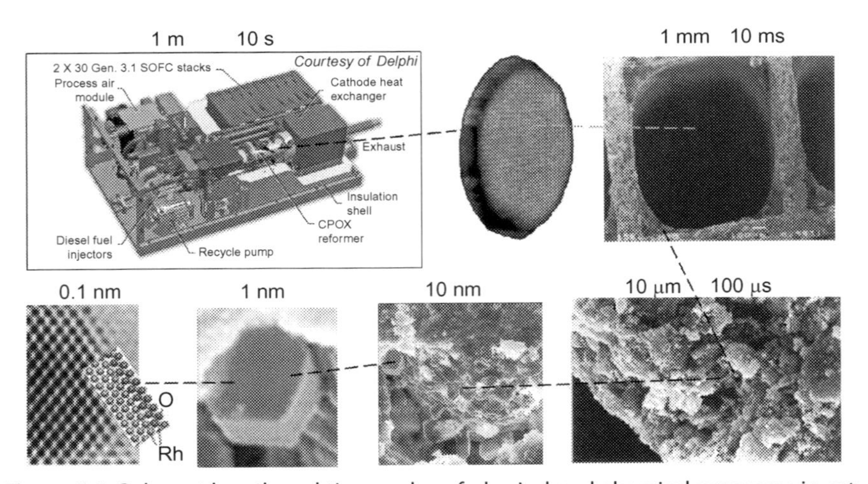

Figure 2.1 Relevant length and time scales of physical and chemical processes in catalytic systems exemplarily shown for the reformer in an auxiliary power unit (APU).

of physical and chemical processes occurring in the catalytic partial oxidation reactor that serves as diesel reformer to produce the hydrogen-rich fuel for an SOFC stack in an auxiliary power unit for on-board production of electricity in heavy duty vehicles (Deutschmann, 2011a; Kaltschmitt and Deutschmann, 2012). In order to design the catalytic reactor and optimize the operating conditions based on a molecular understanding, many different time scales have to be considered.

Multiscale modeling is required from the viewpoint of modeling and numerical simulation. With the increasing power of computers over the last decades and the development of numerical algorithms to solve highly coupled, nonlinear, and stiff equations, modeling, and numerical simulation also have developed into valuable tools in heterogeneous catalysis. These tools have been applied to study the molecular processes in great detail by quantum mechanical computations, density functional theory (DFT), molecular dynamics, and Monte Carlo simulations, but often neglecting the engineering aspects of catalytic reactors such as the interaction of chemistry and mass and heat transfer, on the one hand. On the other hand, mixing, flow structures, and heat transport in technical reactors and processes have been analyzed by computational fluid dynamics (CFD) in great detail, neglecting, however, the details of the microkinetics (Deutschmann, 2011b). In the last years, quite frequently, research proposal and programs in catalysis claim to work on bridging those gaps between surface science and industrial catalysis and indeed some progress has been made. Surface science studies, in experiment, theory, and simulation, include technically relevant conditions more and more. Reaction engineering of technical processes now often tries to understand the underlying molecular processes and even includes quantum-mechanical simulations in the search and development of new catalysts and catalytic reactors. However, convergence here is a slow process. One major reason is the gap between the high complexity of catalysts used in practice and the many approximations still to be made in molecular simulations. Furthermore, using kinetic data derived from numerical simulations in scale-up of technical systems might indeed become risky if the engineer does not take into account the simplifying assumptions and the computational uncertainties of the numerical simulations. Actually, this warning holds for all simulations relevant for heterogeneous catalysis, from DFT to CFD.

Figure 2.2 shows the dream of multiscale modeling of catalytic reactors; here the information would be passed from the molecular level to the technical system through many steps. Starting with DFT, theoretical calculations

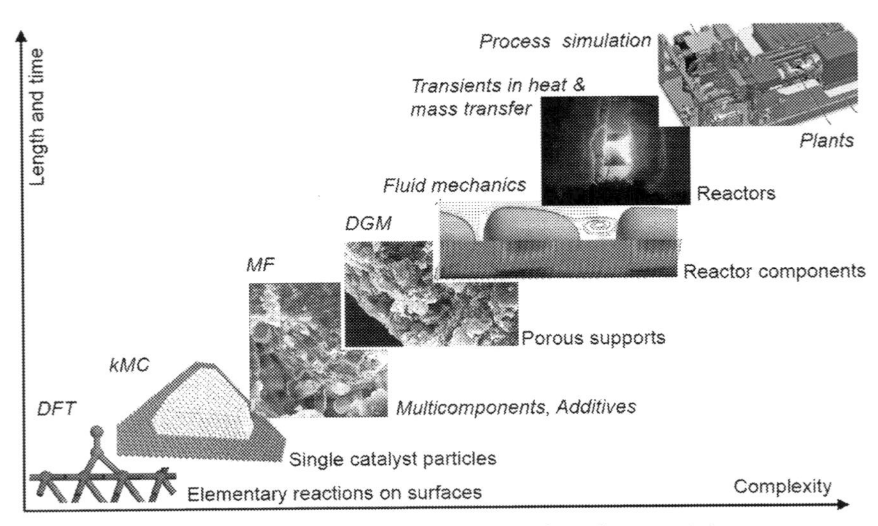

Figure 2.2 Simulation of technical catalytic reactors from first principles.

of surface reactions can lead to the elucidation of elementary reactions and their kinetics. This information could then be passed to kinetic Monte Carlo (kMC) simulations if the interest is rather in the probability whether or not a certain process (adsorption, diffusion, reaction) takes place and how the elementary reaction is coupled with diffusion of adsorbates and adsorbate–adsorbate interaction. Using information derived from DFT, first-principles kMC simulations lead directly to the surface reaction rate as a function of ambient conditions and the state of the catalyst. They combine elementary kinetics with statistics to properly evaluate the kinetics of the catalytic particle. The diverse morphology of catalytic particles in technical systems calls for a simpler approach because MC simulations are too expensive to be directly included in simulation of catalytic reactors including mass and heat transfer. Therefore, MC simulations may be used to derive rate equations to estimate not only surface reaction rates but also homogeneous gas-phase reaction rates that also play a role in many catalytic processes. For modeling surface reactions in CFD simulations, the model of choice is the so-called mean-field approximation, which however cannot take directly into account the details of the molecular process such as diffusion of adsorbates and the dependence of the reaction rate on the crystal phases and defects. We now move from the microscopic to the mesoscopic length scale. The interaction of surface reactions and molecular transport of the reactants and products in the ambience of the catalytic particle needs to be considered covering modeling of processes in catalytic porous media and the interaction of diffusion and reactions in those pore structures. Moving to the macroscopic scale, the molecular-based models, usually expressed by rate equations, have to be coupled with the reactive flow and the surrounding heat transport processes. The comparison of spatially and temporally resolved species profiles in catalytic reactors obtained by sophisticated experimental techniques with the profiles computed by coupling chemistry and mass and heat transport models can be used not only for the evaluation of kinetic schemes developed but also for optimization of reactor design and operating conditions.

1.1. Computational fluid dynamics

CFD numerically simulates the behavior of chemically reactive gas–solid flows with the integration of macro- and microkinetic reaction mechanisms. Macrokinetic reaction mechanisms are usually derived based on a limited range of experimental data and the global rate laws are adjusted to yield the best agreement for the selected conditions. In contrast to them, microkinetic models, which are based on elementary-step reaction mechanisms, give the possibility to investigate the interactions between the reacting species on a molecular level over a wide range of temperature and pressure conditions. Therefore, they are increasingly used in CFD simulations. However, solution of CFD with detailed chemistry is still a challenging task due to large number of species mass conservation equations, their highly nonlinear coupling, and the wide range of time scales introduced by the complex reaction networks (Deutschmann, 2008).

Therefore, mathematical modeling and experimental measurements are needed to go hand in hand for fundamental understanding of catalytic reactors. Since it is challenging to conduct experiments directly in the often porous and tortuous structures of practical reactors, simple configurations for lab-scale reactors are needed. In this respect, stagnation flow reactors (SFRs) and channel reactors offer simple configuration to investigate heterogeneously catalyzed gas-phase reactions under practical operating conditions. The fluid-mechanical properties of stagnation flow and channel reactors enable measuring and modeling the gas-phase boundary layer adjacent to the catalytic surface being zero-dimensional and one-dimensional, respectively. Kinetic measurements along with the coupled model of heterogeneous chemistry with reacting flow facilitate the development of reaction mechanisms for different chemical problems such as heterogeneous catalysis (Deutschmann et al., 1994a, 1996; Ikeda et al., 1995; Karadeniz et al., 2013; Karakaya and Deutschmann, 2013; Mhadeshwar et al., 2002; Vlachos, 1996; Warnatz, 1992) and chemical vapor deposition (CVD) (Ruf et al., 1996, 2000). Physical and chemical steps of heterogeneously catalyzed chemical processes, such as external and internal mass transfer limitations as well as reaction sequences on the catalyst, can be investigated at a fundamental level with the integration of the developed reaction mechanisms into the appropriate numerical models. The fundamental information that is obtained through *in situ* measurements and numerical simulations can be used further for the development and optimization of practical reactors.

1.2. Invasive *in situ* techniques

The experimental analysis of spatial profiles of species concentrations in catalytic reactors has been conducted for many years. Eigenberger et al. for instance used isothermal flat bed reactors to determine species profiles at several axial positions of the honeycomb structured catalysts by sucking a small amount of the gas stream into an analytical device (Brinkmeier et al., 2005; Hauff et al., 2013; Tuttlies et al., 2004). Lyubovsky et al. (2005) presented one of the first studies on CPOX including measurements of concentration and temperature profiles. This technique was based on stacking of microlith screens, in between which 1/16-in. tubing for gas sampling and thermocouples was placed. With this setup, a resolution of 2 mm was obtained.

In general, one can distinguish between invasive and noninvasive techniques to resolve species and temperature concentrations. For the invasive techniques used extensively in the last years, the catalytic system is investigated by inserting a movable probe for concentration and/or temperature measurements along the catalyst length. Horn et al. (2006a,b) have developed an *in situ* sampling technique for foam catalysts. For this technique, a capillary is inserted into a channel, which is drilled through the center of a catalytic foam in axial direction. The capillary is sealed at the upper end and a small orifice at the side of the capillary is used to sample the gases (Horn et al., 2007a) so that the channel in the foam is not left open when the capillary is pulled out. Inside the capillary, a thermocouple is placed to detect temperature profiles simultaneously with the concentration profiles. Many studies have been performed in this setup for partial oxidation over Pt and Rh with mainly methane (Bitsch-Larsen et al., 2008; Donazzi et al., 2008; Horn et al., 2006a, 2007a,b) and also with ethane (Michael et al., 2010). The sampling setup was further improved by Horn et al. (2010), to allow for operation under pressures up to 45 bar and temperatures up to 1573 K (Korup et al., 2011).

Donazzi et al. (2011a,b) have adapted the *in situ* sampling setup for honeycomb catalysts. A very thin capillary (diameter = 340 μm) is used inside the catalytic channel (height ~1 mm). For the temperature profile, two different temperature probes were inserted separately into the catalytic channel. For the gas-phase temperature, a thermocouple was used. The surface temperature was detected by a side-facing optical fiber connected to a pyrometer (Donazzi et al., 2011a). The CPOX of propane was studied over a Rh/Al_2O_3 monolith, which led to the discovery of the combined homogeneous–heterogeneous reaction network for propane (Donazzi et al., 2011b). For methane as well as propane, the *in situ* investigations led to guidelines for the hot spot reduction at the catalyst inlet (Beretta et al., 2011; Livio et al., 2012). The *in situ* sampling technique was modified to allow for operation up to 4 bar (Donazzi et al., 2014).

In parallel, the group of Partridge at Oak Ridge National Laboratory developed an *in situ* capillary sampling technique to study species profiles in automotive catalytic converters. Their technique has meanwhile become commercially available. In this group, also the influence of the probe on the measured concentration profiles was studied for the case of catalytic CO oxidation (Sa et al., 2010). They found a negligible influence of the probe with the capillary in the corner of a channel and an increase of velocity for the channel with probe; the volumetric flux was kept constant for all channels in the simulation.

Similar *in situ* sampling techniques have also recently been used for SFRs to analyze the concentration profiles as a function of the distance from the catalytic surface (Karakaya and Deutschmann, 2013; McGuire et al., 2011), i.e., to resolve the boundary layer over the catalytic plate as discussed below.

1.3. NonInvasive *in situ* techniques

Noninvasive *in situ* analysis of the state of the catalysts became rather popular in the last decade, in particular using X-ray absorption spectroscopy methods (Grunwaldt and Baiker, 2005; Grunwaldt et al., 2006; Hannemann et al., 2007). For instance, Grunwaldt et al. studied CPOX of methane over Rh and Pt–Rh using *in situ* X-ray absorption spectroscopy in an *in situ* spectroscopic cell (Hannemann et al., 2007). The measurements were performed at a synchrotron. However, spatial analysis of the catalyst is beyond the scope of this chapter that focusses on the analysis of the gas phase in a catalytic structure.

In noninvasive techniques for *in situ* investigations of the gas phase, optically accessible reactors offer the possibility to analyze spatial profiles by laser spectroscopy methods (Mantzaras, 2013). The Mantzaras group have developed an optically accessible channel-flow reactor, in which 2D laser-induced fluorescence (LIF) and 1D Raman spectroscopy are employed to measure concentrations of trace species and stable gas phase species, respectively (Mantzaras, 2013; Reinke et al., 2002). The catalytic channel consists of two horizontal plates, which are coated with the catalyst and placed to yield a free height of 7 mm in between. Devices for heating and cooling of the plates are present. Temperature profiles are obtainable from thermocouples which are placed along the channel length beneath the catalytic surface. Additionally, this reactor was also applicable for elevated pressures (Mantzaras, 2013). Employing this reactor, Schneider et al. (2006, 2007) investigated CPOX of methane over Rh/ZrO_2 from 4 to 10 bar with H_2O and CO_2 dilution.

A similar setup for 2D LIF spectroscopy was recently completed in our group at KIT (Zellner et al., 2014), which was used to study the catalytic reduction of NO by hydrogen toward ammonia over a diesel oxidation catalyst consisting of a Pt/Al_2O_3-coated monolith. The combination with *ex situ* analytics and the consideration of quenching effects on the measured signal lead to quantitative NO concentration profiles in the catalytic channel only 2 mm in width. The interaction of diffusion and flow with the surface reaction could be elucidated for this system under different operating conditions (Zellner et al., 2014).

In noninvasive techniques, the catalytic system is not disturbed by a probe as it is the case for invasive techniques. However, the noninvasive methods are very demanding due to their requirement of laser setups or beam time at a synchrotrons. Furthermore, the optical accessibility restricts the operation conditions, which can be investigated. And, in the single catalytic channel setups, the heat management differs, which in particular matters in high-temperature catalysis.

1.4. Objective of this chapter

The understanding of coupling of heterogeneously catalyzed gas-phase reactions with the surrounding flow field is the topic of this chapter. With the experimental capillary techniques discussed, species and temperature profiles are resolved in the submillimeter range, i.e., on the mesoscopic to macroscopic length scales, which is similar of the spatial resolutions usually obtained by CFD simulations. In the CFD modeling discussed, molecular-based multistep reaction schemes are used and internal mass transfer limitations in the porous structure are considered as well. The combination of the *in situ* measurement technique and CFD simulation is discussed for two reactor configurations: a stagnation flow on a catalytically coated plate and the flow through honeycomb structured, catalytically coated monoliths. From a chemical point of view, CO oxidation and partial oxidation of methane over Rh-based catalysts are considered. Most of the results presented originate from studies in recent and running PhD theses in our group.

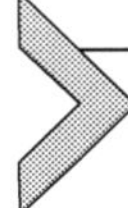

2. FUNDAMENTALS

2.1. Reaction kinetics

A chemical reaction is a series of molecular level processes that lead to transformations of one species to another by rearrangement or exchange of atoms. The general reaction equation can be expressed as

$$\sum_i \nu'_{ik} A_i \rightarrow \sum_i \nu''_{ik} A_i \tag{2.1}$$

with the stoichiometric coefficients of ν'_{ik} and ν''_{ik} for species of reactant or product A_i. For the species i, the rate equation can be expressed as

$$\frac{\mathrm{d}c_i}{\mathrm{d}t} = \left(\nu''_{ik} - \nu'_{ik}\right) k_f \prod_i c_i^{a'_{ik}} \tag{2.2}$$

In this equation, c_i is the concentration of species i, k_f is the reaction rate coefficient, and a'_{ik} is the reaction order with respect to species i.

For elementary-step reactions, a'_{ik} is equal to stoichiometric coefficient of species i (ν'_{ik}), whereas for global rate equations, the order of the reaction may vary. For the forward reaction rate, the rate coefficient k can be expressed by the Arrhenius model:

$$k_{f_k} = A_k T^{\beta_k} \exp\left[\frac{-E_{a_k}}{RT}\right] \tag{2.3}$$

Here, A_k is the preexponential factor, β_k is the temperature exponent, and E_{a_k} is the activation energy.

For surface reactions, pressure has no direct effect on the rate coefficients, whereas for gas-phase reactions, an additional pressure dependency of the rate coefficient is necessary for dissociation and recombination reactions.

Reaction steps occurring on a gas–solid interface at the microscopic level basically involve adsorption, surface reactions, desorption, and diffusion steps. These processes are briefly summarized in Fig. 2.3. The processes are exemplarily shown on a catalyst with pellet structure.

The combination of reaction kinetics and reactor design has been studied as a major subject of catalytic reaction engineering since the 1950s. Early studies used global rate expressions to determine the reaction rate. Purely empirical algebraic expressions were used to express the chemical reaction rate. If a reaction occurs on a molecular level in exactly the way it is described by the reaction equation, it is called an *elementary reaction* (microkinetic model). Otherwise, it is a *global reaction, overall reaction, or net reaction* (macrokinetic) (Deutschmann, 2008).

Macrokinetic models are still widely used for reactor design because they can be derived rather time- and cost-saving with an optimized set of experimental measurements (Deutschmann, 2011b). Even though the models are only valid in the limited range of conditions in which the kinetic data are derived for, they often fit their purpose in particular if the rate expressions reflect the molecular processes in some way and the rate-determining step is adequately taken into account. On the other hand, the microkinetic approach attempts to describe reactions using their most fundamental set

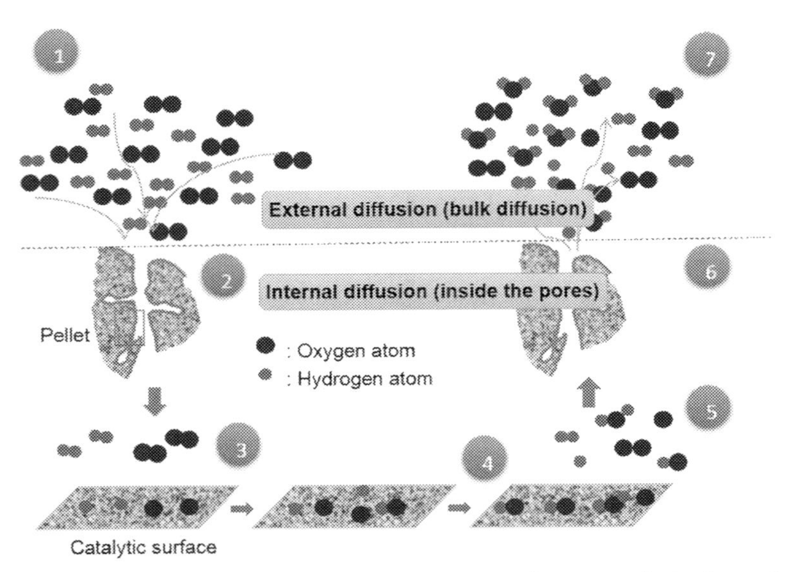

Figure 2.3 Sequence of processes near the catalytic surface exemplarily shown for H_2 oxidation.

of elementary reaction steps. By this, the model can be applied to predict reaction kinetics accurately over a wide range of temperature and pressure. As a result, one model can be used to validate various sets of reaction conditions in a quick and cost-efficient manner, thus allowing the optimal processing conditions to be determined (Dooling et al., 1999). The model can be used to estimate the reactor behavior for varying external conditions such as high pressures and high temperatures which are difficult to realize in laboratory conditions. One drawback of this approach is that large numbers of kinetic parameters are required for a complex reaction mechanism. A detailed description of the different approaches to describe surface reaction kinetics with their individual potentials and limitations can be found elsewhere (Deutschmann, 2011b).

2.2. Surface reactions and mean-field approximation

Although the main goal of microkinetic modeling of heterogeneous catalysts is to describe the system on its most fundamental level, simplifications are necessary because of the complexity of a catalytic reaction which varies in several orders of length and time scales (Deutschmann, 2011b; Dooling et al., 1999).

The mean-field approximation is a popular method for modeling technical chemical reactors. According to the mean-field approximation, the catalytically active surface is associated with a surface site density, Γ, that describes the maximum number of species that can adsorb on a unit surface area of the catalyst, e.g., given in mol/m^2. It should be noted that the reference area here is the active catalytic surface area, reflecting the number of adsorption sites, and not the total area of the porous catalyst structure. The former can be measured by hydrogen and CO chemisorption, at least in the in case of metallic catalysts, and the latter by BET measurements. In the mean-field approximation, adsorbed species are randomly distributed on the surface and lateral interactions can only be taken into account by additional coverage dependencies of the rate expression (Deutschmann, 2008). Hence, on a macroscopic level, the state of the surface is described by the mean surface coverages of the adsorbed species and the temperature. However, this coverage depends on the local position in the reactor. The resolution of the coverage profile is given by the resolution of the computational grid, i.e., the spatial resolution is usually in the micro- to millimeter scale. Since the catalyst surface may change with reaction conditions (Grunwaldt et al., 2000), different catalytic surfaces have to be defined as well as rate expressions for the transition between those surfaces.

The molar net production rate, $\dot{s}_i$, of a gas phase as well as an adsorbed species are defined as

$$\dot{s}_i = \sum_{k=1}^{K_s} \nu_{ik} k_{fk} \prod_{i=1}^{N_g + N_s + N_b} c_i^{\nu'_{ik}} \left(i = 1, \ldots, N_g + N_s + N_b\right) \quad (2.4)$$

The mechanism consists of K_s surface reactions among N_g gas phase, N_s surface, and N_b bulk species. The net production rate $\dot{s}_i$ of a surface species is linked to its coverage θ_i.

$$\dot{s}_i = \frac{\Gamma}{\sigma_i} \frac{d\theta_i}{dt} \left(i = 1, \ldots, N_g + N_s + N_b\right) \quad (2.5)$$

In this equation, σ_i is the coordination number, i.e., the number of adsorption sites covered by each species of i. Additional parameters μ_{i_k} and ε_{i_k} are introduced to account for the coverage dependence of the rate expression:

$$k_{f_k} = A_k T^{\beta_k} \exp\left[\frac{-E_{a_k}}{RT}\right] \prod_{i=1}^{N_s} \theta_i^{\mu_{i_k}} \exp\left[\frac{\varepsilon_{i_k}\theta_i}{RT}\right] \quad (2.6)$$

For adsorption reactions, sticking coefficients are commonly used. The relation between the rate coefficient and the sticking coefficient is shown in Eq. (2.7). S_i^0 denotes the initial sticking coefficient for an uncovered surface:

$$k_{f_k}^{\mathrm{ads}} = \frac{S_i^0}{\Gamma^\tau} \sqrt{\frac{RT}{2\pi M_i}} \text{ and } \tau = \sum_{i=1}^{N_s} \nu'_{ik} \quad (2.7)$$

One of the major issues in developing detailed surface reaction mechanisms is thermodynamic consistency. Even though the recently published reaction mechanisms ensure enthalpic consistency, many of them are not consistent with respect to entropy, which is due to the lack of knowledge about the transition states of the individual reaction steps. Thus, there is not sufficient information for a theory-based determination of pre-exponential factors in the rate equations. However, an independent choice of the rate coefficients causes an inconsistent entropy change in the overall reaction, which leads to an incorrect prediction of equilibrium states. There are two approaches to avoid this inconsistency by adjusting the rate expressions, which are described in the literature (Deutschmann, 2008; Maier et al., 2011; Mhadeshwar et al., 2003).

2.3. Coupling of surface reaction rate and internal mass transfer

Instantaneous diffusion (∞*-approach*) model assumes that the catalyst is virtually distributed at the gas/washcoat interface so that there is infinitely fast mass transport within the washcoat. This model eliminates the washcoat parameters, such as its thickness and porosity, and the diameters of the inner pores. Therefore, ∞*-approach* does not account for internal mass transport limitations that are due to a porous layer. It means that mass fractions of gas-phase species on the surface are obtained by the balance of production or depletion rate with diffusive and convective processes (Deutschmann, 2008; Kee et al., 2001; Warnatz, 1992). Thus, the net production rate of each chemical species due to surface reactions can be balanced with the diffusive flux of that species at the gas-surface boundary, assuming that no deposition or ablation of chemical species occurs on/from the catalyst surface:

$$\vec{j}_i^{\,\mathrm{W}} = F_{\mathrm{cat/geo}} \dot{s}_i W_{i} \vec{n}_i^{\,\mathrm{W}} \tag{2.8}$$

The term $F_{\mathrm{cat/geo}}$ is introduced as a scaling factor; it is a ratio of the active catalytic surface area A_{catalyst} and the geometric surface area $A_{\mathrm{geometric}}$.

$$F_{\mathrm{cat/geo}} = \frac{A_{\mathrm{catalyst}}}{A_{\mathrm{geometric}}} \tag{2.9}$$

Effectiveness factor approach (η*-approach*) accounts for diffusion limitations in the washcoat. η*-approach* is based on the assumption that one target species determines overall reactivity (Deutschmann, 2008). An effectiveness factor for a first-order reaction is calculated for the chosen species based on the dimensionless Thiele modulus (Φ) (Hayes et al., 2007, 2012), and all reaction rates are multiplied by this factor at the species governing equation at the gas–surface interface. Φ is calculated as

$$\Phi = L\sqrt{\frac{\dot{s}_i \gamma}{D_{i,\,\mathrm{eff}}\, c_{i,0}}} \tag{2.10}$$

in which $c_{i,0}$ is the concentration of species i at the gas–washcoat interface, and γ stands for the active catalytic surface area per washcoat volume as

$$\gamma = \frac{F_{\mathrm{cat/geo}}}{L} \tag{2.11}$$

in which L is thickness of the washcoat and $F_{\mathrm{cat/geo}}$ is the ratio of the total catalytically active surface area to the geometric surface area, and the effective diffusion coefficient is $D_{i,\mathrm{eff}}$. The term in the square root in Eq. (2.10) indicates the ratio of intrinsic reaction rate to diffusive mass transport in the washcoat. When Thiele modulus is large, internal mass transfer limits the overall reaction rate; when Φ is small, the intrinsic surface reaction kinetics is usually rate limiting (Fogler, 2006).

Consequently, the effectiveness factor (η) is defined as the ratio of the effective surface reaction rate inside the washcoat to the surface reaction rate without considering the diffusion limitation (Fogler, 2006):

$$\eta = \frac{\dot{s}_{i,\,\mathrm{eff}}}{\dot{s}_i} = \frac{\tanh(\Phi)}{\Phi} \tag{2.12}$$

The zero-dimensional *η-approach* offers a simple and computationally inexpensive solution. However, it might lose the validity in conditions where more than one species' reaction rate and diffusion coefficient determine the overall reactivity (Deutschmann, 2011b).

One-dimensional reaction-diffusion equations (*RD-approach*) offer a more adequate model than the *η-approach* to account for mass transport in the washcoat. The model calculates spatial variations of concentrations and surface reaction rates inside the washcoat. It assumes that the species flux inside the pores is only due to diffusion (Stutz and Poulikakos, 2008). Therefore, it neglects the convective fluid flow inside the porous layer, because of very low permeability assumption (Stutz and Poulikakos, 2008). Eventually, each gas-phase species leads to one reaction-diffusion equation in the *RD-approach*, which is written in the transient form, as (Deutschmann, 2011b; Deutschmann et al., 2014; Karadeniz et al., 2013)

$$\frac{\partial c_{i,\,\mathrm{w}}}{\partial t} = -\frac{\partial \vec{J}_i^{\,\mathrm{w}}}{\partial z} + \gamma \dot{s}_{i,\,\mathrm{w}} \tag{2.13}$$

$$\vec{J}_i^{\,\mathrm{w}} = -D_{i,\mathrm{eff}} \frac{\partial c_{i,\,\mathrm{w}}}{\partial z} \tag{2.14}$$

in which $c_{i,\mathrm{w}}$ is the molar concentration, $\vec{J}_i^{\,\mathrm{w}}$ is the molar diffusion flux, and $\dot{s}_{i,\mathrm{w}}$ is the surface reaction rate of the ith species in the washcoat, all depending on the spatial coordinate z.

2.4. Coupling of surface reactions with external mass and heat transport

The reader is referred to the wide variety of textbooks covering the area of reactive flow simulations in catalytic reactors. Concerning the coupling of mass and heat transport with catalytic surface reactions, the following literature (among many others) is recommended (Deutschmann, 2008, 2011b; Dixon et al., 2008; Kee et al., 2001, 2003; Mantzaras, 2013; Taskin et al., 2008). However, in the remainder of this chapter, the governing equations of the stagnation flow model are explicitly discussed, while for the model used for the CFD simulation of the catalytic monolith, only the special extensions are discussed; otherwise, it is referred to the user manual of the CFD code applied and standard textbooks.

3. STAGNATION FLOW ON A CATALYTIC PLATE

The study of reaction kinetics in flow reactors to derive microkinetic expressions also relies on an adequate description of the flow field and well-defined inlet and boundary conditions. The stagnation flow on a catalytic plate represents such a simple flow system, in which the catalytic surface is zero dimensional and the species and temperature profiles of the established boundary layer depend only on the distance from the catalytic plate. This configuration consequently allows the application of simple measurement and modeling approaches (Sidwell et al., 2002; Warnatz et al., 1994a). SFRs are also of significant technical importance because they have extensively been used for CVD to produce homogeneous deposits. In this deposition technique, the disk is often additionally forced to spin to achieve a thick and uniform deposition across the substrate (Houtman et al., 1986a; Oh et al., 1991).

In research, SFRs have been widely used in combustion research to study the effects of fluid-mechanical strain on flame behavior (Kee et al., 2005a). A number of groups have also studied catalytic chemistry in a stagnation-flow configuration to incorporate detailed chemistry into catalytic combustion simulations (Deutschmann et al., 1994b, 1996; Dupont et al., 2001; Ikeda et al., 1995; Ljungström et al., 1989; McDaniel et al., 2002; McGuire et al., 2009, 2011; Rice et al., 2007; Sidwell et al., 2002; Song et al., 1991; Taylor et al., 2003; Warnatz et al., 1994a). Ljungström et al. have studied the H_2O formation kinetics in H_2/O_2 mixtures on a Pt foil by introducing a laser beam into a stagnation foil surface (Ljungström et al., 1989). Song et al. have studied the ignition criteria of H_2 and propane on stagnation-flow geometry and showed that the SFR geometry provides accurate data on the bifurcation analysis where heterogeneous and homogeneous reactions take place (Song et al., 1991). Many studies have been performed in order to model the ignition of the oxidation of H_2, CH_4, and CO on Pt surfaces (Deutschmann et al., 1996; Fernandes et al., 1999; Försth et al., 1999; Ikeda et al., 1995; Warnatz et al., 1994a). Furthermore, the SFR configuration has been used to investigate the detailed surface reaction kinetics for complex reactions such as methane partial oxidation and steam reforming on noble metals (Dupont et al., 2001; McGuire et al., 2009, 2011; Sidwell et al., 2002, 2003). Recently, McGuire et al. studied dry reforming of methane over Rh-supported strontium-substituted hexaaluminate catalysts in a stagnation flow (McGuire et al., 2011). Their experimental setup is shown in Fig. 2.4. In collaboration with the group of R.J. Kee at Colorado School of Mines, Karakaya and Deutschmann (2013) and Karakaya (2013) set up a very similar system in our group, which is described in detail below.

3.1. Experimental setup

The experimental setup basically consists of the SFR, a gas feeding system and the analytics (Karakaya, 2013; Karakaya and Deutschmann, 2013). The system has been designed to work for a wide range of gases and

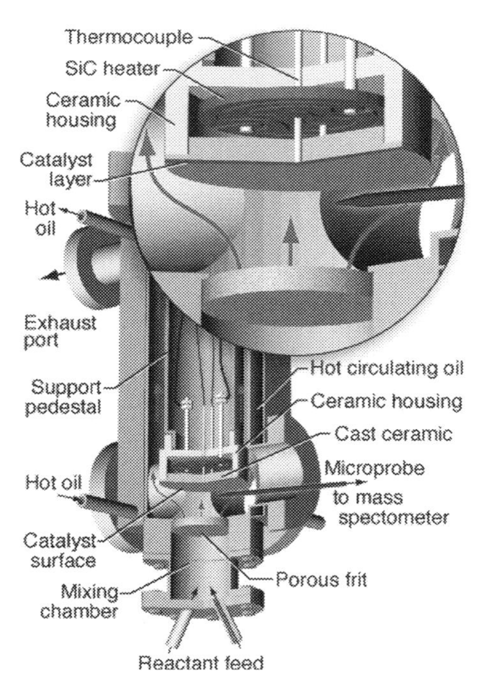

Figure 2.4 Schematic illustration of the stagnation-flow reactor. *Figure taken from McGuire et al. (2011).*

evaporated liquids. The reaction chamber is made of stainless steel, which is isolated from the ambient atmosphere by hot ethylene glycol circulation via a chiller; recirculating water bath (Thermo Neslab RTE7, −298 to +423 K) to keep the reactor wall temperature constant. The SFR setup enables working at 100–1100 mbar and temperatures of 298–1173 K. Gases such as O_2, CO, H_2, CH_4, C_2H_6, and C_3H_8 and vaporized liquids such as water, ethanol, methanol, and iso-octane were already used as reactants. All gases and liquids are dosed via mass flow controllers (MFCs: Bronkhorst). For liquid fuels, a microstructure nozzle evaporation technique is applied (Thormann et al., 2008). These gases are premixed in a mixing unit before they enter the reaction zone. A 2-mm-thick porous sintered metal is placed on top of the glass bead layer serving as mixing chamber. The pore size of 0.1 mm of the sintered metal is chosen to be smaller than the quenching distance of any flammable mixture to prevent flashbacks (Bridges, 2006). A K-type thermocouple is embedded in the center of the mixing chamber to measure the inlet temperature of the gases. The gas mixture is directed to the catalytic surface through a flow straightener (3.75 cm in diameter) made of stainless steel with a 0.8-mm cell size honeycomb structure. The distance between the flow straightener and the catalytic surface is 3.9 cm.

The flow configuration is oriented upward so that the buoyancy effect on the stagnation-flow field is diminished. This configuration provides a stagnation-flow field with a radially uniform velocity profile at the inlet. Gas lines are also heated to prevent the condensation of liquids. The gases are exhausted through an annular pipe and burned in a Bunsen burner which is also housed in the reactor.

The catalyst is coated on a flat stagnation disk which is made of alumina. An R-type (rhodium–platinum) thermocouple with 0.2 mm thickness is embedded in the center of the stagnation surface during casting. The stagnation surface is directly heated by a resistive heater which is located right above the surface (Fig. 2.5).

A chemical ionization mass spectrometer (MS), from V&F named *Airsense 500*, an electron pulse ionization mass spectrometer, from V&F named *H-sense*, and Fourier-transformed infrared spectrometer (FTIR), from *MKS* named *MultiGas*^{TM} *2030 Model*, are used simultaneously to analyze the product composition. H_2O is calculated via the oxygen mass balance.

The concentration profiles of the species within the boundary layer are measured by a microprobe sampling technique. A quartz microprobe with a 50 μm opening and a bend angle of 15° is used. The microprobe has the dimensions of 3 mm OD (outer diameter) and 1 mm ID (inner diameter);

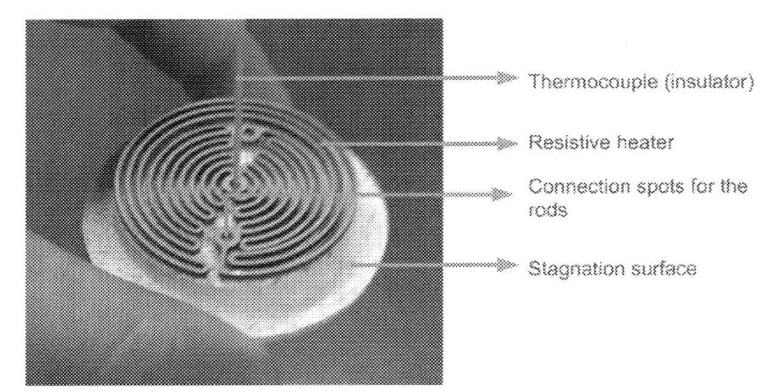

Figure 2.5 Resistive heater to control temperature of the catalytic disc; the catalyst is on the bottom site of the stagnation surface.

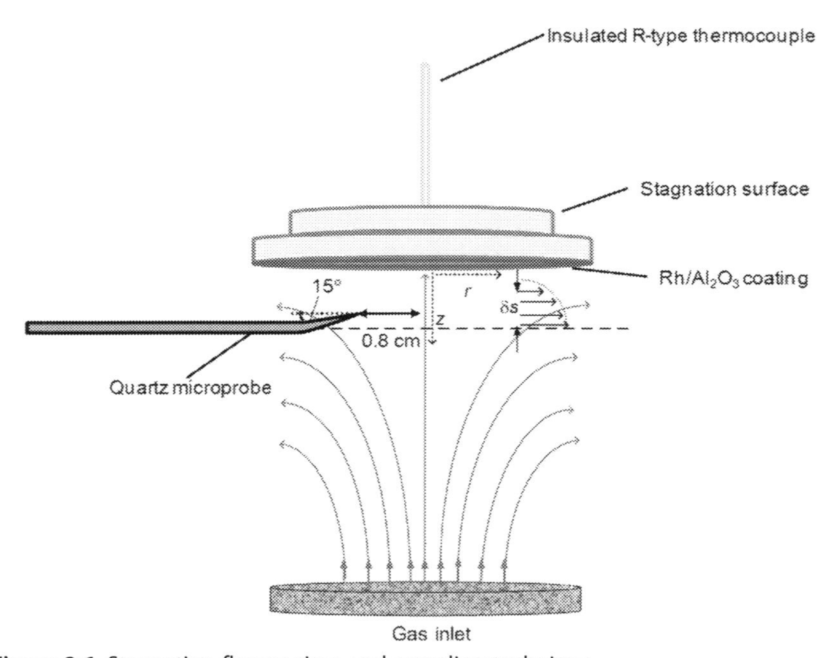

Figure 2.6 Stagnation-flow regime and sampling technique.

however, on the tip where the sampling is performed, the microprobe OD is approximately 0.4 mm. This configuration allows close contact to the catalyst surface. The probe is positioned at a radial distance of nearly 0.8 cm from the center to eliminate the edge effect and provide as little disturbance to the flow field as possible (Fig. 2.6).

The position of the microprobe in the boundary layer is controlled by a step motor controller. Initially, the microprobe is positioned on the surface and the probe-surface contact is determined visually. Starting from the position of this zero point, the probe is moved downward through the boundary layer.

3.2. Modeling the stagnation flow on a catalytic plate

Evans and Greif (1988) formulated a one-dimensional model of the rotating disk/SFR, which is the basis of our modeling approach. They considered two solid disks with a finite distance between them. Both disks had an infinite extent in the r–θ plane. In the rotating disk configuration, one of the disks was rotating, and the other parallel, porous disk was fixed. In the stagnation-point flow, both disks had a zero rotation rate. Gas at ambient temperature was injected through the porous disk normal to its surface. The rotating disk's surface was heated to a constant temperature. Coltrin et al. (1989) extended the model to include the detailed chemical kinetics of species. Therefore, they included a species governing equation for each gas-phase species. These equations account for convective and diffusive transport of species as well as production and consumption of species by elementary chemical reactions (Coltrin et al., 1989). The CHEMKIN SPIN code (Kee et al., 2001), which was developed to solve 1D rotating-disk and SFR models, included an equation for each surface species to consider the effect of surface composition on the system. The CHEMKIN SPIN code solves the models at steady state. Deutschmann et al. (1996) simulated the transient behavior at catalytic ignition with the 1D stagnation-flow model. Raja et al. (2000) formulated the compressible transient stagnation-flow model to study the transient dynamics of catalytic ignition in stagnation flows.

Compressible 1D stagnation-point flow analysis forms the basis of the equation system presented below. It was found that the prediction of the effect of internal mass transfer limitations in the catalytic washcoat of the SFR configuration is crucial to derive microkinetic data from SFR experiments (Karadeniz, 2014; Karadeniz et al., 2013); our model is extended to include the diffusion limitations due to a porous layer. It should be noted that the CHEMKIN code has no ability to account for internal mass transport in the catalytic coating.

Evans and Greif (1988), Houtman et al. (1986b), Kee et al. (2001, 2003), Behrendt et al. (1995), Deutschmann et al. (1996), and Raja et al. (2000)

have formed the continuous development of the simplified formulations of the stagnation flows for semi-infinite and finite domains and steady and transient cases. Kee et al. (2005b) have documented all these cases comprehensively. In this case, axisymmetric flow equations are simplified further based on two principle conjectures. The first conjecture is based on the consideration of the velocity field in terms of a stream function, which has a separable form

$$\psi(z,r) = r^2 U(z) \tag{2.15}$$

where $U(z)$ is an unspecified function of z alone (Kee et al., 2005b). The advantage of the stream function is that it enables defining two different velocity variables in terms of a single variable. In addition, the axial momentum and mass continuity equations are combined into a single equation (Houtman et al., 1986b). The second conjecture is based on presuming the changes in temperature, species composition, and density in the z-coordinate only. Because, in the stagnation flow field, scalar quantities (temperature and species mass fractions) depend only on the distance from the surface, not on the radial position (Behrendt et al., 1995; Coltrin et al., 1989; Warnatz et al., 1994b). This allows us to draw the attention to the center of the catalytic plate, and the system can be modeled by a one-dimensional representation of the Navier–Stokes equations. The flow equations (mass and momentum) are coupled with the thermal energy and species conservation equations in their 1D form. Therefore, their derivatives with respect to r coordinate drop out. The pressure can vary throughout the flow, but the variations of the pressure are assumed to be small comparing to the mean thermodynamic pressure.

After some more mathematical exercises (Karadeniz, 2014; Kee et al., 2005b), the SFR reactor can be described by the following set of equations:

- *Mass continuity*:

$$\frac{\partial \rho}{\partial t} = -2\rho V + \frac{\partial(\rho v_z)}{\partial z} \tag{2.16}$$

- *Axial momentum*:

$$\rho \frac{\mathrm{d}v_z}{\mathrm{d}t} = -\frac{\partial p}{\partial z} - \rho v_z \frac{\partial v_z}{\partial z} + 2\mu \frac{\partial V}{\partial z} + \frac{4}{3}\frac{\partial}{\partial z}\left[\mu \frac{\partial v_z}{\partial z} - \mu V\right] + 2\mu \frac{\partial V}{\partial z} \tag{2.17}$$

- *Scaled radial momentum*:

$$\rho\frac{\partial V}{\partial t}=\rho v_z\frac{\partial V}{\partial z}-\rho V^2-\Lambda+\frac{\partial}{\partial z}\left(\mu\frac{\partial V}{\partial z}\right) \tag{2.18}$$

- *Thermal energy*:

$$\rho c_\mathrm{p}\frac{\partial T}{\partial t}=-\left[\rho c_\mathrm{p}v_\mathrm{z}+\sum_{i=1}^{N_\mathrm{g}}\rho Y_k V_k c_{\mathrm{p},i}\right]\frac{\partial T}{\partial z}-\sum_{i=1}^{N_\mathrm{g}}\dot{\omega}_i M_i h_i+\frac{\partial}{\partial z}\left(\lambda\frac{\partial T}{\partial z}\right) \tag{2.19}$$

- *Species continuity*:

$$\rho\frac{\partial Y_i}{\partial t}=-\rho v_z\frac{\partial Y_i}{\partial z}+\dot{\omega}_i M_i-\frac{\partial \rho Y_k V_k}{\partial z} \tag{2.20}$$

- *Perfect-gas equation*:

$$p=\rho RT\sum_{i=1}^{N_\mathrm{g}}\frac{Y_i}{W_i} \tag{2.21}$$

This simplified system of equations represents the complete 2D Navier–Stokes equations in 1D. In this case, there are also other simplified models such as 1D plug flow and 2D boundary layer equations to predict the behavior of chemically reacting flows. These simplified models neglect some certain physical effects. For instance, plug flow reactor model neglects radial gradients through the reactor (Alopaeus et al., 2008). In addition, convective transport is assumed to dominate over the diffusive transport in the axial direction (Sari et al., 2008). These assumptions lead to a 1D model without considering any diffusive term. Boundary layer approximation ignores the diffusive transport terms along the flow direction and sets all the second derivatives involving the flow direction to zero (Sari et al., 2008). However, the simplified 1D SFR model does not emerge due to neglecting certain physical effects; instead, it emerges due to natural vanishing of some terms because of the mathematical reduction (Kee et al., 2005b). Therefore, it considers all certain physical and chemical effects, and it is convenient to investigate the gas–surface interactions at a detailed fundamental level.

The coupling of the 1D SFR equations with the chemical processes in and on the catalytic plate is straightforward. All models discussed in Section 2.3 can be coupled via the species mass fluxes at the boundary between fluid phase and catalytic plate (Karadeniz, 2014; Karadeniz et al., 2013). Even more sophisticated models for the description of mass transport and chemical reactions in porous media such as the dusty-gas model (DGM) and also energy balances can be implemented into the numerical simulation (Karadeniz, 2014).

The mentioned SFR model is numerically implemented into the software package DETCHEM (Deutschmann et al., 2014) named DETCHEM^{STAG}. The code is validated with experiments for different chemical compositions, reaction mechanisms, temperatures, and flow rates, and an example is given below.

Wehinger et al. (2014) recently studied the influence of fluidic effects on kinetic parameter identifications in lab scale catalysis testing using CFD. The SFR with a detailed surface mechanism was simulated fully in three dimensions. It is shown that the 3D simulations are not advantageous over the commonly used stagnation-flow boundary-layer problem description as we and others use them. They state that the SFR reactor is "setting a valuable example of how fluidic effects on kinetic parameter estimation can be suppressed." In the same study, simplified tubular flow models were analyzed as well, showing significant problems for deriving kinetic data from tubular flow reactors oversimplifying the fluid mechanics (Wehinger et al., 2014).

3.3. Example: CO oxidation on porous Rh/Al_2O_3 plate

In this section, direct oxidation of carbon monoxide (CO) over a porous Rh/Al_2O_3 catalyst is chosen as an example to illustrate the experimental technique and modeling approach for stagnation flows on catalytically coated plates (Karadeniz, 2014; Karadeniz et al., 2013; Karakaya, 2013).

Catalytic CO oxidation on noble metal surfaces is a simple but important reaction because it produces only gaseous CO_2 as the product, which hardly sticks to metal surfaces, but it still exhibits many of the fundamental steps of a heterogeneous catalytic process (Haruta and Tsubota, 2009; McClure and Goodman, 2009). The effect of surface characteristics on reaction kinetics can be investigated at an atomic scale. Therefore, this reaction has been studied extensively in the literature (Anderson, 1991; Bourane and Bianchi, 2004; Hopstaken and Niemantsverdriet, 2000; Karadeniz et al., 2013; McClure and Goodman, 2009; Royer and Duprez, 2011) regarding the heterogeneous catalysis studies, to understand the relation between the fundamental surface science and practical applications. For instance, CO oxidation is an important reaction for the removal of hazardous CO emission in the automotive exhaust catalyst, in which precious noble metals are used. Furthermore, CO is undesirable in ammonia synthesis and fuel cell power generation systems because it reduces the hydrogen productivity and poisons the catalyst in downstream processes. In this case, the undesirable CO content can be removed by using noble metal catalysts. Since the price of the precious noble metals is high, understanding the catalytic CO oxidation at a fundamental level aids optimizing the processes and the catalysts.

3.3.1 Surface reaction mechanism

It is mostly accepted that CO oxidation on noble metals occurs between the CO and O adsorbates (Karadeniz et al., 2013; Karakaya, 2013). The intrinsic kinetics of the CO oxidation over Rh/Al_2O_3 is taken here from the recent study of Karakaya et al. (2014) without any modification. This surface reaction mechanism is a subset of the kinetics of the water–gas shift reaction over Rh/Al_2O_3 catalysts given by Karakaya et al. (2014). This direct oxidation of CO involves 10 elementary-like surface reaction steps among 4 surfaces and 3 gas-phase species. The reaction rates are modeled by a modified Arrhenius expression as given in Eq. (2.6). The nominal values of the preexponential factors are assumed to be $10^{13}N_A/\Gamma$ (cm^2/mol s), where N_A is Avogadro's number (the surface site density was estimated to be 1.637×10^{19} site/cm^2 derived from a Rh(110) surface). The nominal value of 10^{13} is the value calculated from transition state theory ($k_B T/h$) with k_B being Boltzmann's constant and h Plank's constant (Maier et al., 2011).

Exactly, the same kinetics of adsorption and desorption of oxygen as well as the reaction of adsorbed oxygen (O(s)) have also been used before to model hydrogen oxidation (Karakaya and Deutschmann, 2013). The surface reaction kinetics for CO oxidation is given in Table 2.1. The reaction kinetics are thermodynamically consistent at temperatures of 273–1273 K.

3.3.2 Catalyst preparation

The flat stagnation disk was coated with a Rh/Al_2O_3 catalyst, where rhodium particles were distributed in a porous Al_2O_3 washcoat. Appropriate amounts of aqueous solution of rhodium(III) nitrate (Umicore) (9 wt%

Table 2.1 Surface reaction mechanism for CO oxidation on Rh

	Reaction	A (cm mol s)	β(−)	E_a (kJ/mol)
R1	$O_2 + Rh(s) + Rh(s) \rightarrow O(s) + O(s)$	1.000×10^{-2b}	Stick. coeff.	
R2	$CO_2 + Rh(s) \rightarrow CO_2(s)$	4.800×10^{-2b}	Stick. coeff.	
R3	$CO + Rh(s) \rightarrow CO(s)$	4.971×10^{-1b}	Stick. coeff.	
R4	$O(s) + O(s) \rightarrow Rh(s) + Rh(s) + O_2$	5.329×10^{22}	−0.137	387.00
R5	$CO(s) \rightarrow CO + Rh(s)$	1.300×10^{13}	0.295	$134.07\text{-}47\theta_{CO}$
R6	$CO_2(s) \rightarrow CO_2 + Rh(s)$	3.920×10^{11}	0.315	20.51
R7	$CO_2(s) + Rh(s) \rightarrow CO(s) + O(s)$	5.752×10^{22}	−0.175	106.49
R8	$CO(s) + O(s) \rightarrow CO_2(s) + Rh(s)$	6.183×10^{22}	0.034	129.98
R9	$CO(s) + Rh(s) \rightarrow C(s) + O(s)$	6.390×10^{21}	0.000	174.76
R10	$C(s) + O(s) \rightarrow CO(s) + Rh(s)$	1.173×10^{22}	0.000	92.14

The rate constants are given in the form of $k = AT^{\beta} \exp(-E_a/RT)$; adsorption kinetics is given in the form of sticking coefficients; the surface site density is $\Gamma = 2.72 \times 10^{-9}$ mol/cm^2.

Rh) and boehmite (AlOOH) (20% boehmite) were mixed to obtain a 5 wt% Rh/Al_2O_3 composition. The solution was diluted with water and applied to the disk by the spin-spray technique to ensure a homogeneously distributed catalytic layer on the surface. Coating a flat surface with a well-defined particle size and morphology is essential for the SFR application (Thune and Niemantsverdriet, 2009; Vanhardeveld et al., 1995). For this purpose, a simple laboratory-scale spray apparatus was developed. The stagnation surface was heated to 373 K and held on a rotary support which spins at 1000 rpm. The solution was sprayed by compressed air via a spray gun. The surface was dried at 403 K for 10 min and the procedure was repeated until the desired coating thickness of 100–130 μm was achieved. The coated stagnation disk was then calcined at 973 K in air for 2 h. Prior to the measurements, the surface was oxidized by 5 vol% O_2 diluted in Ar at 773 K for 2 h. The resulting rhodium oxide phase was reduced by 5 vol% H_2 diluted in Ar at 773 K for 2 h.

3.3.3 Catalyst characterization

The coating thickness and the homogeneity of the coating layer were investigated by means of light microscopy (LM: Riechert MEF4A). LM investigations showed that there was a uniform ~100 μm catalyst layer on the stagnation surface (Karakaya and Deutschmann, 2013). For the investigation of nanoscale Rh particles and the washcoat structure, scanning electron microscopy (SEM: Hitachi S570) was applied in combination with energy-dispersive X-ray spectroscopy and high-resolution transmission electron microscopy (HR-TEM: Philips CM200 FEG). SEM images indicated a diverse particle size distribution, where as Rh particles of ~100 nm diameter as well as smaller Rh particles of 15–50 nm were also detected in HR-TEM investigations (Karakaya and Deutschmann, 2013). Metal dispersion was measured by the continuous-flow CO chemisorption technique (Karakaya and Deutschmann, 2012). The flat stagnation disk was subjected to the chemisorption measurement before the catalytic measurements. The catalytic surface area was calculated to be 0.21 m^2/g based on the CO chemisorption measurements with the assumption of 1:1 adsorption stochiometry between Rh and CO molecules. With this information, $F_{cat/geo}$ was calculated to be 30, i.e., the total amount active catalytic surface area equals 30 times the geometrical area of the disk surface.

3.3.4 *Kinetic measurements*

CO oxidation measurements were carried out in the SFR at varying CO/O_2 ratios. Ar-diluted gas mixtures were fed to the reactor with a flow rate of 15.5 SLPM (standard liter per minute at 293 K, 1 atm.). The calculated flow velocity and working pressure were 51 cm/s and 500 mbar, respectively. The reactor inlet temperature was 313 K. The reaction was studied at steady-state conditions (Table 2.2). At low temperatures, oxygen-rich conditions were selected to avoid external mass transport limitations and examine the kinetic effects (Case 1). However, for moderate- and high-temperature regimes (Case 2 and Case 3), the reactions were examined under stoichiometric conditions.

The boundary-layer concentration profiles of CO, CO_2, and O_2 were measured by using a chemical ionization mass spectrometer (Airsense

Table 2.2 Stagnation disc temperature and inlet conditions

	T_{disc} (K)	T_{inlet} (K)	CO (% vol.)	O_2 (% vol.)	Ar (carrier gas) (% vol.)	Inlet velocity (cm/s)	Reactor pressure (mbar)
Case 1	521	313	2.67	2.23	95.10	51	500
Case 2	673	313	5.67	2.89	91.44	51	500
Case 3	873	313	5.66	2.83	91.51	51	500

500, V&F) with a quadrupole ion trap. A microprobe sampling technique was used as described in Section 2.1 to measure the gas-phase composition in the boundary layer adjacent to the catalyst surface.

3.3.5 Numerical simulation

The inlet conditions for the numerical simulations are based on the experimental conditions. The simulations are performed with the three different models for internal diffusion as given in Section 2.3 to analyze the effect of internal mass transfer limitations on the system. The thickness (100 μm), mean pore diameter, tortuosity ($\tau = 3$), and porosity ($\varepsilon = 60\%$) of the washcoat are the parameters that are used in the effectiveness factor approach and the reaction-diffusion equations. The values for these parameters are derived from the characterization of the catalyst. The mean pore diameter, which is assumed to be 10 nm, lies in the mesapore range given in the literature (Hayes et al., 2000; Zapf et al., 2003). CO is chosen as the rate-limiting species for the *η-approach* simulations. *η-approach* simulations are also performed with considering O_2 as the rate-limiting species.

3.3.6 Species profiles

The reaction is already active at 521 K (Fig. 2.7); however, total consumption of the reactants is not achieved in the experiment. However, the *∞-approach* predicts complete consumption of CO at the surface, i.e., it strongly overpredicts the overall reaction rate. Simulations with the *η-approach* and *RD-approach* models predict the slow overall reaction rate

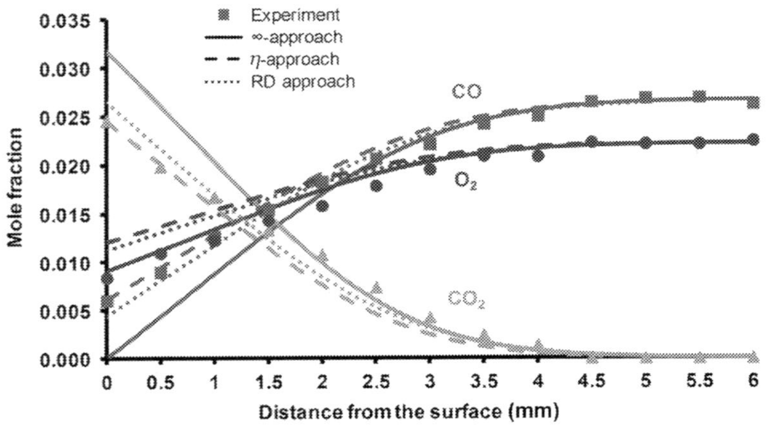

Figure 2.7 Comparison of the experimental and simulation results for the species profiles in catalytic oxidation of CO at 521 K. *Figure taken from Karadeniz et al. (2013).*

of the experiments. The slight deviation for the O_2 consumption might be due to sampling inaccuracies in the experiment.

The *RD approach* predicts the species profiles inside the porous washcoat, for the first case, as given in Fig. 2.8. Species are consumed or produced just within the first 7–7.5 μm of the washcoat. This can be attributed to the fact that surface reactions are very fast even at this low temperature. The rate-limiting process is already diffusion. *η-approach* yields Thiele modulus of $\Phi = 27.44$ and effectiveness factor $\eta = 0.036$, respectively, confirming finite internal diffusion.

In the second case ($T = 673$ K), CO and O_2 concentration at the surface decrease by 82% and 71%, respectively, relative to the inlet conditions (Fig. 2.9). *∞-approach* predicts total consumption for both reactants. Simulations with the *RD approach* estimate results close to the experiments for the consumption of reactants and production of CO_2. There is a relatively good agreement between the experiment and the simulation results with the *η-approach*.

Species profiles inside the washcoat (predicted with *RD-approach*) are similar to Case 1; but the reaction layer decreases from 7.5 to 6.5 μm (Fig. 2.10). For this condition, the dimensionless Φ and η are calculated as 53.7 and 0.019, respectively.

In the last case ($T = 873$ K), CO and O_2 concentration at the surface decrease by 84% and 79%, respectively, relative to the inlet conditions (Fig. 2.11). CO_2 formation has its maximum value, since reaction rate reaches its maximum. *∞-approach* underpredicts consumption of reactants

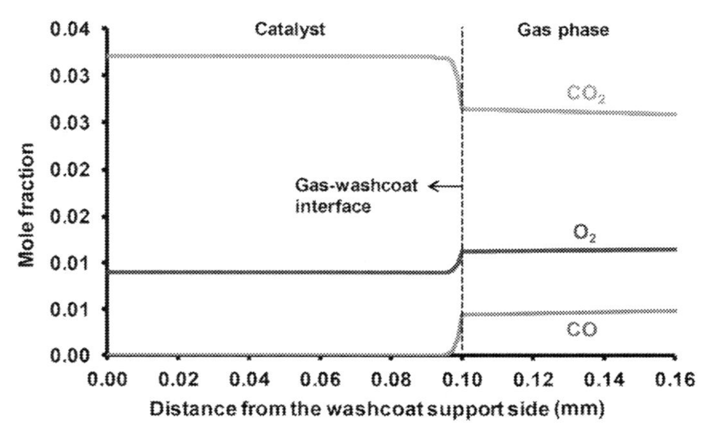

Figure 2.8 Species mole fractions inside the porous washcoat layer at 521 K. *Figure taken from Karadeniz et al. (2013).*

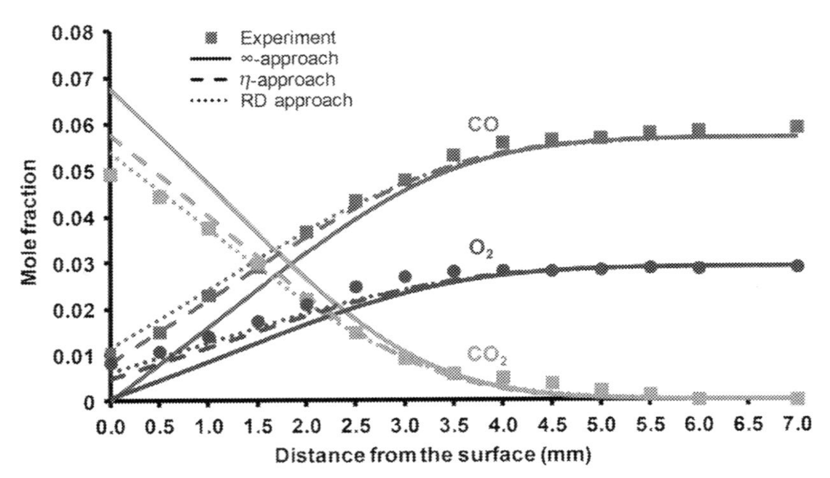

Figure 2.9 Comparison of the experimental and simulation results for the species profiles in catalytic oxidation of CO at 673 K. *Figure taken from Karadeniz et al. (2013).*

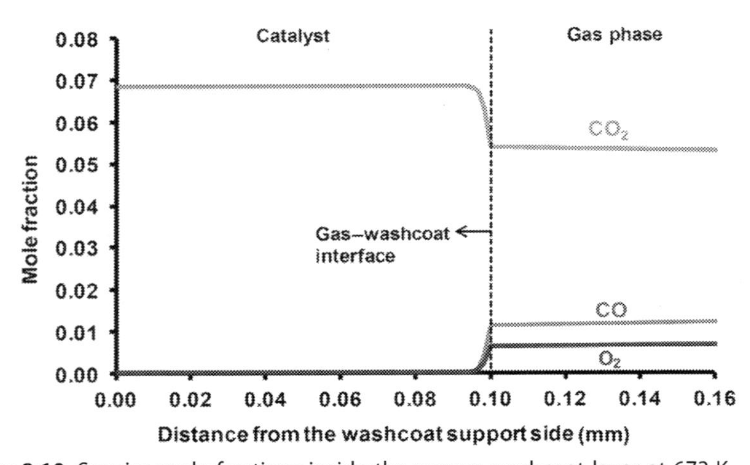

Figure 2.10 Species mole fractions inside the porous washcoat layer at 673 K.

and overpredicts formation of CO_2. Simulation with the *RD-approach* surface model reproduces the experimental data. There is also, again, a relatively good agreement between the experiments and the simulation with the *η-approach*. At this temperature, reactions are even faster, resulting in large concentration gradients within the first 5.5–6 μm in the washcoat (not shown). The Φ and η are 91.75 and 0.011, respectively.

Finally, *η-approach* simulations are performed for considering O_2 instead of CO as the species "rate-limiting species," i.e., the effectiveness factor is

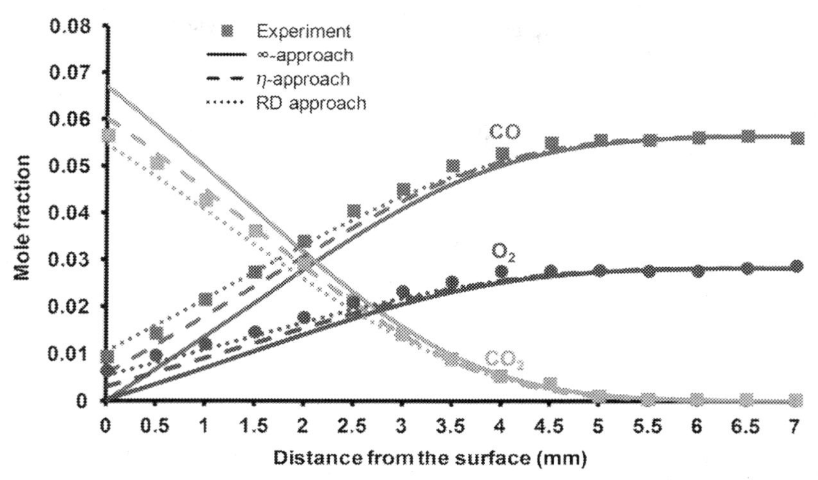

Figure 2.11 Comparison of the experimental and simulation results for the species profiles in catalytic oxidation of CO at 873 K. *Figure taken from Karadeniz et al. (2013).*

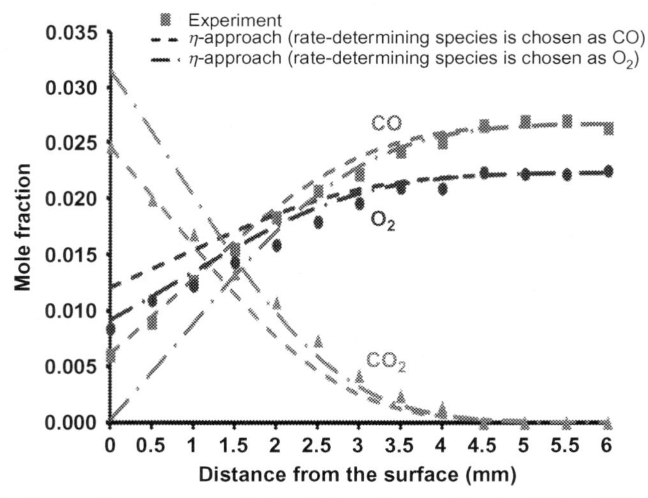

Figure 2.12 Comparing η-approach simulations by considering CO and O_2 as the rate-limiting species at 521 K.

calculated using the diffusion coefficient and reaction rate of CO. Since these parameters depend on the species but mass continuity requires the same effectiveness factor for all species' boundary conditions, it matters which species is considered. Here, using O_2 instead of CO leads to an over-prediction of the consumption of CO and formation of CO_2 for the lean Case 1 (521 K) (Fig. 2.12), however, no difference is observed for the

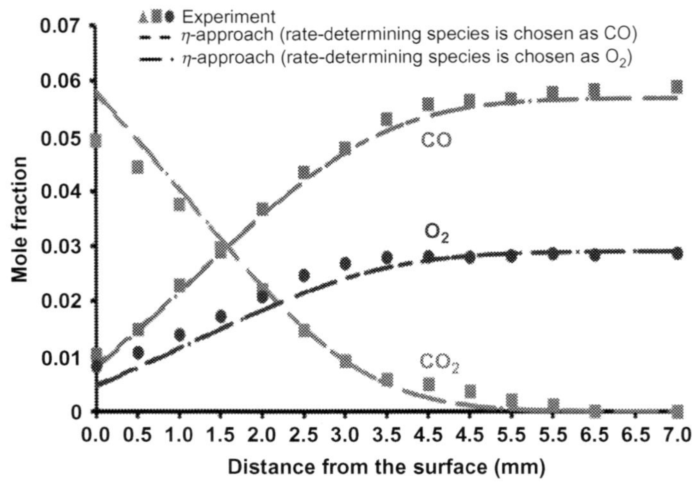

Figure 2.13 Comparing η-approach simulations by considering CO and O_2 as the rate-limiting species at 673 K.

stoichiometric Case 2 (Fig. 2.13) and for Case 3 (not shown). In the latter case, external mass transfer is the rate-limiting step at this high temperature; consequently, the internal diffusion model is not much significant.

3.3.7 Summary

The SFR with catalytically coated plates is an easy setup to investigate heterogeneously catalyzed gas-phase reactions. *In situ* invasive capillary techniques can be used to determine the gas-phase concentration in the one-dimensional boundary layer on top of the catalyst. The measurement species profiles can be compared with numerically predicted profiles to test surface reaction mechanisms, diffusion models but also gas-phase reaction schemes, CVD processes, and others (not discussed here). However, internal diffusion inside the catalytic disc has to be taken into account when thicker catalyst layers are used. Then, the choice of an adequate diffusion model can be crucial for a correct interpretation of the measured data. The computer code DETCHEM^{STAG} offers simulations with the following models to account for internal diffusion in stagnation flows on porous plates with reactions inside: effectiveness factor model, 1D reaction diffusion model (RD-approach), and DGM (not discussed here). While the RD-approach may even play a role in simple cases as discussed here, it is the model of choice when parallel reactions occur (e.g., catalytic partial oxidation, three-way catalysts). In cases with strong variations of the total molar numbers and/or pressure gradients over the porous structures, the DGM is the preferred one.

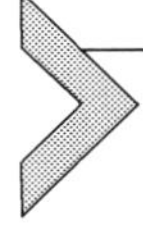

4. CHANNEL REACTORS WITH CATALYTICALLY COATED WALLS

SFRs can easily be used to obtain concentration profiles as a function of the distance from the catalytic surface as described above. The measured concentration profiles above the zero-dimensional catalytic surface can be used to improve our understanding of the reaction kinetics as well as their interaction with external and internal mass transport in perfect laboratory reactors. However, the configuration is rarely used in technical catalytic systems aside from the frequent use in materials synthesis' and coating techniques. Industrial catalytic processes are conducted in flow reactors, characterized by the variation of concentrations in flow directions leading to at least a one-dimensional spatial variation of the gas-phase concentrations, surface coverages, and temperature and, consequently, surface reaction rate. The simplest configuration of a flow system is a tubular reactor, in which the inner tube wall is coated with the catalyst. In environmental catalysis, e.g., automotive catalytic converters, and high-temperature catalysis, e.g., catalytic combustion and reforming, honeycomb-structured catalyst is often applied. The channels of these monolithic reactors can be considered as straight channel reactors, in which the variation of the gas-phase concentration and surface reaction rate primarily depends on the axial (in flow direction) position; hence, the catalytic surface is one dimensional. It should be noted that all remarks in Sections 2 and 3 on internal diffusion also apply for the washcoats in honeycomb structures. In this section, we discuss the potentials and limitation of an invasive *in situ* technique for measuring axial concentration and temperature profiles in straight channels as occurring in catalytically coated honeycomb structures (Fig. 2.1). Exemplarily, we discuss the partial oxidation of methane in Rh/Al_2O_3-coated honeycomb catalysts. The experimental technique described was recently set-up in our laboratory (Diehm, 2013; Diehm and Deutschmann, 2014; Livio et al., 2013). The experimental technique is further critically evaluated by CFD simulations (Diehm, 2013; Hettel et al., 2013).

4.1. Experimental setup

The flow diagram of the experimental setup is shown in Fig. 2.14. For the feeding of the gaseous components, MFCs (Bronkhorst) are used. Up to

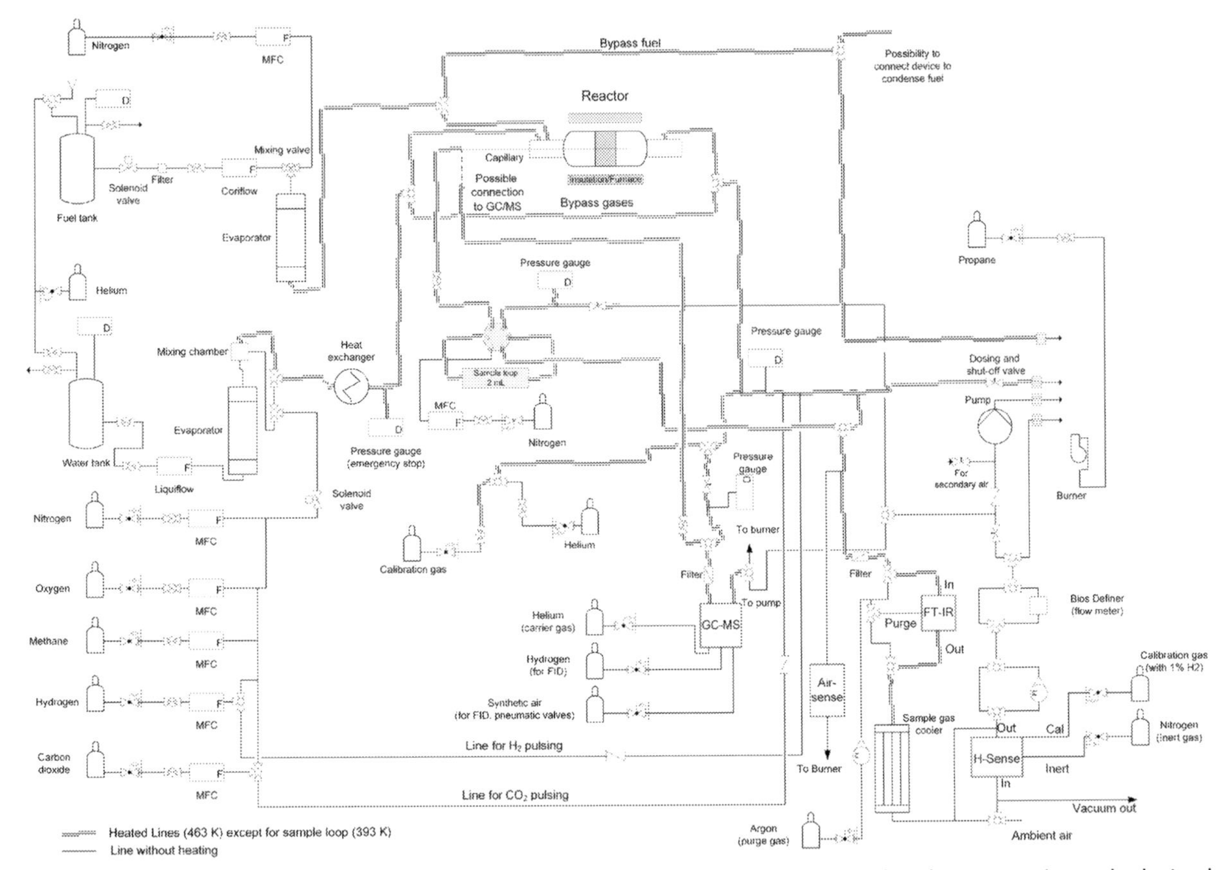

Figure 2.14 Flow diagram of the experimental setup; analytical instruments are highlighted with a green (gray shade in the print version) frame.

10 gaseous components can be fed simultaneously to the reactor. It is possible to feed two different streams of preevaporated components. One is dosed (named "fuel" in Fig. 2.14) from a tank under He pressure with a Cori-Flow (Bronkhorst), a controlled evaporation mixer (Bronkhorst), and an inert N_2 flow. As fuel, liquid hydrocarbons and two-component mixtures of liquid species can be employed. The preevaporated fuel is fed to the reactor inlet through different lines than the gaseous reactants (inlet marked with "fuel" in Fig. 2.14). Water is the second preevaporated component, which is dosed from a second tank under He pressure with a Liqui-Flow (Bronkhorst), a preevaporator, and an inert N_2 flow. Water is fed to the reactor inlet together with the gaseous components (inlet marked with "syn. air" in Fig. 2.14). The reactor inlet (Fig. 2.15) leads to a quick mixing of the components confirmed by CFD simulations.

The reactor consists of a quartz tube (length = 62 cm, ID = 20 mm) placed in an oven for thermal insulation and reaction light-off. The monolithic honeycomb catalyst (diameter = 19 mm, variable length) is placed 230 mm downstream of the reactor inlet. For increased insulation, the part of the reactor tube, in which the catalyst is positioned, is wrapped with a 1-cm-thick layer of quartz wool. Uncoated monoliths up- and downstream of the catalyst are employed as heat shields (front heat shield (FHS) and back heat shield (BHS), respectively). The FHS also serves as support for the probe applied in the sampling technique. To prevent a gas by-pass between monoliths and quartz tube, the heat shields and the catalyst are wrapped with

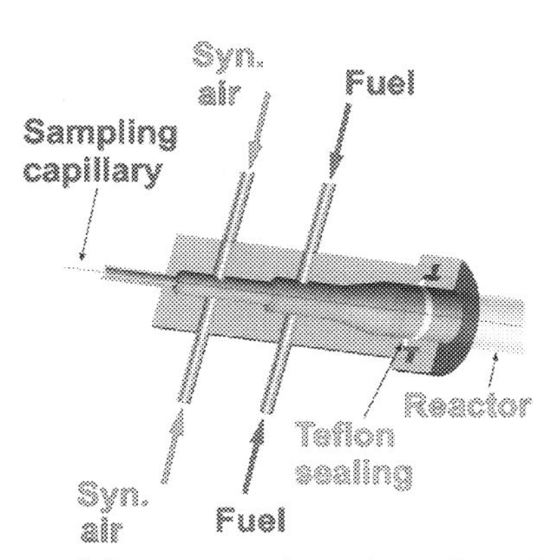

Figure 2.15 Cross section of the reactor inlet with capillary. As fuel, a preevaporated liquid fuel in N_2, such as ethanol, can be supplied. If a gaseous reactant, such as methane, is fed, it will be fed through the "syn. air" inlet together with the other gaseous components.

a 1-mm-thick layer of ceramic fiber paper. The catalysts—used for the results presented here—were supported on a honeycomb monolith made out of cordierite.

The total gas flow from the reactor outlet is monitored on-line by an FTIR analyzer (MKS Multigas 2030) and an EI-MS (V&F, H-sense). The FTIR is operated at a cell temperature of 463 K. A sample gas cooler is installed between FTIR and EI-MS to protect the EI-MS from excessive amounts of water in the gas composition. The sample gas cooler is set to 278 K to remove water and other species, e.g., unconverted hydrocarbons, which condense at these conditions. Additionally, an IMR-MS (V&F, Air-sense) and a GC-MS (6890 N, Agilent) can be used. The total gas flow is fully oxidized in the flame of a bunsen burner after it leaves the analytics. To quantify the total gas flow that reaches the FTIR, an adapted external standard method is used (Hartmann, 2009; Hartmann et al., 2009; Kaltschmitt, 2012).

In Fig. 2.16, a close-up picture of the reactor with the capillary for temperature profiles is displayed. The FHS, the capillary inserted into one channel and the ceramic fiber paper covering FHS, catalyst, and BHS are visible. The setup can be used for *in situ* measurements, where axial temperature and concentration profiles can be detected from one channel of the monolithic catalyst. Figure 2.17 shows the microvolume tee mounted onto the motorized linear stage used for moving the capillary. The setup is similar to the one introduced for foam monoliths by Horn et al. (2006a,b) and then adapted to honeycomb monoliths by Donazzi et al. (2011a,b). A novel feature of our

Figure 2.16 Picture of the reactor with capillary for temperature profile measurements. *Figure taken from Diehm and Deutschmann (2014).*

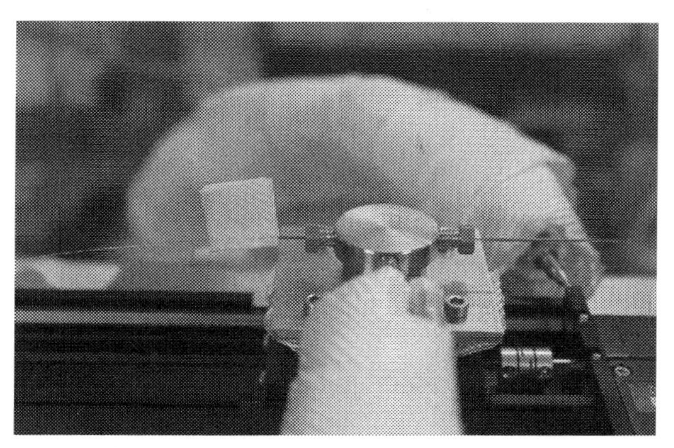

Figure 2.17 Picture of the microvolume tee, with the connected capillary, mounted onto the miniature motorized linear stage (black part below). The thin yellow (gray shade in the print version) fiber coming out of the back of the capillary is the VIS–IR optical fiber.

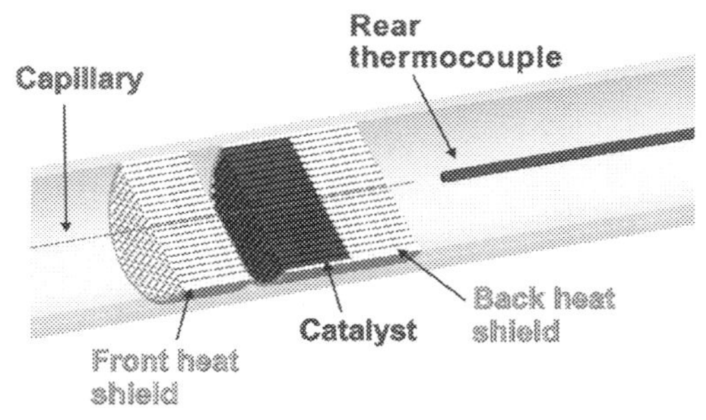

Figure 2.18 Cross section of reactor with front heat shield, catalyst, back heat shield, sampling probe, and rear thermocouple. *Figure taken from Livio et al. (2013).*

sampling system is the ability to maintain spatial concentration and temperature profiles in channels differently positioned across the monolith, e.g., a channel close to the periphery of the monolith. The cross section of the reactor setup used for the sampling technique is shown in Fig. 2.18 for sampling from a central channel. A void space of 5 mm is kept between the front inert monolith and the catalyst to avoid flow disturbance at the catalyst entrance due to partial occlusion of the channels by misalignment of the FHS. The temperature probes consist of a thin K-type thermocouple (Omega Engineering Inc., diameter = 250 μm) and a VIS-IR optical fiber (Leoni, core diameter = 200 μm, clad diameter = 220 μm, OD = 245 μm) connected to an infrared narrow band pyrometer (LumaSense Technologies GmbH,

IGA LO-50 plus). The IR detector is an InGaAs photodiode with a spectral range of 1.45–1.8 μm. With this spectral range, temperatures between 623 and 2073 K can be detected. The tip of the optical fiber is industrially polished at an angle of 45° in order to guarantee the local measurement of the temperature (Donazzi et al., 2011a). The connector of the optical fiber to the pyrometer is equipped with an anti-twist protection to avoid measurement bias due to fiber misalignment. The temperature measured by thermocouple and the optical fiber were representing the temperature of the gas phase and the surface, respectively (Donazzi et al., 2010, 2011a). Both temperature probes are inserted into a deactivated, fused silica capillary (ID = 570 μm, OD = 630 μm), which serves as a protection sheath. The distal end of the capillary is sealed by melting the capillary to isolate the temperature probe from the reacting system, while the proximal end is connected to a microvolume tee and is moved with a constant velocity of 0.009 cm/s along the monolith channels (channel diameter = 876 μm) through a miniature motorized linear stage (Zaber Technologies, T-LSM100A) with very high spatial resolution. As a sealing between reactor inlet, which ends in a 1/16-in. connection, and capillary, a graphite ferrule is used inside the stainless steel nut. The tightening of the sealing system with ferrule and nut is limited as the capillary still needs to be movable. Small leakages at this point of the reactor are possible. For the temperature profiles, the measurements can be performed during both pulling the capillary out of the reactor and pushing it back in. There is no visible difference in the arising profiles.

To measure the axial profile of the gas composition, an analogous sampling probe to the one used for the temperature profiles is applied. The dimension of the capillary (ID = 100 μm, OD = 170 μm) was chosen so that the disturbance of the probe on the flow field inside the channel (channel diameter = 880 μm) is as small as possible and the capillary is relatively simple to handle. The capillary is flexible and bends due to gravity. The radial position of the capillary inside the channel is not controllable, but the capillary will most likely be leaning on a channel wall. The gas is sampled from the capillary tip at low sucking rate and is sent to the analytics section via a 6-port valve and a pump (Diehm, 2013). During a measurement, the pressure is kept constant at 610 mbar corresponding to a sucking flow rate of approximately 1 ml/min, which was chosen, so that the contact time inside the channel with the probe is very close to that inside a channel without a probe. Besides, it was considered to be a good compromise between the influence of a high sucking rate on the flow field and the occurrence of homogeneous reactions inside the capillary at a low sucking rate. For the same reason, the gas is drawn from the reactor inlet (cold side) countercurrently to the reacting flow. However, gas-phase reactions in the probe cannot be completely ruled out. The capillary that is used for the measurement of the concentration profiles is very thin (OD = 170 μm) and bends easily. For this reason, this capillary can only be pulled out of the reactor.

The sampling system is capable of the simultaneous detection of a wide spectrum of gaseous species potentially produced in the oxidation and reforming of higher hydrocarbons. The experiments discussed below were performed at four SLPM (standard liters per minute, standard refers to standard conditions, $T = 298$ K and $p = 1.013$ bar), yielding a gas hourly space velocity of 120,000 h^{-1} (referred to open volume of the catalyst, calculated for standard conditions, $T = 298$ K and $p = 1.013$ bar) for a typical catalyst (diameter = 19 mm, length = 11 mm, 600 channels per square inch (cpsi)). Very short residence times on the order of 10 ms were reached in a channel of the catalyst. The experiments were carried out at 80% N_2 dilution. The gases were preheated to 463 K. For light-off, the furnace was heated to 523 K and a diluted mixture of H_2 and O_2 was fed until a temperature of 563 K was measured downstream of the BHS. After this temperature was reached, the inlet composition was switched to fuel and O_2 in nitrogen. The furnace was kept at 523 K to maintain a steady temperature around the reactor tube, leading to quasi-autothermal operation.

Figures 2.19 and 2.20 exemplarily show the measured temperature and concentration profiles in a single channel located in the center of the catalytic monolith. In Fig. 2.19, the axial temperature profiles for the gas phase

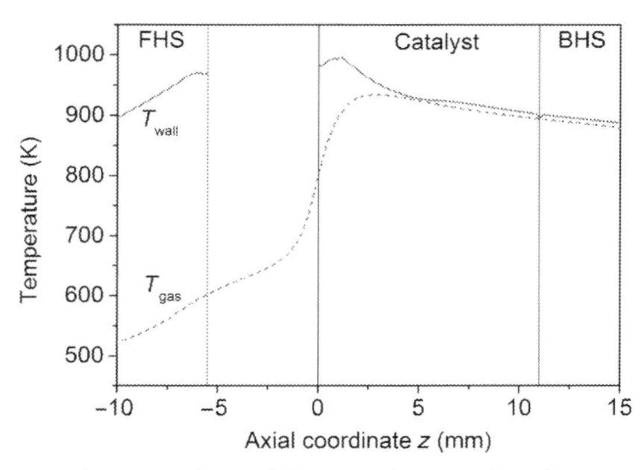

Figure 2.19 Measured temperature of the gas phase and surface as a function of the axial coordinate for CPOX of CH_4 over Rh/Al_2O_3 at molar C/O ratio of unity. *Figure taken from Hettel et al. (2013).*

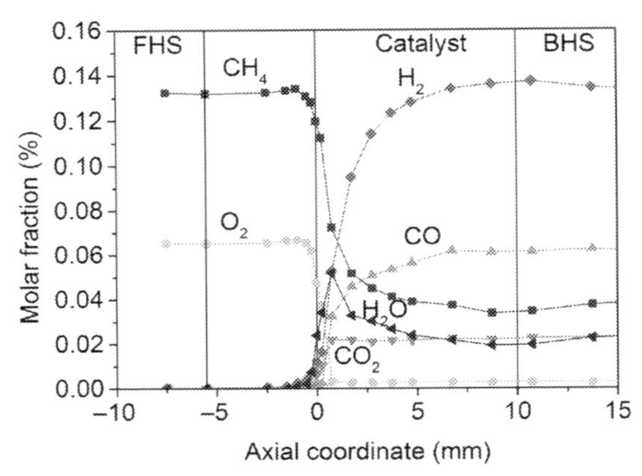

Figure 2.20 Measured mole fractions as a function of the axial coordinate for CPOX of CH_4 over Rh/Al_2O_3 at molar C/O ratio of unity. *Figure taken from Hettel et al. (2013).*

and the solid phase in a central channel of the monolith are shown. Within the first millimeter of the catalyst-coated channel, the surface temperature shows a hot spot of 990 K. This maximum is followed by a steep decrease in temperature to 922 K downstream of half of the channel length ($z = 5$ mm) and further downstream by a much slower decline to 903 K at the end of the channel ($z = 11$ mm). For the gas phase, the temperature rises from 799 K at the catalyst entrance ($z = 0$ mm) to 928 K at 2.6 mm and declines to 891 K at the outlet of the catalyst ($z = 11$ mm). The highest temperature of the gas phase is shifted further downstream in the channel compared to the hot spot on the surface. Inside the catalyst, a thermal equilibrium between gas and solid is reached toward the end of the channel.

The concentration profiles of the reactants and products are displayed in Fig. 2.20. The temperature and concentration profiles clearly indicate that the CPOX proceeds in two steps. The first zone of the catalyst (0–1 mm) is the oxy-reforming zone. In the first part of the channel, exothermic total oxidation is the determining process, leading to the temperature hot spot at the channel wall. Oxygen is completely converted after 1 mm. The main consumption of methane occurs upstream of an axial position of 3.5 mm, followed by a much slower consumption at the end of the catalyst. For water, a maximum is detected at $z = 0.5$ mm. Simultaneously, partial oxidation and reforming reactions take place, leading to the formation of hydrogen and carbon monoxide and the consumption of water. Carbon dioxide is formed within the first millimeter and nearly no change in mole fraction is observed throughout the rest of the channel of the catalyst. CO_2 is mainly formed in total oxidation in the oxy-reforming zone and is neither consumed in dry reforming or WGS nor formed in reverse WGS or other reactions as also suggested in the literature (Horn et al., 2006b, 2007b; Schwiedernoch et al., 2003). The second zone ($z = 1$–11 mm) is the reforming zone. The endothermic process predominates after all oxygen has been consumed, thus leading to the steep decrease in surface temperature and a further consumption of methane and water as well as formation of hydrogen and carbon monoxide. These findings support former studies of CPOX of CH_4 over Rh (Donazzi et al., 2011a; Horn et al., 2006b; Schwiedernoch et al., 2003). The formation of syngas seems to occur on Rh by a combination of direct and indirect routes as already proposed in the literature (Horn et al., 2006b). Nevertheless, it is not possible to infer the concentrations close to the catalytic surface from the gas composition measured with the *in situ* technique.

4.2. CFD modeling

The *in situ* technique presented above is an invasive technique, in which the probe inside the channel will influence the velocity and species profiles. To be able to evaluate the measured profiles, CFD simulations were performed (Hettel et al., 2013). The CFD code FLUENT (ANSYS Fluent) was applied for the numerical simulation of the three-dimensional flow field in the monolithic structure. Transport processes in the solid structure were not included in the computations.

A detailed surface-reaction mechanism was used for the example system presented here, the partial oxidation of methane over rhodium (Deutschmann et al., 2001). The mechanism includes 20 species and 38 reactions. It can be imported into FLUENT in the format of the CHEMKIN database. The mean-field approximation was applied for modeling the surface chemistry (Section 2). In these simulations, the influence of the internal mass transfer—in contrast to the SFR models discussed above—was not directly covered by the employed CFD code but estimated by adapting the active catalytic surface area by an effectiveness factor. No gas-phase reaction mechanism was employed, as it was already shown in the literature that reactions in the gas phase can be neglected for the given operational conditions (Beretta et al., 2011; Bitsch-Larsen et al., 2008; Deutschmann and Schmidt, 1998; Veser and Frauhammer, 2000).

In addition to the transport equations for mass, momentum, and enthalpy, a transport equation for six of the seven gas-phase species CH_4,

O_2, N_2, H_2O, CO_2, H_2, CO has to be solved. To analyze the residence time τ of the fluid, an additional transport equation was implemented into the CFD code by implementing user-defined functions. The appropriate transport equation reads (Habisreuther et al., 2005):

$$\frac{\partial(\rho \cdot \tau)}{\partial t} + \text{div}\,(\rho \cdot u \cdot \tau) = \text{div}\left(\frac{\mu}{Sc} \cdot \text{grad}\,\tau\right) + \rho \cdot S_\tau \tag{2.22}$$

τ may be regarded as the mean value of the residence times of molecules which are transported and exchanged according to convection and diffusion. Sc is the Schmidt number of the mixture ($Sc = 0.7$). The source term S_τ is set to unity (one second residence time per second physical time). To integrate the residence time inside the channels, the source term was only active downstream of $z = 0$ mm (begin of the catalytic zone).

The computations were performed for the monolithic honeycomb used in the experiments described above. For simplicity, the rounding of the channel edges due to the washcoat was neglected. The free height of a quadratic single channel is $h = 795$ μm. The calculation domain comprises only a part of the monolith, including one channel with the probe and several channels without a probe. Figure 2.21 shows a part of the monolith for the configuration in which the probe is positioned in the center of a channel and in which the tip of the probe is at $z_{\text{probe}} = 5$ mm. In this case, the flow domain includes nine partially sliced channels. It was verified that the amount of calculated channels is large enough to guarantee a negligible influence of the size of the calculation domain on the results. The solid was not included in the calculations. The fluid enters on the left side (magenta), 5 mm upstream of the monolith, and leaves the channels on

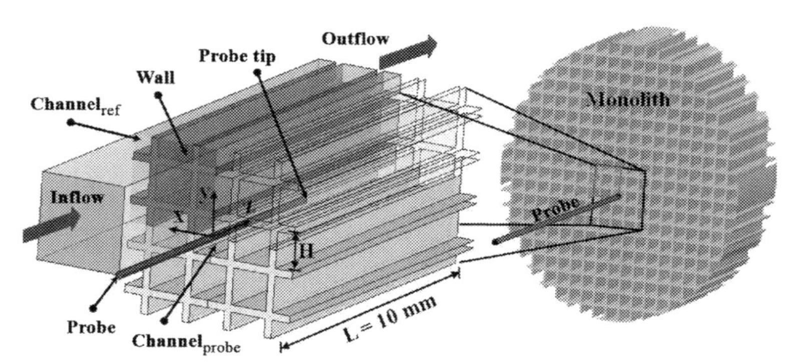

Figure 2.21 Calculation domain and boundary condition. *Figure taken from Hettel et al. (2013).*

the right side. Part of the fluid is sucked in at the tip of the probe which is colored blue. The probe points into the z-direction and is perpendicular to the xy plane.

Different positions of the probe in the xy plane were examined, while the probe was fully inserted into the channel. Furthermore, for the case in which the probe is in the center of a channel, nine axial positions of the probe tip were explored. For each configuration, a block-structured grid with a typical size of 700,000 prism cells was generated.

In the discussion below, the conditions in the channel with the probe (subscript "probe") and in a reference channel without a probe (subscript "ref") are compared. The reference channel is the channel without a probe, which is the farthest from the channel with the probe and, therefore, the influence of the probe was expected to be the smallest. However, the difference of the volume fluxes in the various empty channels is very small (below 1%).

The conditions at the flow inlet were according to the experimental conditions: velocity $u_{in} = 0.86$ m/s, pressure $p_{in} = 1$ bar, temperature $T_{in} = 873$ K, mole fraction of methane in air 0.133 with a molar C/O ratio of 1.0, i.e., stoichiometric for partial oxidation to hydrogen and carbon monoxide. At the channel walls, the experimentally determined wall temperature was used as boundary condition. The Reynolds number was always lower than 40 and, thus, the flow was laminar.

4.3. CFD evaluation of the sampling technique

The sampling capillary is placed inside the channel of the catalyst from the upstream side and thus obviously will affect the gas flux into the channel to a certain degree. As discussed above, this placement was chosen to avoid gas-phase reactions inside the sampling capillary. The colder area of the reactor is reached after a much shorter distance than for sampling from the downstream side of the catalyst. Nevertheless, even if gas-phase reactions are avoided in the capillary, the impact of the sampling probe in the investigated channel needs to be considered. For this evaluation, CFD simulations have been performed using the software packages FLUENT (Hettel et al., 2013), DETCHEM (Deutschmann et al., 2014), and OpenFOAM coupled with DETCHEM (Diehm, 2013). Here, we discuss only the CFD simulations using FLUENT.

4.3.1 Impact on flow field without chemical conversion

First, the impact of the capillary on the volume flux for different axial and radial positions (xy plane = plane perpendicular to the flow direction z) was investigated. Four different positions of the capillary inside the channel are compared; the bending of the capillary, however, is neglected. Figure 2.22 (left) shows the velocity in the z-direction at the outlet (xy plane) of the reference channel (Ref) without probe taken from the CFD simulation for case A, where the capillary is positioned straight in the center axis of the channel probed (A). In configuration B, the probe is placed midway between channel center and a corner. In C and D, the probe is positioned in the center of a wall and in a channel corner, respectively. For the last two configurations, the shape of the probe was varied to guarantee a block-structured grid with good quality. This variation slightly increases the cross-sectional area of the probe, which is a conservative approximation with respect to the volume flux reduction in channel$_{probe}$. In all configurations, the probe tip is at $z_{probe} = 10$ mm, which means that the probe is completely launched into the monolith (Fig. 2.23).

The area-averaged velocity at the outlet of the channel with a probe is strongly dependent on the probe position and decreased from D to A. The area-averaged velocity at the outlet of a reference channel slightly changes vice versa because the reduction of volume flux in the channel with the probe is distributed toward the neighboring empty channels. From configuration A over B and C to D, the probe is positioned closer and closer to the corner of the channel. As the velocity profiles in the channels develop fast, the channels are "filled" with boundary. Hence, the probe lies more and more in the area of influence of the channel walls where the velocity

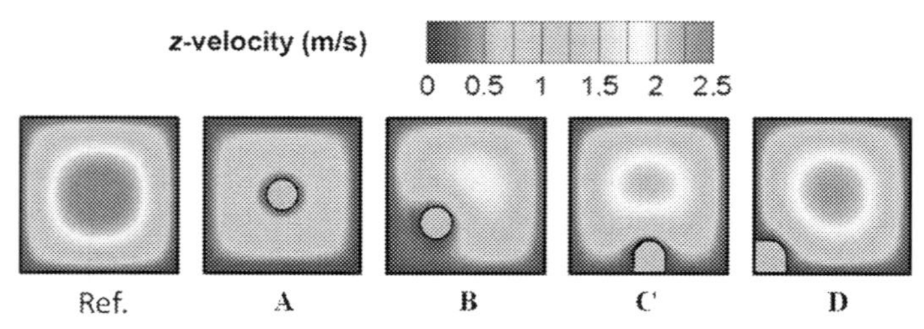

Figure 2.22 *z*-velocity in the *xy* plane at the outlet of the reference channel and four different configurations of the probe. *Figure taken from Hettel et al. (2013).*

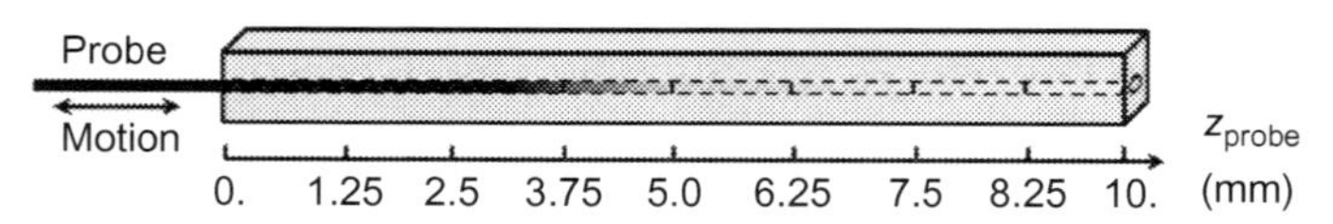

Figure 2.23 Positions of the probe tip z_{probe} used in the calculations. *Figure taken from Hettel et al. (2013).*

decreases to zero. The smaller the original velocity inside the reference channel, the smaller is the effect of the blocking and of the additional adhesion at the wall of a capillary, which is inserted into this region. The ratio of the volume flux in the channel with probe to the volume flux in the reference channel is 0.46 in case A and 0.91 in case D. Consequently, the capillary can modify the residence time in the channels by more than a factor of 2 (!) in the worst case (capillary located central location and moved throughout the entire monolith), even though the cross-sectional area of the capillary covers approximately 3.5% of the cross-sectional area of open channel only. In the experiment, the thin capillary bends and often touches the surface. The smallest measurement impact is in case D with less than 9% concerning the residence time. However, positioning the capillary on the walls or even in the corners of the channel may cause trouble when moving it through the channel.

For the case in which the probe is positioned in the center of the *xy* plane (configuration A in Fig. 2.22), the influence of the position of the probe tip in the *z*-direction was evaluated. Nine different positions of the probe tip were explored being in the range of $z_{probe} = 0–10$ mm (Fig. 2.23). In Fig. 2.24, the field of *z*-velocity for the tip position $z_{probe} = 5$ mm is shown. Launching the probe into the channel implies that a part of the volume flux of the formerly empty channel cannot flow through this channel anymore because it is blocked by the probe decreasing velocity in this channel (left bottom channel of the monolith section shown in Fig. 2.24). As a consequence, the velocity in all other channels increases slightly compared to a

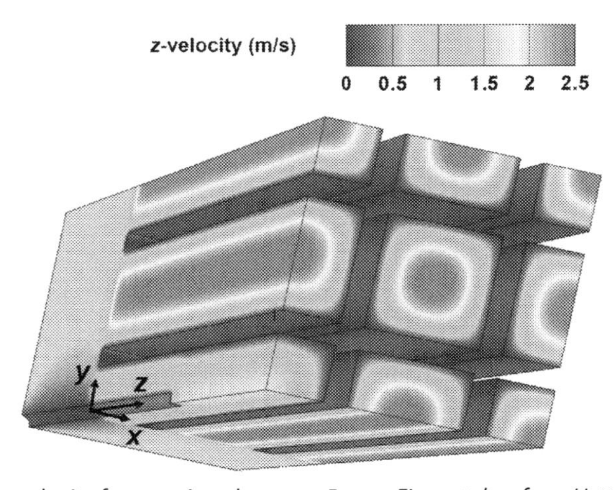

Figure 2.24 *z*-velocity for case A and $z_{probe} = 5$ mm. *Figure taken from Hettel et al. (2013).*

calculation where no probe would be considered. However, the velocity distribution in all channels without a probe is basically the same and not dependent from the distance to channel with the probe. This indicates that the volume flux blocked by the probe is equally distributed to all other empty channels.

Figure 2.25 shows isoplots of the velocity in the z-direction in the yz plane through the center of the channels including a region of 2 mm upstream of the monolith for three different z positions of the probe tip. The velocity profile in the channels is completely developed after a distance of approximately 0.5 mm. As expected, the smaller the velocity in the channel with the probe, the deeper the probe is launched into the channel. It should be noted that even if the probe tip is at $z_{\text{probe}}=0$ mm, the volume flux in channel with the probe is already reduced by 4%. It is furthermore observed that the ratio of the volume fluxes in the channel with probe and in the reference channel is rather independent from the inflow velocity and the wall temperature. This means that for a larger flow rate at the inlet, the velocities in all channels increase, but the ratio remains the same. Also, the suction at the probe tip has a negligible influence, even if the volume flux sucked is increased to 5 ml/min. The reason for this is that only a small portion of the volume flux of the entire channel is sucked in.

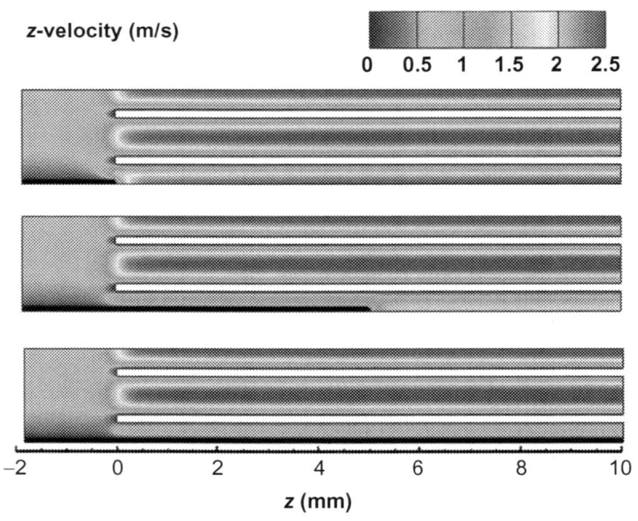

Figure 2.25 z-velocity for case A in the yz plane for three different positions $z_{\text{probe}}=0, 5$, and 10 mm (from top to bottom). The fluid flows from left to right; the white blocks indicate the solid parts of the monolith. *Figure taken from Hettel et al. (2013).*

4.3.2 Impact on species profiles (with chemical conversion)

Figure 2.26 shows distributions of concentrations, residence time, and temperature for configuration A for $z_{\text{probe}} = 5$ mm in the same yz plane as in Fig. 2.25. The concentrations of methane (A) and oxygen (B) decrease relatively fast due to total oxidation of methane, which forms water (C) with a maximum at ca. 2 mm. Afterward, the water is consumed by endothermic steam reforming, which yields hydrogen (D) and carbon monoxide (not shown). The comparison of the concentrations in the different channels shows that the entire process takes place earlier (in terms of the z position) in the channel with the probe than in a channel without a probe. The reference to the real time can be seen in the plot of the residence time (E). As the velocity in channel with probe is smaller than that in the reference channel, the residence time increases faster in the z-direction. Therefore, at the same z position, the reaction progress in the channel with probe is advanced further than in the reference channel. The entire process is dominated by the given wall temperature (F), as the velocity is relatively small. Except for the first millimeter, all quantities are mainly homogeneous in the cross-sectional

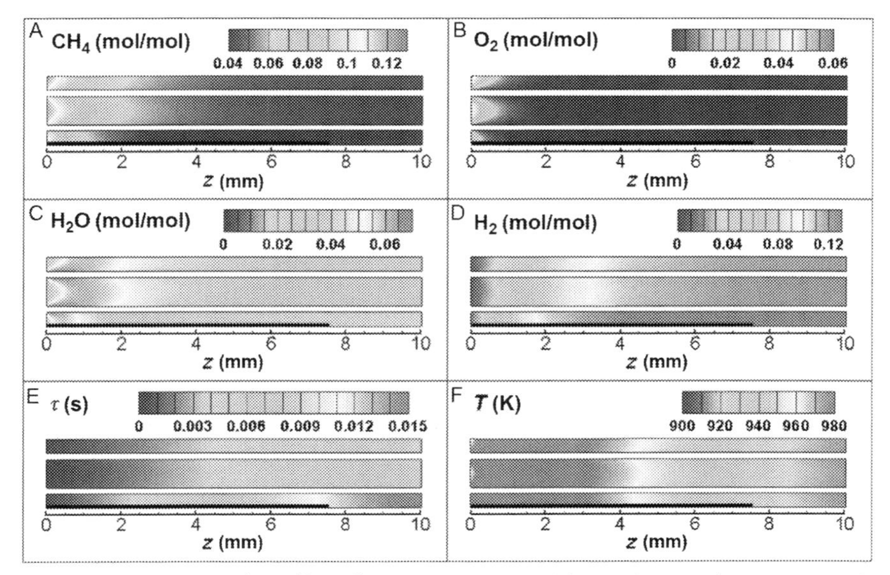

Figure 2.26 Computed profiles of concentrations, residence time, and temperature in the yz plane through channel centers (case A, $z_{\text{probe}} = 7.5$ mm). *Figure taken from Hettel et al. (2013).*

area of each channel, i.e., that strong radial mass-transfer limitation of the conversion rate occurs only in the entrance zone.

Figure 2.27 shows an enlargement of the region near the probe tip in Fig. 2.26. The reduced volume flux in the channel with probe leads to a residence time of 7 ms for the fluid at the probe tip ($z=5$ mm) versus a residence time of 4 ms at the same axial position in the reference channel. Therefore, the concentration of the hydrogen sucked in at the probe tip is higher than that in the reference channel (8% vs. 6% molar percentage).

The probe was launched into a channel located in the center of the monolith. The position of the probe in the *xy* plane cannot be controlled exactly but was expected to be near the central axis of the channel. The detailed surface reaction mechanism from the literature as discussed above was used without any modification. No single parameter was fitted to the experiment, neither in the mechanism nor in any other physical submodel used. The comparison between experimental and computed data for the reactant CH_4 and O_2 as well as the products H_2O and H_2 is shown along the channel axis in Fig. 2.28. The calculations shown here used exactly the parameters of the experimental data presented before. The solid lines present the CFD computed profiles in the reference channel, the dashed line the CFD computed profiles in the channel with the probe, and the symbols are the measured with the probe. Inside the channel of the catalyst ($z=0$–10 mm), the values for the empty reference channel (solid lines) differ from the measured data (square symbols). In contrast, the concentrations for

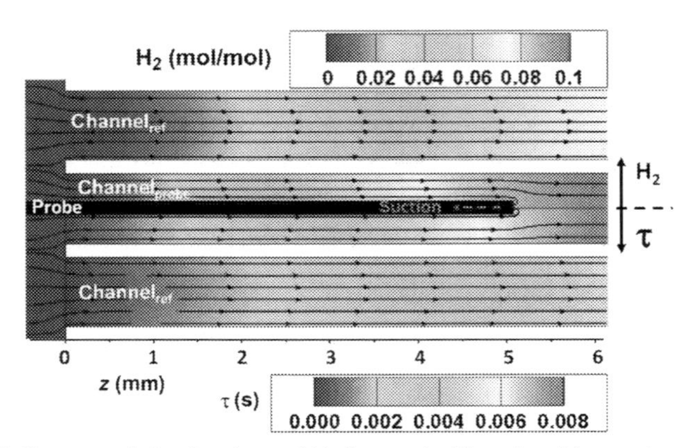

Figure 2.27 Computed distributions of H_2 (upper half) and residence time (lower half) in the *yz* plane through channel centers (case A, $z_{probe}=5$ mm). The flow is from left to right. The streamlines depict the direction of the flow and show the suction of the capillary. *Figure taken from Hettel et al. (2013).*

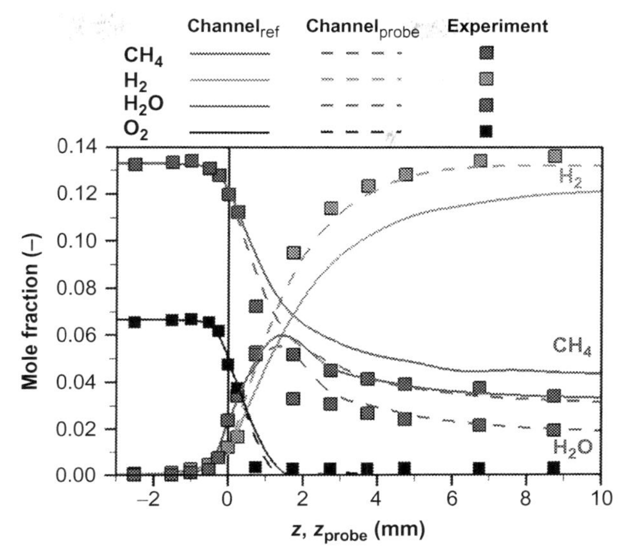

Figure 2.28 Comparison of computed (with and without probe) and measured concentration profiles as function of axial coordinate (case A). *Figure taken from Hettel et al. (2013).*

the different positions of the probe tip in the channel with the probe (dashed lines) agree very well with the measured values. Only the position of the maximum of H_2O is a little bit further downstream in the calculation compared to the measurement. The important statement that can be gained from this plot is the general influence of a probe, which is positioned in the channel center, on the data. CFD simulations are able to govern this influence. The difference between the experimental data (symbols) and the solid line represents the experimental error introduced by the modification of the flow field, and consequently the residence time, by the capillary. Even though we have a quantitative error only, i.e., the capillary technique will not change the reaction sequence itself, the error becomes significant when the measured data are used to derive kinetic expressions or to evaluate catalytic kinetics. It is highly recommended to account for this error in any case. However, the deviations between channel with capillary and without very much depend on the position of the capillary (axial and radial) and the chemical system. Hence, *a priori* quantification of the experimental uncertainty is a formidable task, which can be solved for quasi-isothermal systems (such as in exhaust-gas after-treatment) but not for systems determined by very exothermic reactions and strong temperature gradients (high-temperature catalysis) without three-dimensional CFD simulations.

In the region upstream of the catalyst, the influence of the capillary on the concentration profiles is negligible. Nevertheless, an interesting shape of the concentration profiles can be observed. For the calculations as well as for the experiments, the concentrations of oxygen and methane start to show a decrease already at $z = -0.5$ mm and the concentrations of water and hydrogen begin to increase from $z = -1$ mm. In the CFD simulations, potential (but unlikely) gas-phase reactions are not accounted for. This leaves diffusion as source for the decrease in reactants and increase in products upstream of the catalyst in the simulations. The operating conditions in the region between $-1 < z < 1$ mm, e.g., high temperature ($T = 690$ K) and large concentration gradients, favor diffusion effects. In this region, the diffusive flux of the species is in the same order of magnitude than the convective flux. The very good agreement between measured and calculated data upstream of the catalyst is a piece of evidence that also in the experiments, no gas-phase reactions take place. Additionally, simulations including a gas-phase mechanism (Golovitchev et al., 1999) using DETCHEM^{BATCH} (Deutschmann et al., 2014) did not show a conversion of gas phase species for the reaction conditions given in this region. Consequently, upstream diffusion matters even in these CPOX reactors of very high flow rates, which has been neglected so far, ever since L.D. Schmidt made short contact time reactors popular in the early 1990s.

5. CONCLUSIONS

Measurement techniques for the resolution of concentration and temperature profiles in chemical reactors with heterogeneously catalyzed gas-phase reactions are a very useful tool not only for a better understanding of the reaction sequence and derivation of reaction kinetics but also for the elucidation of the coupling between catalytic reaction kinetics and mass and heat transport. The combination of numerical simulations of the reactive flow in catalytic reactors incorporating microkinetic reaction schemes and those recently developed invasive and noninvasive *in situ* techniques can today support the optimization of reactor design and operating conditions in industrial applications.

In this chapter, we presented and, by discussing CFD simulations, critically analyzed two reactor configurations equipped with capillary technique for the resolution of species and temperature profiles, which have gained immense interest in the community recently: (1) the SFR, in which the reactants flow is directed on a catalytic plate and (2) catalytically coated monoliths. For both systems, the potentials and limitations have been discussed.

The SFRs can be used for laboratory-scale analysis of catalytic reaction schemes under technical operating conditions in an idealized flow regime, in which the surface can be considered zero dimensional. Hence, the entire catalytic surface is exposed to the same gas-phase environment; a homogeneous boundary layer is formed in radial direction on the catalytic plate. This configuration consequently allows one-dimensional measurements, e.g., by a movable capillary technique, to resolve the species concentration profiles in the boundary layer. The CFD simulation of this configuration as presented here and elsewhere can not only help to interpret the experimental data and analyze reaction kinetics but also to investigate the limitations of the technique. While CFD simulations showed that the general flow field assumption, i.e., reduction to a 1D problem, holds true, they also showed that internal mass transfer inside the catalytic coating easily becomes significant when a porous catalyst coating is used. Hence, the interpretation of the kinetic data derived from the experimentally determined species profiles needs to account for internal mass transfer aside from the externally formed boundary layer. Here, we presented several approaches to account for internal mass transport in SFRs, and also a recently developed computer code, DETCHEM^{STAG}, in which these models are implemented.

The resolution of species profiles within honeycomb foam structured catalysts by a capillary sampling technique has become rather popular in the last years, helping to understand chemical reaction kinetics quantitatively in particular in the field of exhaust-gas after-treatment and high-temperature catalysis. A recent experimental realization of this technique has been presented in this chapter. Very valuable information can be gained by those measurements, which again can be used to derive and evaluate detailed kinetic schemes. The derivation of the kinetics requires at least some kind of flow field models. Here, again, CFD can help to interpret the data by accounting for external and internal mass and heat transport. However, CFD can also be used to critically evaluate the invasive technique. In this chapter, we exemplarily showed how CFD simulations allow a quantifying of the impact of the capillary on the flow field, residence time, and species profiles measured and a computation of the uncertainty introduced. The study presented clearly reveals the necessity to consider the effect of the technique for the derivation and evaluation of microkinetic schemes.

As final conclusion, we would like to state that experiment and modeling have to go hand in hand to interpret the observed reactor behavior and to develop reliable models that can be used in up-scaling and optimizing the technical process.

ACKNOWLEDGMENTS

The work presented includes studies of former and current MS and PhD students and postdocs in our group at the Karlsruhe Institute of Technology; in particular, we would like to mention the PhD theses by Canan Karakaya, Claudia Diehm, and Hüsyein Karadeniz as well as the MS thesis by Bentolhoda Torkashvand and would also like to thank Lubow Maier and Steffen Tischer. We very much appreciate the collaboration with Colorado School of Mines (Robert J. Kee) and Politecnico di Milano (Dario Livio, Alessandro Donazzi, Alessandra Beretta) in setting up the stagnation flow and catalytic honeycomb reactors, respectively. Thanks to Yvonne Dedecek (KIT) for editorial assistance. Financial support by the German Research Foundation (DFG), Helmholtz Association via the Helmholtz Research School Energy-Related Catalysis, the European Union, Steinbeis GmbH für Technologietransfer, and many industrial partners is gratefully acknowledged.

REFERENCES

Alopaeus V, Laavi H, Aittamaa J: A dynamic model for plug flow reactor state profiles, *Comput Chem Eng* 32:1494–1506, 2008.

Anderson JA: Co oxidation on Rh/Al_2O_3 catalysts, *J Chem Soc Faraday Trans* 87:3907–3911, 1991.

ANSYS® FLUENT Academic Research, Release 13.0, ANSYS, Inc.

Behrendt F, Deutschmann O, Maas U, Warnatz J: Simulation and sensitivity analysis of the heterogeneous oxidation of methane on a platinum foil, *J Vac Sci Technol A* 13:1373–1377, 1995.

Beretta A, Donazzi A, Livio D, et al: Optimal design of a CH4 CPO-reformer with honeycomb catalyst: combined effect of catalyst load and channel size on the surface temperature profile, *Catal Today* 171:79–83, 2011.

Bitsch-Larsen A, Horn R, Schmidt LD: Catalytic partial oxidation of methane on rhodium and platinum: spatial profiles at elevated pressure, *Appl Catal A Gen* 348:165–172, 2008.

Bourane A, Bianchi D: Oxidation of CO on a Pt/Al_2O_3 catalyst: from the surface elementary steps to light-off tests: V. Experimental and kinetic model for light-off tests in excess of O-2, *J Catal* 222:499–510, 2004.

Bridges CL: An experimental study of catalytic partial oxidation of CH_4 on rhodium-strontium hexaaluminate in a stagnation-flow boundary layer. MS Thesis, Division of Engineering, Colorado School of Mines, Golden, 2006.

Brinkmeier C, Opferkuch F, Tuttlies U, Schmeisser V, Bernnat J, Eigenberger G: Car exhaust fumes purification—a challenge for procedure technology, *Chem Ing Tech* 77:1333–1355, 2005.

Coltrin ME, Kee RJ, Evans GH: A mathematical-model of the fluid-mechanics and gas-phase chemistry in a rotating-disk chemical vapor-deposition reactor, *J Electrochem Soc* 136:819–829, 1989.

Deutschmann O: Computational fluid dynamics simulation of catalytic reactors. In Ertl HKG, Schüth F, Weitkamp J, editors: *Handbook of heterogeneous catalysis*, Weinheim, 2008, Wiley-VCH, pp 1811–1828.

Deutschmann O: High temperature catalysis: efficient way for chemical conversion of logistic fuels, *Chem Ing Tech* 83:1954–1964, 2011a.

Deutschmann O, editor: *Modeling and simulation of heterogeneous catalytic reactions: from the molecular process to the technical system*, Weinheim, 2011b, Wiley-VCH.

Deutschmann O, Schmidt LD: Modeling the partial oxidation of methane in a short-contact-time reactor, *AIChE J* 44:2465–2477, 1998.

Deutschmann O, Behrendt F, Warnatz J: Modeling and simulation of heterogeneous oxidation of methane on a platinum foil, *Catal Today* 21:461–470, 1994a.
Deutschmann O, Behrendt F, Warnatz J: Modelling and simulation of heterogeneous oxidation of methane on a platinum oil, *Catal Today* 21:461–470, 1994b.
Deutschmann O, Schmidt R, Behrendt F, Warnat J: Numerical modeling of catalytic ignition, *Symp Combust* 26:1747–1754, 1996.
Deutschmann O, Schwiedernoch R, Maier LI, Chatterjee D: Natural gas conversion in monolithic catalysts: interaction of chemical reactions and transport phenomena. In Iglesia JJSE, Fleisch TH, editors: *Studies in surface science and catalysis*, Amsterdam, 2001, Elsevier, pp 251–258.
Deutschmann O, Tischer S, Kleditzsch S, Janardhanan VM, Correa C, Chatterjee D, Mladenov N, Minh HD, Karadeniz H, Hettel M: DETCHEM Software package, 2.5 ed. Karlsruhe, Germany, 2014.
Diehm C: *Catalytic reforming of fuels over noble metal-coated honeycomb monoliths: capillary-based in-situ sampling technique,* Dissertation (Ph.D. Thesis), 2013, Faculty of Chemistry and Biosciences, Karlsruhe Institute for Technology.
Diehm C, Deutschmann O: Hydrogen production by catalytic partial oxidation of methane over staged Pd/Rh coated monoliths: spatially resolved concentration and temperature profiles, *Int J Hydrogen Energ*, 2014. http://dx.doi.org/10.1016/j.ijhydene.2014.06.094 (Article In Press).
Dixon AG, Taskin ME, Nijemeisland M, Stitt EH: Wall-to-particle heat transfer in steam reformer tubes: CFD comparison of catalyst particles, *Chem Eng Sci* 63: 2219–2224, 2008.
Donazzi A, Michael BC, Schmidt LD: Chemical and geometric effects of Ce and washcoat addition on catalytic partial oxidation of CH4 on Rh probed by spatially resolved measurements, *J Catal* 260:270–275, 2008.
Donazzi A, Maestri M, Michael BC, et al: Microkinetic modeling of spatially resolved autothermal CH_4 catalytic partial oxidation experiments over Rh-coated foams, *J Catal* 275:270–279, 2010.
Donazzi A, Livio D, Beretta A, Groppi G, Forzatti P: Surface temperature profiles in CH_4 CPO over honeycomb supported Rh catalyst probed with in situ optical pyrometer, *Appl Catal A Gen* 402:41–49, 2011a.
Donazzi A, Livio D, Maestri M, et al: Synergy of homogeneous and heterogeneous chemistry probed by in situ spatially resolved measurements of temperature and composition, *Angew Chem Int Ed Engl* 50:3943–3946, 2011b.
Donazzi A, Livio D, Diehm C, Beretta A, Groppi G, Forzatti P: Effect of pressure in the autothermal catalytic partial oxidation of CH_4 and C_3H_8: spatially resolved temperature and composition profiles, *Appl Catal A Gen* 469:52–64, 2014.
Dooling DJ, Rekoske JE, Broadbelt LJ: Microkinetic models of catalytic reactions on nonuniform surfaces: application to model and real systems, *Langmuir* 15:5846–5856, 1999.
Dupont V, Zhang SH, Williams A: Experiments and simulations of methane oxidation on a platinum surface, *Chem Eng Sci* 56:2659–2670, 2001.
Evans GH, Greif R: Forced flow near a heated rotating disk: a similarity solution, *Numer Heat Transfer* 14:373–387, 1988.
Fernandes NE, Park YK, Vlachos DG: The autothermal behavior of platinum catalyzed hydrogen oxidation: experiments and modeling, *Combust Flame* 118:164–178, 1999.
Fogler SH: *Elements of chemical reaction engineering*. Harlow, England, 2006, Pearson Education Limited. pp 645–756.
Försth M, Gudmundson F, Persson JL, Rosén A: The influence of a catalytic surface on the gas-phase combustion of $H_2 + O_2$, *Combust Flame* 119:144–153, 1999.
Golovitchev VI, Tao F, Chomiak J: Numerical evaluation of soot formation control at diesel-like conditions by reducing fuel injection timing, SAE Technical Paper 1999-01-3552, 1999.

Grunwaldt JD, Baiker A: Axial variation of the oxidation state of Pt-Rh/Al_2O_3 during partial methane oxidation in a fixed-bed reactor: an in situ X-ray absorption spectroscopy study, *Catal Lett* 99:5–12, 2005.
Grunwaldt JD, Molenbroek AM, Topsøe NY, Topsøe H, Clausen BS: In situ investigations of structural changes in Cu/ZnO catalysts, *J Catal* 194:452–460, 2000.
Grunwaldt JD, Hannemann S, Schroer CG, Baiker A: 2D-mapping of the catalyst structure inside a catalytic microreactor at work: partial oxidation of methane over Rh/Al_2O_3, *J Phys Chem B* 110:8674–8680, 2006.
Habisreuther P, Philipp M, Eickhoff H, Leuckel W: Mathematical modeling of turbulent swirling flames. In *High intensity combustors—steady isobaric combustion*, Weinheim, Germany, 2005, Wiley-VCH Verlag, pp 156–175.
Hannemann S, Grunwaldt J-D, van Vegten N, Baiker A, Boye P, Schroer CG: Distinct spatial changes of the catalyst structure inside a fixed-bed microreactor during the partial oxidation of methane over Rh/Al_2O_3, *Catal Today* 126:54–63, 2007.
Hartmann M: *Erzeugung von Wasserstoff mittels katalytischer Partialoxidation höherer Kohlenwasserstoffe an Rhodium,* Dissertation (PhD thesis), 2009, Department of Chemistry and Biosciences, Karlsruhe Institute of Technology.
Hartmann M, Lichtenberg S, Hebben N, Zhang D, Deutschmann O: Experimental investigation of catalytic partial oxidation of model fuels under defined constraints, *Chem Ing Tech* 81:909–919, 2009.
Haruta M, Tsubota S: Catalysis and electrocatalysis at nanoparticle surfaces. In Andrzej W, Savinova ER, Vayenas Constantino G, editors: *Effects of size and contact structure of supported noble metal catalysts in low temperature CO oxidation*, New York, 2009, Taylor & Francis, pp 645–666.
Hauff K, Boll W, Chan D, et al: Macro- and microkinetic simulation of DOC: effect of aging, noble metal loading and platinum oxidation, *Chem Ing Tech* 85:673–685, 2013.
Hayes RE, Kolaczkowskib ST, Li PKC, Awdry S: Evaluating the effective diffusivity of methane in the washcoat of a honeycomb monolith, *Appl Catal B Environ* 25:93–104, 2000.
Hayes RE, Mok PK, Mmbaga J, Votsmeier M: A fast approximation method for computing effectiveness factors with non-linear kinetics, *Chem Eng Sci* 62:2209–2215, 2007.
Hayes RE, Fadic A, Mmbaga J, Najafi A: CFD modelling of the automotive catalytic converter, *Catal Today* 188:94–105, 2012.
Hettel M, Diehm C, Torkashvand B, Deutschmann O: Critical evaluation of in situ probe techniques for catalytic honeycomb monoliths, *Catal Today* 216:2–10, 2013.
Hopstaken MJP, Niemantsverdriet JW: Structure sensitivity in the CO oxidation on rhodium: effect of adsorbate coverages on oxidation kinetics on Rh(100) and Rh(111), *J Chem Phys* 113:5457–5465, 2000.
Horn R, Degenstein NJ, Williams KA, Schmidt LD: Spatial and temporal profiles in millisecond partial oxidation processes, *Catal Lett* 110:169–178, 2006a.
Horn R, Williams KA, Degenstein NJ, Schmidt LD: Syngas by catalytic partial oxidation of methane on rhodium: mechanistic conclusions from spatially resolved measurements and numerical simulations, *J Catal* 242:92–102, 2006b.
Horn R, Williams KA, Degenstein NJ, Schmidt LD: Mechanism of and CO formation in the catalytic partial oxidation of on Rh probed by steady-state spatial profiles and spatially resolved transients, *Chem Eng Sci* 62:1298–1307, 2007a.
Horn R, Williams KA, Degenstein NJ, et al: Methane catalytic partial oxidation on autothermal Rh and Pt foam catalysts: oxidation and reforming zones, transport effects, and approach to thermodynamic equilibrium, *J Catal* 249:380–393, 2007b.
Horn R, Korup O, Geske M, Zavyalova U, Oprea I, Schlögl R: Reactor for in situ measurements of spatially resolved kinetic data in heterogeneous catalysis, *Rev Sci Instrum* 81:064102, 2010.

Houtman C, Graves DB, Jensen KF: CVD in a stagnation point flow—an evaluation of the classical 1D-treatment, *J Electrochem Soc* 133:961–970, 1986a.

Houtman C, Graves DB, Jensen KF: CVD in stagnation point flow—an evaluation of the classical 1D-treatment, *J Electrochem Soc* 133:961–970, 1986b.

Ikeda H, Sato J, Williams FA: Surface kinetics for catalytic combustion of hydrogen-air mixtures on platinum at atmospheric-pressure in stagnation flows, *Surf Sci* 326:11–26, 1995.

Kaltschmitt T: *Catalytic partial oxidation of higher hydrocarbon fuels for hydrogen production—process investigation with regard to the concept of an auxiliary power unit,* Dissertation (PhD thesis), 2012, Department of Chemistry and Biosciences, Karlsruhe Institute of Technology.

Kaltschmitt T, Deutschmann O: Chapter 1—fuel processing for fuel cells. In Kai S, editor: *Advances in chemical engineering*, San Diego, California, USA, 2012, Academic Press, pp 1–64.

Karadeniz H: *Numerical modeling of stagnation flows over porous catalytic surfaces,* Dissertation (Ph.D. Thesis), 2014, Fakultät für Maschinenbau, Karlsruher Institut für Technologie, submitted.

Karadeniz H, Karakaya C, Tischer S, Deutschmann O: Numerical modeling of stagnation-flows on porous catalytic surfaces: CO oxidation on Rh/Al_2O_3, *Chem Eng Sci* 104:899–907, 2013.

Karakaya C: *A novel, hierarchically developed surface kinetics for oxidation and reforming of methane and propane over Rh/Al_2O_3,* Dissertation (Ph.D. Thesis), 2013, Faculty of Chemistry and Biosciences, Karlsruhe Institute for Technology.

Karakaya C, Deutschmann O: A simple method for CO chemisorption studies under continuous flow: adsorption and desorption behavior of Pt/Al_2O_3 catalysts, *Appl Catal A Gen* 445:221–230, 2012.

Karakaya C, Deutschmann O: Kinetics of hydrogen oxidation on Rh/Al_2O_3 catalysts studied in a stagnation-flow reactor, *Chem Eng Sci* 89:171–184, 2013.

Karakaya C, Otterstätter R, Maier L, Deutschmann O: Kinetics of the water-gas shift reaction over Rh/Al_2O_3 catalysts, *Appl Catal A Gen* 470:31–44, 2014.

Kee RJ, Rupley FM, Miller JA, et al: *CHEMKIN Collection, Release 3.6,* San Diego, 2001, Reaction Design, Inc.

Kee RJ, Coltrin ME, Glarborg P: *Chemically reacting flow, theory and practice,* Hoboken, New Jersey, USA, 2003, John Wiley & Sons.

Kee RJ, Coltrin ME, Glarborg P: *Chemically reacting flow, theory and practice,* Hoboken, New Jersey, USA, 2005a, John Wiley & Sons.

Kee RJ, Coltrin ME, Glarborg P: *Chemically reacting flow, theory and practice*, Hoboken, New Jersey, USA, 2005b, John Wiley & Sons, Inc., pp 249–308

Korup O, Mavlyankariev S, Geske M, Goldsmith CF, Horn R: Measurement and analysis of spatial reactor profiles in high temperature catalysis research, *Chem Eng Process Process Intensif* 50:998–1009, 2011.

Livio D, Donazzi A, Beretta A, Groppi G, Forzatti P: Experimental and modeling analysis of the thermal behavior of an autothermal C_3H_8 catalytic partial oxidation reformer, *Ind Eng Chem Res* 51:7573–7583, 2012.

Livio D, Diehm C, Donazzi A, Beretta A, Deutschmann O: Catalytic partial oxidation of ethanol over Rh/Al_2O_3: spatially resolved temperature and concentration profiles, *Appl Catal A Gen* 467:530–541, 2013.

Ljungström S, Kasemo B, Rosen A, Wahnström T, Fridell E: An experimental study of the kinetics of OH and H_2O formation on Pt in the $H_2 + O_2$ reaction, *Surf Sci* 216:63–92, 1989.

Lyubovsky M, Roychoudhury S, LaPierre R: Catalytic partial "oxidation of methane to syngas" at elevated pressures, *Catal Lett* 99:113–117, 2005.

Maier L, Schadel B, Delgado KH, Tischer S, Deutschmann O: Steam reforming of methane over nickel: development of a multi-step surface reaction mechanism, *Top Catal* 54:845–858, 2011.

Mantzaras J: New directions in advanced modeling and in situ measurements near reacting surfaces, *Flow Turbul Combust* 90:681–707, 2013.
McClure SM, Goodman DW: New insights into catalytic CO oxidation on Pt-group metals at elevated pressures, *Chem Phys Lett* 469:1–13, 2009.
McDaniel AH, Lutz AE, Allendorf MD, Rice SF: Effects of methane and ethane on the heterogeneous production of water from hydrogen and oxygen over platinum in stagnation flow, *J Catal* 208:21–29, 2002.
McGuire NE, Sullivan NP, Kee RJ, et al: Catalytic steam reforming of methane using Rh supported on Sr-substituted hexaaluminate, *Chem Eng Sci* 64:5231–5239, 2009.
McGuire NE, Sullivan NP, Deutschmann O, Zhu H, Kee RJ: Dry reforming of methane in a stagnation-flow reactor using Rh supported on strontium-substituted hexaaluminate, *Appl Catal A Gen* 394:257–265, 2011.
Mhadeshwar AB, Aghalayam P, Papavassiliou V, Vlachos DG: Surface reaction mechanism development for platinum-catalyzed oxidation of methane, *Proc Combust Inst* 29:997–1004, 2002.
Mhadeshwar AB, Wang H, Vlachos DG: Thermodynamic consistency in microkinetic development of surface reaction mechanisms, *J Phys Chem B* 107:12721–12733, 2003.
Michael BC, Nare DN, Schmidt LD: Catalytic partial oxidation of ethane to ethylene and syngas over Rh and Pt coated monoliths: spatial profiles of temperature and composition, *Chem Eng Sci* 65:3893–3902, 2010.
Oh IH, Takoudis CG, Neudeck GW: Mathematical-modeling of epitaxial silicon growth in pancake chemical vapor-deposition reactors, *J Electrochem Soc* 138:554–567, 1991.
Raja LL, Kee RJ, Serban R, Petzold LR: Computational algorithm for dynamic optimization of chemical vapor deposition processes in stagnation flow reactors, *J Electrochem Soc* 147:2718–2726, 2000.
Reinke M, Mantzaras J, Schaeren R, Bombach R, Kreutner W, Inauen A: Homogeneous ignition in high-pressure combustion of methane/air over platinum: comparison of measurements and detailed numerical predictions, *Proc Combust Inst* 29:1021–1029, 2002.
Rice SF, McDaniel AH, Hecht ES, Hardy AJJ: Methane partial oxidation catalyzed by platinum and rhodium in a high-temperature stagnation flow reactor, *Ind Eng Chem Res* 46:1114–1119, 2007.
Royer S, Duprez D: Catalytic oxidation of carbon monoxide over transition metal oxides, *ChemCatChem* 3:24–65, 2011.
Ruf B, Behrendt F, Deutschmann O, Warnatz J: Simulation of homoepitaxial growth on the diamond (100) surface using detailed reaction mechanisms, *Surf Sci* 352:602–606, 1996.
Ruf B, Behrendt F, Deutschmann O, Kleditzsch S, Warnatz J: Modeling of chemical vapor deposition of diamond films from acetylene-oxygen flames, *Proc Combust Inst* 28:1455–1461, 2000.
Sa J, Fernandes DLA, Aiouache F, et al: SpaciMS: spatial and temporal operando resolution of reactions within catalytic monoliths, *Analyst* 135:2260–2272, 2010.
Sari A, Safekordi A, Farhadpour FA: Comparison and validation of plug and boundary layer flow models of monolithic reactors: catalytic partial oxidation of methane on Rh coated monoliths, *Int J Chem React Eng* 6, 2008, Article A73.
Schneider A, Mantzaras J, Jansohn P: Experimental and numerical investigation of the catalytic partial oxidation of mixtures diluted with and in a short contact time reactor, *Chem Eng Sci* 61:4634–4649, 2006.
Schneider A, Mantzaras J, Bombach R, Schenker S, Tylli N, Jansohn P: Laser induced fluorescence of formaldehyde and Raman measurements of major species during partial catalytic oxidation of methane with large H_2O and CO_2 dilution at pressures up to 10 bar, *Proc Combust Inst* 31:1973–1981, 2007.
Schwiedernoch R, Tischer S, Correa C, Deutschmann O: Experimental and numerical study on the transient behavior of partial oxidation of methane in a catalytic monolith, *Chem Eng Sci* 58:633–642, 2003.

Sidwell RW, Zhu H, Kee RJ, Wickham DT, Schell C, Jackson GS: Catalytic combustion of premixed methane/air on a palladium-substituted hexaluminate stagnation surface, *Proc Combust Inst* 29:1013–1020, 2002.

Sidwell RW, Zhu H, Kee RJ, Wickham DT: Catalytic combustion of premixed methane-in-air on a high-temperature hexaaluminate stagnation surface, *Combust Flame* 134:55–66, 2003.

Song X, Schmidt LD, Aris R: The ignition criteria for stagnation-point flow: SemenovFrank-Kamenetski or van't Hoff, *Combust Sci Technol* 75:311–331, 1991.

Stutz MJ, Poulikakos D: Optimum washcoat thickness of a monolith reactor for syngas production by partial oxidation of methane, *Chem Eng Sci* 63:1761–1770, 2008.

Taskin ME, Dixon AG, Nijemeisland M, Stitt EH: CFD study of the influence of catalyst particle design on steam reforming reaction heat effects in narrow packed tubes, *Ind Eng Chem Res* 47:5966–5975, 2008.

Taylor JD, Allendorf MD, McDaniel AH, Rice SF: In situ diagnostics and modeling of methane catalytic partial oxidation on Pt in a stagnation-flow reactor, *Ind Eng Chem Res* 42:6559–6566, 2003.

Thormann J, Pfeifer P, Schubert K, Kunz U: Reforming of diesel fuel in a micro reactor for APU systems, *Chem Eng J* 135:S74–S81, 2008.

Thune PC, Niemantsverdriet JW: Surface science models of industrial catalysts, *Surf Sci* 603:1756–1762, 2009.

Tuttlies U, Schmeisser V, Eigenberger G: A mechanistic simulation model for NO_x storage catalyst dynamics, *Chem Eng Sci* 59:4731–4738, 2004.

Vanhardeveld RM, Gunter PLJ, Vanijzendoorn LJ, Wieldraaijer W, Kuipers EW, Niemantsverdriet JW: Deposition of inorganic salts from solution on flat substrates by spin-coating—theory, quantification and application to model catalysts, *Appl Surf Sci* 84:339–346, 1995.

Veser G, Frauhammer J: Modelling steady state and ignition during catalytic methane oxidation in a monolith reactor, *Chem Eng Sci* 55:2271–2286, 2000.

Vlachos DG: Homogeneous-heterogeneous oxidation reactions over platinum and inert surfaces, *Chem Eng Sci* 51:2429–2438, 1996.

Warnatz J: Resolution of gas phase and surface combustion chemistry into elementary reactions, *Symp Combust* 24:553–579, 1992.

Warnatz J, Allendorf MD, Kee RJ, Coltrin ME: A model of elementary chemistry and fluid-mechanics in the combustion of hydrogen on a platinum surfaces, *Combust Flame* 96:393–406, 1994a.

Warnatz J, Allendorf MD, Kee RJ, Coltrin ME: A model of elementary chemistry and fluid mechanics in the combustion of hydrogen on platinum surfaces, *Combust Flame* 96:393–406, 1994b.

Wehinger GD, Eppinger T, Kraume M: Fluidic effects on kinetic parameter estimation in lab-scale catalysis testing—a critical evaluation based on computational fluid dynamics, *Chem Eng Sci* 111:220–230, 2014.

Zapf R, Becker-Willinger C, Berresheim K, et al: Detailed characterization of various porous alumina-based catalyst coatings within microchannels and their testing for methanol steam reforming, *Chem Eng Res Des* 81:721–729, 2003.

Zellner A, Suntz R, Deutschmann O: In situ investigations of catalytic NO reduction inside an optically accessible flow reactor, *Chem Ing Tech* 86:538–543, 2014.

CHAPTER THREE

Catalytic Combustion of Hydrogen, Challenges, and Opportunities

John Mantzaras
Paul Scherrer Institute, Combustion Research, Villigen, Switzerland

Contents

Abstract

The fundamentals of hydrogen hetero-/homogeneous combustion and its applications to power generation systems are reviewed. Multidimensional numerical modeling and *in situ* spatially resolved measurements of gas-phase thermoscalars over the catalyst boundary layer have allowed investigation of the hydrogen catalytic and gaseous kinetics at industrially relevant conditions. Combination of advanced numerical modeling and *in situ* near-wall species and velocity measurements is used to firstly validate hydrogen kinetics and then to address the interplay between interphase fluid transport (laminar or turbulent) and hetero-/homogeneous chemistry. For the hydrogen fuel, key

Advances in Chemical Engineering, Volume 45
ISSN 0065-2377
http://dx.doi.org/10.1016/B978-0-12-800422-7.00003-0

parameters are the low-temperature homogeneous ignition kinetics and its intricate pressure dependence, flame propagation characteristics, diffusional imbalance of hydrogen that leads to superadiabatic surface temperatures, and fuel leakage through the gaseous combustion zone. Methodologies for thermal management of hydrogen catalytic reactors are presented, and the impact of hydrogen on the catalytic oxidation of other commercial fuels is finally outlined.

NOMENCLATURE

A cross-sectional flow area
A_s cross-sectional area of solid wall
B ratio of catalytically active to geometrical surface area
c_p, c_s specific heat of gas at constant pressure, specific heat of solid
$D_{k\ell}$ multicomponent diffusion coefficient between species k and ℓ
D_k^T thermal diffusion coefficient of kth species
D_{km} mixture average species diffusion coefficient of kth species
E activation energy
f fanning friction factor
h total enthalpy
$\underline{\underline{I}}$ unity matrix
k_s thermal conductivity of solid
K_g total number of gas-phase species
M sum of surface and bulk species for a given surface phase
Le Lewis number (thermal over species diffusivity)
p pressure
P channel wetted perimeter
Q heat of combustion per unit fuel mass
q_{rad} surface radiative flux
R^o universal gas constant
Re Reynolds number
$\dot{s}_k$ heterogeneous molar production rate of kth species
S_L laminar flame speed
T temperature
T_{ad} adiabatic equilibrium temperature
t_{ig}, t_{st} ignition time, steady-state time
u, v streamwise and transverse velocity
U_{IN} inlet streamwise velocity
$\vec{V}_k$ diffusion velocity vector for kth species
W_k molecular weight of kth species
$\bar{W}$ average mixture molecular weight
x, y, z streamwise, transverse, and lateral coordinate
X_k, $[X_k]$ mole fraction and concentration of kth gaseous species
Y_k mass fraction of kth gaseous species

GREEK LETTERS

α_k mass transfer coefficient of *k*th gas-phase species
α_T heat transfer coefficient
γ_k sticking coefficient of *k*th gas-phase species
Γ surface site density
δ_s channel wall thickness
ε surface emissivity
θ_m coverage of *m*th surface species
λ_g thermal conductivity of the gas
μ dynamic viscosity
ν''_{ki}, ν'_{ki} stoichiometric coefficients of *k*th species in reaction *i*
ρ, ρ_s gas density, solid density
σ Stefan–Boltzmann constant
σ_m site occupancy of *m*th surface species
φ fuel-to-oxidizer equivalence ratio
$\dot{\omega}_k$ molar production rate of *k*th gas-phase species

SUBSCRIPTS

∞ free-stream properties in flow over catalytic flat plate
ig ignition
IN inlet
k*, *m index for gas-phase species, index for surface species
w wall
s solid, surface
st steady state
x*, *y streamwise and transverse components

ABBREVIATIONS

CPO catalytic partial oxidation
CST catalytically stabilized thermal combustion
i-CST inverse catalytically stabilized thermal combustion
IGCC integrated gasification combined cycle

1. INTRODUCTION

Catalytic combustion of hydrogen and hydrogen-rich fuels has attracted increased attention in numerous engineering applications, which include large thermal power plants, microreactors for portable power generation, household burners, and recombiners in nuclear power plants for deflagration mitigation in the reactor containment during severe accidents. Moreover, heterogeneous processes involving hydrogen are of fundamental interest in many fuel-processing systems such as catalytic partial oxidation (CPO) of hydrocarbons and methanation of biogas, and also in polymer electrolyte fuel cells.

A recent approach for controlling greenhouse gas emissions is the precombustion CO_2 capture, applied mainly within the integrated gasification combined cycle (IGCC) power plants (Daniele et al., 2013; Nord et al., 2009). Via gasification, IGCC concepts convert solid and liquid hydrocarbons (biomasses, oil, coal, tars, etc.) to a gaseous syngas fuel, containing mostly hydrogen and carbon monoxide. Water gas shift (WGS) further converts most of CO to CO_2, which is subsequently captured prior to combustion. Apart from solid or liquid fuels, precombustion CO_2 capture technologies can also be economically applied to natural-gas-fueled turbine burners by reforming the natural gas to syngas (Tock and Marechal, 2012).

In addition to the aforementioned precombustion CO_2 capture approach that leads to hydrogen-rich fuels, hydrogen-containing fuels are also relevant for the latest postcombustion CO_2 capture concepts in natural-gas-fired power plants. Therein, the reactants are diluted with large exhaust gas recycle so as to increase the CO_2 content in the flue gases and thus to facilitate its subsequent capture (Cormos, 2011; Winkler et al., 2009). In such approaches, addition of hydrogen in the natural gas is highly advantageous as it increases the combustion stability of the heavily diluted reactive mixture (Griffin et al., 2004). Hence, combustion of hydrogen-rich syngas fuels is of prime interest for both pre- and postcombustion CO_2 capture methods in power generation. Besides large-scale power generation, hydrogen and hydrogen-rich fuels are also of interest in microreactors of small ($\sim$100 W_e) portable power units (Karagiannidis and Mantzaras, 2012; Norton et al., 2004). In such systems, hydrogen can be produced from methane or high hydrocarbons in suitable onboard microreformers (Schneider et al., 2008; Stefanidis et al., 2009).

Even though lean premixed combustion constitutes the mainstream approach in gas-fired power plants (Daniele et al., 2011), hybrid methodologies involving catalytic (heterogeneous) and gas-phase (homogeneous) combustion are intensively investigated in the last years as a means to reduce NO_x emissions and improve combustion stability (Carroni and Griffin, 2010; Schlegel et al., 1996). While hetero-/homogeneous combustion is an option for large power plants, it is usually the preferred method for microreactors. Prime reasons for this choice are the large surface-to-volume ratios of microreactors that favor surface catalytic over volumetric gaseous combustion, the occurrence of many undesirable homogeneous combustion instabilities driven by flame–wall interactions (Evans and Kyritsis, 2009; Kurdyumov et al., 2009; Maruta, 2010; Pizza et al., 2008a, 2008b, 2010a, 2012) and the efficient suppression of such instabilities by coating the reactor walls with an active catalyst (Pizza et al., 2009, 2010b).

In the conventional catalytically stabilized thermal (CST) combustion approach (Beebe et al., 2000; Carroni and Griffin, 2010; Carroni et al., 2003) shown in Fig. 3.1A, fractional fuel conversion is achieved in a catalytic honeycomb reactor operated at fuel-lean stoichiometries, while the remaining fuel is combusted in a follow-up gaseous combustion zone, again at fuel-lean stoichiometries. Nonetheless, for diffusionally imbalanced limiting reactants with Lewis numbers (*Le*) less than unity (such as H_2 whereby $Le_{H_2} \sim 0.3$ at fuel-lean stoichiometries in air), CST is compounded by the

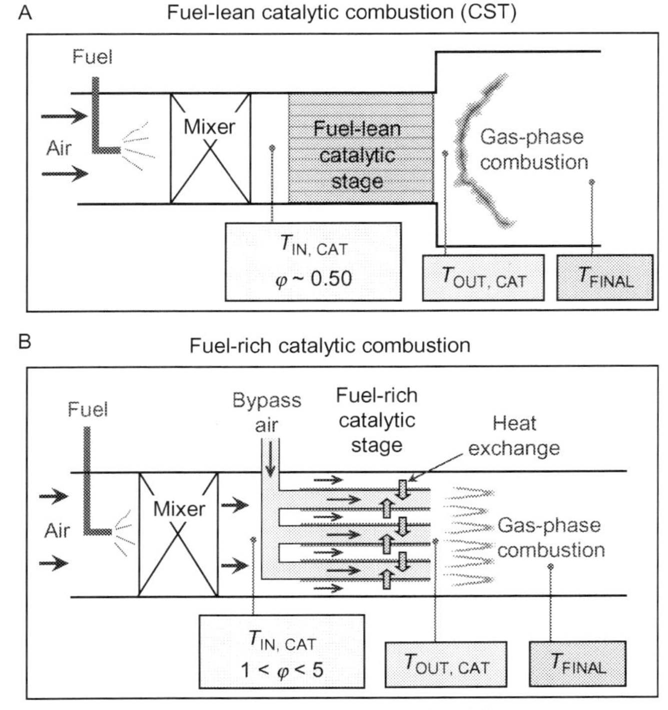

Figure 3.1 Hetero-/homogeneous combustion methodologies in power generation: (A) catalytically stabilized thermal combustion (CST) and (B) fuel-rich catalytic/gaseous-lean combustion.

attainment of superadiabatic surface temperatures (Appel et al., 2002a; Ghermay et al., 2010; Pfefferle and Pfefferle, 1986) that endanger the catalyst and reactor integrity.

Alternative to the fuel-lean CST concept in Fig. 3.1A is the more recent catalytic-rich/gaseous-lean combustion in Fig. 3.1B, initially proposed for natural-gas-fired plants (Schneider et al., 2007; Smith et al., 2006). In this concept, part of the air and all of the natural gas undergo CPO at fuel-rich stoichiometries to produce syngas. To this purpose, catalyst coating is applied to the outer surfaces of a bundle of tubes through which bypass air flows (Fig. 3.1B). The catalyst surfaces are cooled by the bypass air, thus avoiding reactor overheating. Syngas and unconverted reactants are subsequently mixed with the bypass air, forming an overall fuel-lean homogeneous combustion zone. The fuel-rich catalytic/fuel-lean gaseous combustion methodology has several advantages compared to the conventional fuel-lean CST, which include extended catalytic extinction limits (Schneider et al., 2008; Smith et al., 2006), control of catalytic combustion by the availability of oxygen that in turn prevents complete fuel consumption inside the catalytic stage even in the event of accidental gas-phase ignition, and enhanced stability of the follow-up flame due to the presence of highly reactive hydrogen contained in the syngas (Griffin et al., 2004).

The catalytic-rich/gaseous-lean combustion method in Fig. 3.1B is applicable not only to natural gas but also to pure hydrogen and syngas fuels. In this case, the catalytic stage does not have a prime CPO function, but rather acts as a preheater and stabilizer for the subsequent homogeneous combustion zone. This method is suitable for a wide range of fuels that include low calorific value biofuels, whereby flame stability is an issue, and also for hydrogen-rich syngases for which the conventional lean premixed gaseous combustion entails the risk of flame flashback. The aforementioned advantages have led to the development of integrated hetero-/homogeneous combustors based on the catalytic-rich/gaseous-lean approach for coal-derived syngas and also for pure hydrogen (Alavandi et al., 2012; Bolaños et al., 2013; Smith et al., 2006). In fuel-rich combustion of hydrogen or syngas, thermal management is not an overriding issue, as the catalyst temperature does not exceed the adiabatic equilibrium temperature. Catalysts in power generation systems are usually supported noble metals (Pd, Pt, or Rh). The support is a porous metal oxide layer (e.g., Al_2O_3, ZrO_2) that furnishes high surface area and aids the uniform dispersion of the precious metal.

Under the high-pressure operation of power systems, a complication is that homogeneous combustion may occur not only at the designated flame burnout stages in Fig. 3.1 but also inside the catalytic stages themselves (Ghermay et al., 2011; Karagiannidis et al., 2009, 2011; Reinke et al., 2005a). This happens despite the typically large surface-to-volume ratios of modern catalytic reactors (hydraulic diameters ~1 mm) that in turn favor catalytic over gaseous chemical reactions. Thus, this chapter emphasizes on the combined hetero-/homogeneous combustion of hydrogen rather than on sole catalytic combustion.

Further improvements in catalytic combustion technologies require the development of catalysts with enhanced activity and thermal stability, understanding the fundamental surface processes including catalyst–substrate interactions, knowledge of low-temperature homogeneous kinetics and its coupling with the corresponding heterogeneous kinetics. Moreover, the combustor geometries in Fig. 3.1 call for the development of multi-dimensional numerical tools with capabilities of detailed hetero-/homogeneous chemistry, transport, flow, and heat transfer in the solid wall.

This chapter is organized as follows. First, hydrogen heterogeneous chemistry (focusing on noble metal catalysts) and gaseous kinetics are elaborated in Section 2. State-of-the art numerical models for hetero-/homogeneous reactive flows are introduced in Section 3, while the impact of preferential hydrogen diffusion on the attained surface and gas-phase temperatures is discussed in Section 4. In Section 5, experiments and modeling lead to validation of hetero-/homogeneous chemical reaction schemes at fuel-lean and fuel-rich hydrogen/air stoichiometries. The coupling of hydrogen hetero-/homogeneous chemistry at high pressures and the effect of fluid turbulence is addressed in Section 6, while conventional and new hybrid combustion methodologies for hydrogen are introduced in Section 7. Finally, the impact of hydrogen on the catalytic combustion of other key fuels in power generation (methane and carbon monoxide) is reviewed in Section 8.

2. HYDROGEN HETERO-/HOMOGENEOUS CHEMISTRY

Essential inputs in any numerical model are salient heterogeneous and low-temperature homogeneous chemical reaction mechanisms. This section summarizes the current understanding of heterogeneous kinetics and its recent development, and briefly addresses the more mature field of homogeneous kinetics.

2.1. Heterogeneous Chemistry

Turnover frequencies depend on the local gas-phase species concentrations above the catalyst surface, the temperature, and the surface coverage. The surface reaction rate is specific to the catalyst formulation, such that there is a unique rate expression for every catalyst which depends not only on the active material but also on the support, the washcoat type and structure, and the preparation method. Different catalyst surface structures generally vary in their reaction pathways and their corresponding kinetic expressions.

Global catalytic chemical steps for hydrogen (Schefer, 1982) or low hydrocarbons (Song et al., 1990) were the preferred modeling choice for many years, with the reaction rate expressed in various ways (catalyst mass, catalyst volume, reactor volume, or catalyst external surface area). A commonly used approach for kinetic rates was the Langmuir–Hinshelwood–Hougen–Watson (LHHW) methodology (Hougen and Watson, 1943). LHHW is based on Langmuir adsorption, surface reaction between adsorbed intermediates and desorption, under the key assumption that one of these steps is rate limiting. Such an approach cannot account for the complex variety of all aforementioned surface phenomena, such that rate parameters must be evaluated experimentally for each new catalyst and different operating conditions.

Current kinetic studies focus on the development of detailed reaction mechanisms based on the elementary steps on the catalyst surface. Mean field is the most popular description, treating heterogeneous reactions in a way similar to gas-phase reactions. It entails the assumption that all adsorbates are randomly distributed on the catalyst surface, which is further considered uniform. The state of the catalytic surface is described by the temperature and a set of surface coverages θ_i, both depending on the macroscopic position in the reactor, which is an average over microscopic fluctuations. Reaction rates depend on gas phase as well as on surface and bulk species. Each surface structure (there can be several surface structures representing different materials, crystal structures, or reconstructions of the same crystal structure) has a surface site density Γ, defining the maximum number of species that can adsorb per unit catalyst area. The surface site density can be computed from the molecular structure of the catalyst material and should not be confused with the active catalytic surface area that is related to the catalyst loading. Surface site densities are of the order of 10^{-9} mol/cm^2, corresponding roughly to 10^{15} adsorption sites per cm^2. A coordination number σ_m is further associated with each surface species m, accounting for the number of surface sites it occupies. Under the aforementioned mean-field assumptions, multistep reaction mechanisms can be constructed (Kee et al., 2003). The net production rate $\dot{s}_k$ of the kth gaseous species due to adsorption and desorption on the catalyst is then described in a manner analogous to gas-phase chemical reactions:

$$\dot{s}_k = \sum_{i=1}^{N_R} \nu_{ki} \left\{ k_{fi} \prod_{j=1}^{k} [X_j]^{\nu'_{ji}} - k_{ri} \prod_{j=1}^{k} [X_j]^{\nu''_{ji}} \right\} \quad (3.1)$$

with N_R be the total number of surface reactions including adsorption and desorption, K be the number of all species (gas, surface, and bulk), ν''_{ji} and ν'_{ji} be the stoichiometric coefficients of species j on the product and reactant sides of reaction i, respectively, k_{f_i} and k_{r_i} be the forward and reverse rate coefficients for the ith reaction, and $[X_j]$ be the concentration of species j. Lateral interactions between surface species on the catalytic surface are ignored in the mean-field approach. This is justified since the computational cells used in a reactor simulation are typically much larger than the range of lateral surface interactions. Thus, for any reactor configuration, changes of surface coverage do not include spatial gradients:

$$\frac{d\theta_m}{dt} = \sigma_m \frac{\dot{s}_m}{\Gamma} - \frac{\theta_m}{\Gamma}\frac{d\Gamma}{dt}, \quad m = 1, \ldots, M_s \quad (3.2)$$

with M_s be the number of surface species. The second term in the right side of Eq. (3.2) denotes a nonconserved surface site density, although in most catalytic reaction schemes Γ is usually conserved. Because the adsorption binding energies on the surface vary with surface coverage, the rate coefficients in Eq. (3.1) become complicated functions (Kee et al., 2003):

$$k_{f_k} = A_k T^{\beta_k} \exp\left[\frac{-E_{a_k}}{R^{\circ} T}\right] \prod_{i=1}^{M_s} \theta_i^{\mu_{i_k}} \exp\left[\frac{\varepsilon_{i_k}\theta_i}{R^{\circ} T}\right] \quad (3.3)$$

The parameters μ_{i_k} and ε_{i_k} describe the dependence of the rate coefficients on the surface coverage of the ith species. For adsorption reactions, sticking coefficients are commonly used and are converted to conventional rate coefficients as follows:

$$k_{f_i}^{\text{ads}} = \left(\frac{\gamma_i}{1-\gamma_i/2}\right)\frac{1}{\Gamma^m}\sqrt{\frac{R^{\circ} T}{2\pi W_i}} \quad (3.4)$$

with γ_i be the sticking coefficient of the ith gaseous species and m be the number of sites occupied by the adsorbing species. The denominator

$(1-\gamma_i/2)$ in Eq. (3.4) is the Motz–Wise correction (Motz and Wise, 1960), accounting for non-Maxwellian velocity distribution of the molecules when the sticking coefficient is large. This factor has recently been updated to $(1-\gamma_i\Theta_{\text{free}}/2)$, with$\Theta_{\text{free}}$ be the free-site coverage, in Dogwiler et al. (1999).

Effects resulting from lateral adsorbate interactions are either neglected or incorporated by mean-rate coefficients and mean-coverage dependencies. However, a variety of adsorbate–adsorbate interactions have been experimentally observed, which include two-dimensional (2D) phase transitions depending on temperature and coverage, island formation of adsorbed species, dependence of sticking probability, etc. When the specific surface interaction mechanisms are known, Monte Carlo simulation of the surface chemistry can be carried out at the nanometer scale (Kissel-Osterrieder et al., 2000). This kinetic modeling is, however, prohibitive for engineering system sizes.

Detailed (mean-field) heterogeneous chemical reaction schemes have relied primarily on ultra-high vacuum surface science experiments. Notwithstanding recent advances in *in situ* diagnostics, surface science measurements at practical conditions are still not feasible. Therefore, extension of reaction schemes to realistic pressures and technical catalysts has necessitated appropriate kinetic rate modifications in order to bridge the well-known "pressure and materials gap." These modifications were aided by kinetic measurements in a variety of laboratory reactors. Two basic reactor configurations have been singled out in the last years for kinetic measurements: (a) the nearly isothermal, low-temperature ($T \leq 600$ °C), gradientless in the radial direction tubular or annular flow reactors (Ciambelli et al., 1999; Groppi et al., 2001), which are fed with highly diluted fuel/oxidizer mixtures so as to maintain a small temperature rise, and (b) the stagnation-flow reactor (Deutschmann et al., 1996; Park et al., 1999). Although modeling of the former reactor is straightforward, the low-to-moderate temperature (necessary to avoid transport limitations) constrains its wide applicability. Stagnation-flow configurations, on the other hand, have provided a wealth of data on catalytic ignition/extinction, steady fuel conversion, and product selectivity under realistic temperatures and mixture compositions. These measurements, in conjunction with numerical predictions from well-established one-dimensional (1D) stagnation-flow codes (Coltrin et al., 1991), have aided considerably the refinement of surface reaction mechanisms.

More recently, a new methodology involving *in situ* spatially resolved Raman measurements of major gas-phase species concentrations across the boundary layer of a channel-flow catalytic reactor has been introduced (Appel et al., 2002a; Ghermay et al., 2011; Reinke et al., 2004; Schneider et al., 2007; Zheng et al., 2014), providing a direct way to assess the catalytic reactivity—as well as the gaseous reactivity when combined with laser-induced fluorescence (LIF) of radical species. The test rig includes a rectangular reactor and a cylindrical steel tank providing pressurization up to 20 bar (see Fig. 3.2A). The reactor comprises two horizontal Si[SiC] nonporous ceramic plates (300 mm long, 104 mm wide, placed 7 mm apart) and two 3-mm-thick vertical quartz windows. The inner Si[SiC] surfaces are coated with the catalyst of interest, while the surface temperatures are monitored with thermocouples (12 per plate) positioned along the x–y streamwise symmetry plane and embedded 0.9 mm beneath the catalyst surfaces. To establish finite-rate catalytic reactant conversion and onset of homogeneous ignition, the surface temperatures are controlled by a cooling/heating arrangement. The front faces of the ceramic plates are contacted to a water-cooled support metal frame, while two resistive heaters are positioned above the outer ceramic plate surfaces (see Fig. 3.2B). Optical accessibility is maintained from both reactor sides via two 35-mm-thick quartz windows on the tank; counterflow streamwise optical access is also available through a rear tank flange. *In situ*, laser-based spectroscopic techniques are applied, shown in their latest configuration in Fig. 3.2A (Ghermay

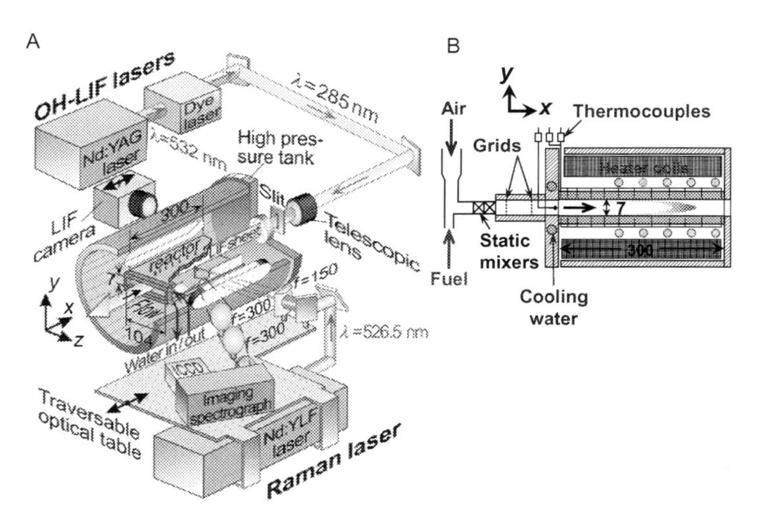

Figure 3.2 (A) Schematic of catalytic channel flow reactor used for kinetic evaluation and the optical arrangement of the Raman/OH-LIF setups (Ghermay et al., 2011; Zheng et al., 2014), (B) reactor detail. All distances and focal lengths are in mm.

et al., 2011; Zheng et al., 2014). The spectroscopic techniques include 1D spontaneous Raman of major gaseous species concentrations and temperature across the entire 7-mm channel height using an Nd:YLF laser and a planar LIF of trace species (typically OH or hot O_2 for hydrogen combustion) along the x–y symmetry plane using a tunable dye laser pumped by an Nd: YAG laser. All kinetic investigations are performed at steady and laminar flow conditions. Application of this methodology to hydrogen hetero-/homogeneous combustion over noble metals will be presented in Section 5 and for syngas heterogeneous combustion in Section 8.

Ultimately, the development of kinetic data involves optimization of the reaction parameters, many obtained with *ab initio* calculations (Czekaj et al., 2013; Kacprzak et al., 2012; Ludwig and Vlachos, 2008; Maestri and Reuter, 2011), through comparisons with experimental data obtained in the aforementioned reactor configurations. Detailed (mean field) kinetic schemes for the oxidation of H_2 over noble metals were thus constructed (Aghalayam et al., 2000; Deutschmann et al., 1996, 2000; Fridell et al., 1994; Hellsing et al., 1991; Kramer et al., 2002; Maestri et al., 2008; Park et al., 1999; Williams et al., 1992). The hydrogen oxidation mechanism on platinum from Deutschmann et al. (2000), which will be used in Sections 5–7, is provided in Table 3.1.

2.2. Homogeneous Chemistry

Although elementary homogeneous reaction schemes for hydrogen and low hydrocarbons have a much longer development period compared to catalytic kinetics, uncertainties still exist in the low-temperature gas-phase chemistry associated with catalytic applications. Moreover, the coupling between catalytic and gaseous chemical reaction pathways (via radical adsorption/desorption reactions and especially via catalytically formed major products, Appel et al., 2002a; Bui et al., 1996; Reinke et al., 2005a) necessitates simultaneous validation of heterogeneous and homogeneous chemical reaction schemes.

Gaseous reaction mechanisms in hetero-/homogeneous combustion can be assessed with homogeneous ignition studies and measurements of radical concentrations above the catalyst surface. Radical concentrations above the catalyst have been routinely measured with LIF, point, or planar (Appel et al., 2002a; Dogwiler et al., 1998; Forsth et al., 1999; Marks and Schmidt, 1991; Pfefferle et al., 1989; Reinke et al., 2005a, 2005b; Schneider et al., 2007). Using planar LIF of the OH or CH_2O radical in

Table 3.1 Catalytic Reaction Scheme for H_2 Oxidation on Pt[a]

		A (γ)	b	E
Adsorption reactions				
S1	$O_2 + 2Pt(s) \rightarrow 2O(s)$	0.023	0.0	0.0
S2	$O_2 + 2Pt(s) \rightarrow 2O(s)$	1.8×10^{21}	−0.5	0.0
S3	$H_2 + 2Pt(s) \rightarrow 2H(s)$	4.5×10^{10}	0.5	0.0
S4	$H + Pt(s) \rightarrow H(s)$	1.0	0.0	0.0
S5	$O + Pt(s) \rightarrow O(s)$	1.0	0.0	0.0
S6	$H_2O + Pt(s) \rightarrow H_2O(s)$	0.75	0.0	0.0
S7	$OH + Pt(s) \rightarrow OH(s)$	1.0	0.0	0.0
Surface reactions				
S8	$H(s) + O(s) = OH(s) + Pt(s)$	3.7×10^{21}	0.0	11.5
S9	$H(s) + OH(s) = H_2O(s) + Pt(s)$	3.7×10^{21}	0.0	17.4
S10	$OH(s) + OH(s) = H_2O(s) + O(s)$	3.7×10^{21}	0.0	48.2
Desorption reactions				
S11	$2O(s) \rightarrow O_2 + 2Pt(s)$	3.7×10^{21}	0.0	$213.2\text{–}60\theta_O$
S12	$2H(s) \rightarrow H_2 + 2Pt(s)$	3.7×10^{21}	0.0	$67.4\text{–}6\theta_H$
S13	$H_2O(s) \rightarrow H_2O + Pt(s)$	1.0×10^{13}	0.0	40.3
S14	$OH(s) \rightarrow OH + Pt(s)$	1.0×10^{13}	0.0	192.8

[a]The surface site density is $\Gamma = 2.7 \times 10^{-9}$ mol/cm^2. In the surface and desorption reactions, the reaction rate coefficient is $k = AT^b \exp(-E/RT)$, A (mole-cm-Kelvin-s) and E (kJ/mol). In all adsorption reactions, except S2 and S3, A denotes a sticking coefficient (γ). Reactions S1 and S2 are duplicate. Reaction S3 has an order of one with respect to platinum. The suffix (s) denotes a surface species and θ_i the coverage of surface species i.
From Deutschmann et al. (2000).

the catalytic reactor of Fig. 3.2 along with detailed numerical simulations, validated gas-phase mechanisms have been provided for fuel-lean and fuel-rich CH_4/air combustion at pressures up to 16 bar (Reinke et al., 2005a, 2007; Schneider et al., 2007). Similar investigations were performed for fuel-lean H_2/air at pressures up to 15 bar and fuel-rich H_2/air at pressures up to 5 bar, using OH-LIF and Raman measurements (Appel et al., 2002a; Ghermay et al., 2010, 2011; Mantzaras et al., 2009). For fuel-rich H_2/CO mixtures, a novel combination of OH-LIF at low pressures (up to 5 bar) and O_2-LIF (from 5 to 14 bar) has recently been applied (Schultze et al., 2014).

Validated heterogeneous and homogeneous reaction schemes can be reduced jointly, thus accounting for their coupling: reduction methodologies for combined hetero-/homogeneous reaction schemes have also advanced in the last years (Bui et al., 1997; Reinke et al., 2004; Yan and Maas, 2000).

The elementary H_2/O_2 reaction mechanism from Princeton University (Li et al., 2004), which is suitable for hetero-/homogeneous combustion of hydrogen as will be shown in Section 5, is provided in Table 3.2.

3. NUMERICAL MODELING OF HETEROGENEOUS AND HOMOGENEOUS COMBUSTION

State-of-the-art numerical models are reviewed, in this section, with emphasis on those employed in Sections 5–8. 1D Models with lumped heat and mass transport coefficients (Nusselt and Sherwood numbers) were initially used in catalytic combustion research (Cerkanowicz et al., 1977; Tien, 1981). These models are appealing for their simplicity, however, unique lumped heat and mass transport coefficients for the entry channel-flow problem pertinent to catalytic reactors are not guaranteed in the presence of hetero-/homogeneous chemical reactions. Although problem-specific transport correlations, some with dependence on surface Damköhler numbers (Groppi et al., 1995), may yield acceptable predictions for catalytic combustion alone (no presence of gaseous chemistry) and steady-state applications, a spatial dimensionality of at least two is required to correctly describe homogeneous combustion. This is because gas-phase combustion is strongly dependent on the reactant and temperature boundary layer profiles (Mantzaras and Appel, 2002; Mantzaras and Benz, 1999). It is emphasized that the contribution of the gaseous reaction pathway cannot be neglected in the catalytic combustion of hydrocarbons at elevated pressures (Karagiannidis et al., 2009; Mantzaras, 2006; Reinke et al., 2002, 2005a; Zheng et al., 2013a). The same applies for hydrogen, albeit its gaseous reactivity is a complicated function of pressure as will be discussed in Section 6.

Two-dimensional models realistically reproduce the in-channel fluid mechanical transport and, when coupled with detailed hetero-/homogeneous chemistry, provide a powerful tool for both reactor design and fundamental research. Three-dimensional (3D) reactive flow models are necessary in complex geometries with nonstraight channels and cross-flow between channels used for achieving reactor radial uniformity. Although 3D effects do play a role even for straight-channel honeycomb

Table 3.2 Homogeneous Chemical Reaction Mechanism for H_2[a]

		A	*b*	*E*
H_2/O_2 reactions				
R1	$H + O_2 = O + OH$	3.55×10^{15}	−0.41	69.45
R2	$O + H_2 = H + OH$	5.08×10^{4}	2.67	26.32
R3	$H_2 + OH = H_2O + H$	2.16×10^{8}	1.51	14.35
R4	$O + H_2O = OH + OH$	2.97×10^{6}	2.02	56.07
H_2/O_2 dissociation-recombination				
R5	$H_2 + M = H + H + M$	4.58×10^{19}	−1.40	436.73
R6	$O + O + M = O_2 + M$	6.16×10^{15}	−0.50	0.00
R7	$O + H + M = OH + M$	4.71×10^{18}	−1.0	0.00
R8	$H + OH + M = H_2O + M$	3.80×10^{22}	−2.00	0.00
HO_2 formation–consumption				
R9	$H + O_2 + M = HO_2 + M$	1.48×10^{12}	0.60	0.00
	$H + O_2 + M = HO_2 + M$	6.37×10^{20}	−1.72	2.18
R10	$HO_2 + H = H_2 + O_2$	1.66×10^{13}	0.00	3.43
R11	$HO_2 + H = OH + OH$	7.08×10^{13}	0.00	1.26
R12	$HO_2 + O = O_2 + OH$	3.25×10^{13}	0.00	0.00
R13	$HO_2 + OH = H_2O + O_2$	2.89×10^{13}	0.00	−2.09
H_2O_2 formation–consumption				
R14	$HO_2 + HO_2 = H_2O_2 + O_2$	4.22×10^{14}	0.00	50.12
R15	$HO_2 + HO_2 = H_2O_2 + O_2$	1.30×10^{11}	0.00	−6.82
R16	$H_2O_2 + M = OH + OH + M$	2.95×10^{14}	0.00	202.63
	$H_2O_2 + M = OH + OH + M$	1.20×10^{17}	0.00	190.37
R17	$H_2O_2 + H = H_2O + OH$	2.41×10^{13}	0.00	16.61
R18	$H_2O_2 + H = H_2 + HO_2$	4.81×10^{13}	0.00	33.26
R19	$H_2O_2 + O = OH + HO_2$	9.55×10^{6}	2.00	16.61
R20	$H_2O_2 + OH = H_2O + HO_2$	1.00×10^{12}	0.00	0.00
R21	$H_2O_2 + OH = H_2O + HO_2$	5.80×10^{14}	0.00	40.00

[a]Reaction rate $k = AT^b \exp(-E/RT)$, A (mole-cm-Kelvin-s), E (kJ/mol). Third body efficiencies in reactions R5-R8 and R16 are $\omega(H_2O) = 12.0$, $\omega(H_2) = 2.5$, $\omega(CO_2) = 3.8$; in reaction R9 $\omega(H_2O) = 11.0$, $\omega(H_2) = 2.0$, $\omega(O_2) = 0.78$, $\omega(CO_2) = 3.8$. Reaction pairs (R14, R15) and (R20, R21) are duplicate. Reactions R9 and R16 are Troe reactions centered at 0.8 and 0.5, respectively (second entries are the low-pressure limits).
From Li et al. (2004).

reactors, despite the rounding of the sharp corners in individual channels during catalyst coating (Hayes and Kolaczkowski, 1994) that in turn largely suppresses secondary flows, the associated computational effort is large and not always justified.

Navier-Stokes (elliptic) 2D models with detailed hetero-/homogeneous chemistry have nowadays become common research tools (Deutschmann et al., 2000; Dogwiler et al., 1999; Karagiannidis and Mantzaras, 2010). On the other hand, parabolic (boundary layer) 2D models (Coltrin et al., 1996; Markatou et al., 1993), again with detailed hetero-/homogeneous chemistry, are computationally very efficient by utilizing an axially marching solution algorithm. The applicability of parabolic models has been demonstrated for pure catalytic combustion at inlet Reynolds numbers as low as 20 (Raja et al., 2000). Comparisons between elliptic and parabolic solvers have also delineated the regimes of applicability of the parabolic approach in combined hetero-/homogeneous combustion (Mantzaras et al., 2000a). It was shown that for laminar propagation speeds small compared to the flow velocities, the parabolic model could adequately describe homogeneous combustion.

3.1. One-Dimensional Channel-Flow Models

One-dimensional channel models with lumped heat and mass transport coefficients have been extensively used in the literature, initially with simplified chemistry (Cerkanowicz et al., 1977) and recently with detailed kinetics (Kramer et al., 2002). The time-dependent 1D governing equations for a channel become:

Continuity

$$\frac{\partial \rho}{\partial t} + \frac{\partial(\rho u)}{\partial x} = 0 \tag{3.5}$$

Momentum

$$\rho\frac{\partial u}{\partial t} + \rho u\frac{\partial u}{\partial x} + \frac{\partial p}{\partial x} + \frac{P\cdot f}{2A}(\rho u)u = 0 \tag{3.6}$$

Gas-phase species

$$\rho\frac{\partial Y_k}{\partial t} + \rho u\frac{\partial Y_k}{\partial x} + \frac{P}{A}\alpha_k(Y_k - Y_{k,s}) - \dot{\omega}_k W_k = 0, \quad k = 1, \ldots, K_g \tag{3.7}$$

Energy

$$\rho \frac{\partial (c_p T)}{\partial t} + \rho u \frac{\partial (c_p T)}{\partial x} + \frac{P}{A} \alpha_T (T - T_s) + \sum_{k=1}^{K_g} h_k \dot{\omega}_k W_k = 0 \quad (3.8)$$

with P and A be the perimeter and area of the channel cross section. The interfacial boundary conditions for the gas-phase species and energy are as follows:

$$\alpha_k (Y_k - Y_{k,s}) + B \dot{s}_k W_k = 0, \quad k = 1, \ldots, K_g \text{ and} \quad (3.9)$$

$$A_s \frac{\partial (\rho_s c_s T_s)}{\partial t} - A_s k_s \frac{\partial^2 T_s}{\partial x^2} - P \alpha_T (T - T_s) - P \dot{q}_{rad} + P \sum_{k=1}^{K_g + M} h_k B \dot{s}_k W_k = 0 \quad (3.10)$$

The surface coverage is given by simplifying Eq. (3.2) for a conserved site density Γ:

$$\frac{\partial \theta_m}{\partial t} = \sigma_m \frac{B \dot{s}_m}{\Gamma}, \quad m = 1, \ldots, K_s \quad (3.11)$$

Finally, the ideal and caloric gas laws close the system of equations:

$$p = \frac{\rho R^{\circ} T}{\bar{W}}, \quad h_k = h_k^{o}(T_o) + \int_{T_o}^{T} c_{p,k} \mathrm{d}T, \quad k = 1, \ldots, K_g \quad (3.12)$$

In Eqs. (3.9) and (3.10), α_k and α_T are suitable mass and heat transport lumped coefficients and A_s is the cross section of the solid wall. In Eqs. (3.9)-(3.11), the parameter B is the ratio of the catalytically active to the geometrical surface area. This parameter cannot account for porous (intraphase) diffusion in the washcoat of technical catalysts but only for geometrical surface area increase. Intraphase transport modeling may be required for technical catalysts with sufficiently thick washcoats and well-dispersed catalyst loading within the washcoat volume. High temperatures, in particular, enhance diffusion limitations through the porous catalyst structure. For intraphase diffusion, various modeling approaches have been proposed (Mladenov et al., 2010). Intraphase diffusion will not be elaborated in this chapter.

In most cases, the flow is considered isobaric, such that Eq. (3.6) reduces to $p = p(x = 0, t)$. This is a key simplification that removes the strong pressure–velocity coupling and the time integration limitations associated with the compressible flow equations. The isobaric assumption is in most cases adequate considering the small pressure drop (around 1%) in straight catalytic channels with laminar flows and length-to-hydraulic diameter ratios typically less than 100. In Eq. (3.10), uniform properties were considered for the solid over its cross-sectional area A_s, such that a 1D model for the solid was used. In addition, thermal radiation exchange between the solid wall elements was accounted for in Eq. (3.10). Surface radiation has been shown to moderate the steep axial solid temperature gradients created upon light-off (Boehman, 1998). At both solid wall ends ($x=0$ and $x=L$), radiative boundary conditions can be used:

$$k_s \frac{\partial T_s}{\partial x}\bigg|_{x=0} = \varepsilon\sigma\left[T_s^4(x=0) - T_{\text{IN}}^4\right], \; -k_s \frac{\partial T_s}{\partial x}\bigg|_{x=L} = \varepsilon\sigma\left[T_s^4(x=L) - T_{\text{OUT}}^4\right] \quad (3.13)$$

with T_{IN} and T_{OUT} be appropriate radiation exchange temperatures and ε be the emissivity of the solid wall. Eq. (3.13) further considers black body inlet and outlet radiation enclosures (Karagiannidis and Mantzaras, 2010). The radiative boundary condition at the entry face can be further improved by adding convective cooling due to flow impingement (Kramer et al., 2002).

Given the long timescale for the solid substrate heat-up compared to the relevant chemical, convective, and diffusive timescales in a catalytic monolith reactor, the quasisteady assumption can be applied for the gas phase (Tien, 1981), whereby the only retained transient term in Eqs. (3.5)–(3.10) is that of the solid energy Eq. (3.10). Although this assumption does not necessarily hold at all spatial locations and/or times during the reactor response history, it was nevertheless employed in all early 1D transient studies (Ahn et al., 1986; Tien, 1981). Rules for using the quasisteady-state assumption have been recently laid down in Karagiannidis and Mantzaras (2010) within the context of 2D models, but the results are equally applicable to 1D models. More recently, 1D simulations with detailed surface chemistry and without invoking the quasisteady assumption have also been reported (Kramer et al., 2002; Robbins et al., 2003).

3.2. One-Dimensional Stagnation-Flow Models

Stagnation point flow configurations are common in heterogeneous kinetic studies. The flow of reactants issuing from a nozzle impinges on a catalytically active surface positioned at a distance L from the nozzle exit. The stagnation point flow reduces mathematically, via the von Karman similarity transformation, to a spatially 1D problem. The time-dependent governing equations in cylindrical coordinates are (Egolfopoulos, 1994) as follows:

Continuity

$$\frac{\partial \rho}{\partial t} + \frac{\partial (\rho u)}{\partial x} + 2\rho G = 0 \tag{3.14}$$

Radial momentum

$$\rho \frac{\partial G}{\partial t} + \rho u \frac{\partial G}{\partial x} + \rho G^2 + H - \frac{\partial}{\partial x}\left(\mu \frac{\partial G}{\partial x}\right) = 0 \tag{3.15}$$

Gas-phase species

$$\rho \frac{\partial Y_k}{\partial t} + \rho u \frac{\partial Y_k}{\partial x} + \frac{\partial (\rho Y_k V_k)}{\partial x} - \dot{\omega}_k W_k = 0 \tag{3.16}$$

Energy

$$\rho c_p \frac{\partial T}{\partial t} + \rho c_p u \frac{\partial T}{\partial x} - \frac{\partial p}{\partial t} - \frac{\partial}{\partial x}\left(\lambda_g \frac{\partial T}{\partial x}\right) + \rho \sum_{k=1}^{K_g} c_{p,k} Y_k V_k \frac{\partial T}{\partial x} + \sum_{k=1}^{K_g} \dot{\omega}_k h_k = 0 \tag{3.17}$$

where $G = v/r$ and $H = (1/r)\mathrm{d}p/\mathrm{d}r$ is the pressure-gradient eigenvalue (Kee et al., 1988). A more accurate transient formulation has also been presented (Raja et al., 1998), which includes gas-dynamic effects in the boundary layer (including temporal pressure variations) and retains the axial pressure gradient. The surface coverage is provided by Eq. (3.11) and the gas laws by Eq. (3.12).

At the nozzle exit ($x = L$), constant parameters independent of time are imposed:

$$T = T_L, \quad Y_k = Y_{k,L}, \quad G = 0, \quad u = U_L \tag{3.18}$$

At the catalytic surface ($x = 0$):

$$\rho Y_k (V_{k,x} + u_{st}) = W_k B \dot{s}_k \tag{3.19}$$

with the Stefan velocity defined as:

$$u_{st} \equiv (1/\rho) \sum_{k=1}^{K_g} W_k B \dot{s}_k \tag{3.20}$$

For all steady-state applications, the Stefan velocity is identically zero (there is no etching or deposition in steady-state catalytic combustion). Given the 1D dimensionality of the stagnation-flow problem, a full multicomponent transport approach for the diffusion velocities $V_{k,x}$ is computationally manageable:

$$\nabla X_k = \sum_{\ell=1}^{K_g} \frac{X_k X_\ell}{D_{k\ell}} \left(\vec{V}_\ell - \vec{V}_k \right) + (Y_k - X_k) \frac{\nabla p}{p} + \sum_{\ell=1}^{K_g} \frac{X_k X_\ell}{\rho D_{k\ell}} \left(\frac{D_\ell^{\mathrm{T}}}{Y_\ell} - \frac{D_k^{\mathrm{T}}}{Y_k} \right) \frac{\nabla T}{T} \tag{3.21}$$

The surface temperature is either provided (measured) or solved via a corresponding 1D solid energy equation:

$$\delta_s \frac{\partial(\rho_s c_s T_s)}{\partial t} + \sum_{k=1}^{K_g+M} h_k B \dot{s}_k W_k - \lambda_g \left(\frac{\partial T}{\partial x} \right)_{x=0} - \dot{q}_{\mathrm{rad}} - \dot{P} = 0 \tag{3.22}$$

with δ_s be the wall thickness, $\dot{P}$ be the external power (e.g., via a resistive heater) to the surface, and T_s be the spatially uniform surface temperature.

Transient stagnation-flow models have been used to determine catalytic ignition of CH_4/air, CO/air, and H_2/air mixtures over noble metals (Deutschmann et al., 1996; Raja et al., 1998). Steady-state catalytic stagnation simulations will be presented in Section 5.1, when discussing the validation of surface chemistry.

3.3. Multidimensional Models

The governing equations are presented in vectorial form, appropriate for 2D and 3D geometries in any coordinate system. These models resolve all relevant spatiotemporal scales.

Continuity equation:

$$\frac{\partial \rho}{\partial t} + \nabla \cdot \left(\rho \vec{u} \right) = 0 \tag{3.23}$$

Momentum equations:

$$\frac{\partial (\rho \vec{u})}{\partial t} + \nabla \cdot \left(\rho \vec{u}\, \vec{u} \right) + \nabla p - \nabla \cdot \mu \left[\nabla \vec{u} + \left(\nabla \vec{u} \right)^{\mathrm{T}} - \frac{2}{3} \left(\nabla \cdot \vec{u} \right) \underline{\underline{I}} \right] = 0 \tag{3.24}$$

Total enthalpy equation:

$$\frac{\partial(\rho h)}{\partial t} + \nabla\cdot\left(\rho \vec{u} h\right) + \nabla\cdot\left(\sum_{k=1}^{K_g} \rho Y_k h_k \vec{V}_k - \lambda_g \nabla T\right) = 0 \quad (3.25)$$

Gas-phase species equations:

$$\frac{\partial(\rho Y_k)}{\partial t} + \nabla\cdot\rho Y_k\left(\vec{u} + \vec{V}_k\right) - \dot{\omega}_k W_k = 0, \quad k = 1, \ldots, K_g \quad (3.26)$$

The surface coverage is given by Eq. (3.11) and the gas laws by Eq. (3.12). Buoyancy is usually neglected in Eq. (3.24), given the small channel hydraulic diameters and the horizontal flow direction in most industrial applications.

The diffusion velocities $\vec{V}_k$ in Eqs. (3.25) and (3.26) are generally computed using mixture average diffusion, including thermal diffusion for the light species (Kee et al., 1996), rather than from the full multicomponent approach of Eq. (3.21):

$$\vec{V}_k = -D_{km}\nabla[\ln(Y_k \bar{W}/W_k)] + \left[D_k^{\mathrm{T}} W_k/(\rho Y_k \bar{W})\right]\nabla(\ln T) \quad (3.27)$$

The interfacial gas-phase species boundary conditions become

$$\left[\rho Y_k\left(\vec{V}_k + \vec{u}_{\mathrm{st}}\right)\right]_+ \cdot \vec{n}_+ = W_k B \dot{s}_k \quad (3.28)$$

with $\vec{n}_+$ be the outward-pointing normal to the catalytic walls, $\vec{u}_{\mathrm{st}}$ be the Stefan velocity given by Eq. (3.20), and the "+" symbol denoting gas properties just above the gas–wall interface. Even though the Stefan velocity involves rearrangement of mass only in an atomic monolayer (catalyst surface), it can reach during the fast light-off event of hydrogen over platinum appreciable values (~10 cm/s as reported in Brambilla et al. (2014)). A multidimensional model for the solid wall heat transfer is also considered, given the rise of significant temperature gradients in the normal-to-surface direction during transient reactor operation:

$$\frac{\partial(\rho_s c_s T_s)}{\partial t} - \nabla\cdot(k_s \nabla T_s) = 0 \quad (3.29)$$

with interfacial energy boundary condition:

$$\left[-\lambda_g \nabla T + \vec{\dot{q}}_{\mathrm{rad}}\right]_+ \cdot \vec{n}_+ + (k_s \nabla T_s)_- \cdot \vec{n}_+ + \sum_{k=1}^{K_g+M} h_k B \dot{s}_k W_k = 0 \quad (3.30)$$

For the laboratory channel reactor in Fig. 3.2, the numerical model is simplified since kinetic studies are performed at steady state and the interfacial energy boundary condition is prescribed wall temperatures. Using the coordinate notation shown in the channel of Fig. 3.3:

$$T(x, y=-b) = T_{\mathrm{w,l}}(x), \quad T(x, y=b) = T_{\mathrm{w,u}}(x) \quad (3.31)$$

with $T_{\mathrm{w,u}}(x)$ and $T_{\mathrm{w,l}}(x)$ be the temperature profiles of the upper and lower wall, respectively, fitted through the 12 thermocouple measurements of each catalytic plate (Fig. 3.2B).

A complete flow model considers the full Navier-Stokes, i.e., elliptic description, with axial diffusion in Eqs. (3.24)-(3.26). As discussed in the beginning of Section 3, boundary layer (parabolic) models are an option for channel flows with hetero-/homogeneous reactions, at least under certain operating conditions (Mantzaras et al., 2000a). However, the presence of heat conduction in the solid, Eq. (3.29), and of surface radiation heat transfer, Eq. (3.30), negates the advantages of a parabolic forward-marching solution, since for such heat transfer modes information flows both upstream and downstream. The parabolic solver retains its full computational advantages only when the surface temperature is prescribed (e.g., boundary conditions in Eq. 3.31), thus eliminating the need for modeling of solid heat conduction and surface radiation.

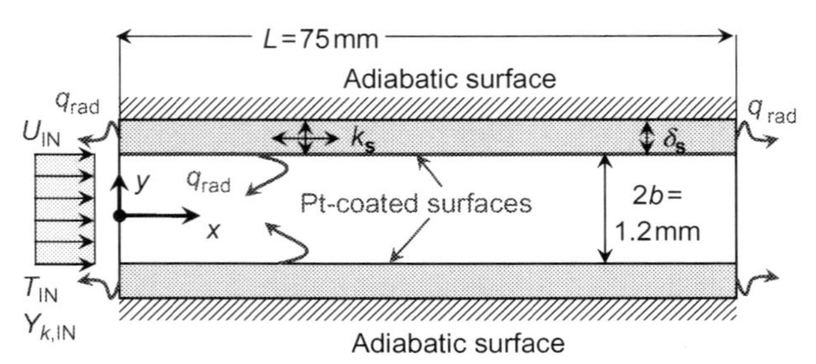

Figure 3.3 Channel domain for the simulations in Fig. 3.4. The catalytic planar channel has a length $L = 75$ mm, a half-height $b = 0.6$ mm, wall thickness $\delta_s = 0.05$ mm, and solid wall thermal conductivity $k_s = 16$ W/m/K.

The quasisteady-state approach can be applied in transient simulations, in a similar fashion to the 1D models discussed in Section 3.1. The only retained transient term is that of solid heat conduction in Eq. (3.29). Quasisteady-state and fully transient 2D channel simulations have been recently compared for fuel-lean hydrogen hetero-/homogeneous combustion over Pt in Brambilla et al. (2014). Requirements for the applicability of quasisteady-state approach have been elaborated in Schneider et al. (2008) and Karagiannidis and Mantzaras (2010).

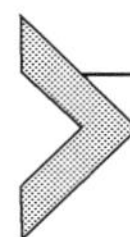

4. IMPACT OF HYDROGEN MOLECULAR TRANSPORT ON REACTOR THERMAL MANAGEMENT

The attained surface temperature is a very important design parameter, as it directly impacts the catalyst thermal stability and the overall reactor integrity. Even for an adiabatic catalytic reactor, the coupling of transport and catalytic chemistry can give rise to surface temperatures markedly different from the corresponding adiabatic equilibrium temperatures. Reasons for this behavior are the diffusional imbalance in heat and mass transport (i.e., nonunity Lewis number of the limiting reactant, $Le \neq 1$), finite-rate surface chemistry, and nonequilibration of the combustion products due to different timescales of the involved reaction pathways (Mantzaras, 2006). The first two factors are present in either fuel-lean or fuel-rich H_2/air combustion as both stoichiometries are predominantly controlled by the total oxidation reaction $H_2 + ½O_2 \rightarrow H_2O$. The third factor (chemical nonequilibration due to multiple reaction pathways) is irrelevant for hydrogen and is of interest only in fuel-rich hydrocarbon/air catalytic combustion, wherein total oxidation, partial oxidation, steam/dry reforming, and WGS reaction pathways are present (Eriksson et al., 2006).

Excursions of the catalyst surface temperature above or below the adiabatic equilibrium temperature (with the latter based on the incoming mixture stoichiometry and temperature) were reported in early catalytic combustion studies (Hegedus, 1975; Satterfield et al., 1954). Such temperature excursions actually reflect the specific nature of the catalytic reaction zone. In heterogeneous combustion, a degenerate diffusion reaction sheet is formed (Williams, 1985), with this terminology reflecting the fixed location of the reaction zone (catalyst surface). The inability of the catalytic reaction zone to adjust its position (in contrast to a freely moving gaseous diffusion flame) leads to temperatures markedly different form the adiabatic equilibrium temperature. For a catalytic flat plate, it can be shown by means of analytical similarity solutions that under infinitely fast catalytic chemistry, the surface temperature is (Zheng and Mantzaras, 2014) as follows:

$$T_w = T_\infty + (\Delta T)_c\, Le^{-2/3} \tag{3.32}$$

$(\Delta T)_c$ is the adiabatic temperature rise:

$$(\Delta T)_c = T_{ad} - T_\infty = QY_{F,\infty}/c_p \tag{3.33}$$

with T_∞ be the free-stream temperature, $Y_{F,\infty}$ be the free-stream mass fraction of the fuel, Q be the heat release per unit mass of fuel, c_p be the mixture heat capacity, and Le be the Lewis number of the limiting reactant (either fuel, F, in fuel-lean combustion or oxidizer, O, in fuel-rich combustion). Eq. (3.32) indicates that the wall temperature is constant along a catalytic flat plate and that it differs from the adiabatic equilibrium temperature, $T_{ad} = T_\infty + (\Delta T)_c$, such that $T_w > T_{ad}$ when $Le < 1$ and $T_w < T_{ad}$ when $Le > 1$.

In catalytic channels, the flat plate surface temperature in Eq. (3.32) is attained at the channel entry ($x \to 0$). As the catalytic channel is not amenable to analytical solutions, simulations are provided next for the channel geometry shown in Fig. 3.3. A planar channel is considered in Fig. 3.3, with a length $L = 75$ mm, height $2b = 1.2$ mm, and a wall thickness $\delta_s = 50\,\mu$m. A 2D steady model for the gas and solid (described in Section 3.3) is used. The solid thermal conductivity is $k_s = 16$ W/m/K referring to FeCr alloy, a common material for catalytic honeycomb reactors in power generation (Carroni et al., 2003). Surface radiation heat transfer was accounted for, with an emissivity $\varepsilon = 0.6$ for each discretized catalytic surface element, while the inlet and outlet sections were treated as black bodies ($\varepsilon = 1.0$). To illustrate differences between the surface temperatures of fuel-lean and fuel-rich hydrogen/air catalytic combustion, computed axial temperature profiles at the gas–wall interface ($y = b$ in Fig. 3.3) are shown in Fig. 3.4 for a lean ($\varphi = 0.3$) and a rich ($\varphi = 6.9$) equivalence ratio, $p = 1$ bar, inlet temperature, and velocity $T_{IN} = 300$ K and $U_{IN} = 10$ m/s, respectively. The two selected equivalence ratios have the same adiabatic equilibrium temperature, $T_{ad} = 1189$ K.

In fuel-lean H_2/air combustion, superadiabatic surface temperatures are attained as illustrated by three computations in Fig. 3.4 (Cases 1–3). Case 1 (marked $T_{w,1}$) was computed using an infinitely fast catalytic step $H_2 + ½O_2 \to H_2O$ (i.e., transport-limited catalytic hydrogen conversion), without gas-phase chemistry and without inclusion of heat conduction in

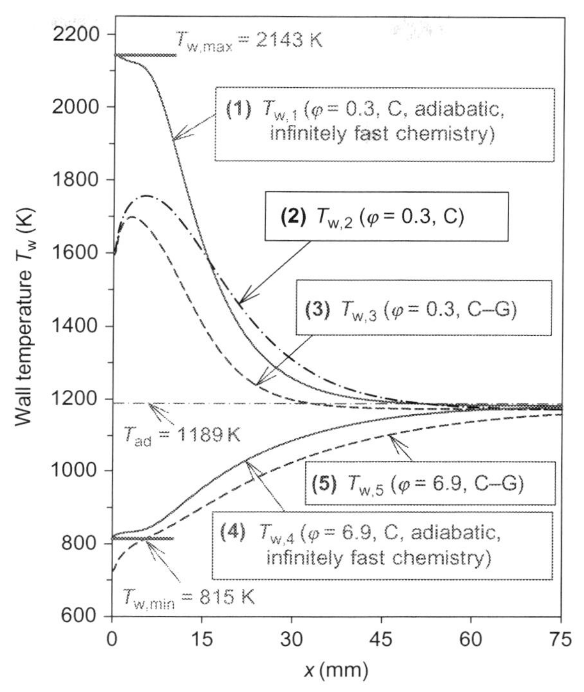

Figure 3.4 Computed wall temperature profiles for the catalytic channel geometry in Fig. 3.3 and two H_2/air stoichiometries ($\varphi = 0.3$ and 6.9) having the same adiabatic equilibrium temperature T_{ad}. In the fuel-lean $\varphi = 0.3$, results are shown for: Case 1 which is adiabatic and transport-limited, Case 2 with only catalytic reactions (C), and Case 3 with both catalytic and gas-phase reactions (C–G). In the fuel-rich $\varphi = 6.9$, results are shown for: Case 4 which is adiabatic and transport-limited, and Case 5 with catalytic and gas-phase reactions (C–G). *Adapted from Schultze and Mantzaras (2013) (with permission).*

the solid or surface radiation heat transfer. The maximum temperature was obtained at the channel entry ($x \rightarrow 0$) and was $T_{w,1,max} = 2143\,K$, i.e., 954 K above the adiabatic equilibrium temperature $T_{ad} = 1189\,K$. As the transport-limited catalytic channel solution at $x \rightarrow 0$ is equivalent to the corresponding transport-limited solution over a catalytic flat plate, the maximum computed temperature $T_{w,1,max}$ could be compared against the analytical solution in Eq. (3.32). The computed $T_{w,1,max} = 2143\,K$ is in excellent agreement with the theoretical transport-limited solution, $T_{w,max} = 2143\,K$, obtained from Eq. (3.32):

$$T_{w,\,max} = T_{IN} + (\Delta T)_c \, Le_{H_2}^{-2/3} \tag{3.34}$$

or equivalently,

$$T_{w,max} - T_{ad} = (\Delta T)_c \left(Le_{H_2}^{-2/3} - 1 \right) \quad (3.35)$$

with the adiabatic combustion temperature rise $(\Delta T)_c = T_{ad} - T_{IN} = 889\,K$ and $Le_{H_2} = 0.335$. The Lewis number of the limiting hydrogen reactant in Eq. (3.34) was computed for a $\varphi = 0.3$ H_2/air mixture using mixture average diffusion. In Fig. 3.4, the wall temperature $T_{w,1}$ dropped monotonically with increasing axial distance to the adiabatic equilibrium temperature $T_{ad} = 1189\,K$ upon complete conversion of hydrogen.

Additional results are shown in Fig. 3.4 for the $\varphi = 0.3$ H_2/air mixture, using finite-rate chemistry, and the full numerical model of Section 3.3, with the inclusion of solid heat conduction and surface radiation heat transfer. Case 2 ($T_{w,2}$) refers to predictions using only the detailed catalytic reaction mechanism (C) from Deutschmann et al. (2000) (Table 3.1) and Case 3 ($T_{w,3}$) to simulations with both the catalytic and gaseous (C–G) reaction mechanisms (Deutschmann et al., 2000; Li et al., 2004) of Tables 3.1 and 3.2. Simulations with only catalytic chemistry, $T_{w,2}$, yielded a maximum wall temperature of 1756 K, i.e., 567 K above the adiabatic equilibrium temperature. Although this temperature was appreciably lower than that obtained from the transport-limited adiabatic solution, $T_{w,1}$, it was still high and posed great concerns for the reactor thermal management and catalyst stability. The reduced degree of superadiabaticity in Case 2 was a combined result of radiation heat losses toward the cold entry, redistribution of energy in the wall via solid heat conduction, and finite-rate surface chemistry. Simulations with combined catalytic and gas-phase chemistry (C–G) at $\varphi = 0.3$ (Case 3 in Fig. 3.4) yielded an even lower superadiabaticity compared to the transport-limited Case 1 and the only catalytic chemistry (C) Case 2, with a maximum wall temperature of 1698 K. Reason for this quite interesting behavior was the presence of gaseous combustion, which partially shielded the catalyst from the hydrogen-rich channel core and hence reduced the catalytically driven superadiabaticity (Appel et al., 2002a; Ghermay et al., 2010). This issue will be further discussed in Section 6.

For the fuel-rich stoichiometry $\varphi = 6.9$, the predictions in Fig. 3.4 refer to adiabatic, transport-limited conditions (Case 4, $T_{w,4}$) and to finite-rate chemistry with detailed hetero-/homogeneous reaction mechanisms and the full model of Section 3.3 (Case 5, $T_{w,5}$). As in matter of fact, the results of Case 5 were the same in the absence of gaseous chemistry (the surface temperatures in the $\varphi = 6.9$ simulations were too low for homogeneous combustion to play a role). With increasing axial distance, the transport-limited solution $T_{w,4}$ approached the adiabatic equilibrium temperature $T_{ad} = 1189\,K$ from underadiabatic values. The Lewis number of the deficient reactant, calculated using a mixture average transport model was $Le_{O_2} = 2.268$. Substituting Le_{O_2} to Eq. (3.32), the resulting theoretical minimum temperature (flat plate solution) was $T_{w,min} = 815$ K, again in excellent agreement with the computed $T_{w,4,min} = 815$ K at $x \to 0$ (see Fig. 3.4). Computations using finite-rate chemistry with surface radiation and heat conduction ($T_{w,5}$ in Fig. 3.4) yielded temperatures 20–88 K lower than $T_{w,4}$. These modest differences (compared to the larger differences in the fuel-lean Cases 1–3) were attributed to the less significant radiation heat losses, a result of the substantially lower wall temperatures at the reactor entry.

It was evident from Fig. 3.4 that fuel-lean catalytic combustion of hydrogen posed severe challenges due to the superadiabatic surface temperatures at the channel entry that could lead to reactor meltdown and/or catalyst deactivation. The foregoing analysis exemplified the impact of lean/rich hydrogen stoichiometries on the surface temperatures. However, very recent theoretical studies (Zheng and Mantzaras, 2014) have shown that the surface temperature underadiabaticity in the fuel-rich H_2/air mixtures is accompanied by an excess total energy (chemical and sensible) in the gas phase, which, upon homogeneous ignition, would manifest itself with superadiabatic flame temperatures. The gas-phase superadiabaticity in channels is a monotonically increasing function of the Lewis number (for $Le > 1$) reaching a peak of 20.8% above the adiabatic equilibrium temperature of the fresh reactants as $Le \to \infty$. For fuel-rich H_2/air mixtures, where the Lewis number of the limiting reactant is $Le_{O_2} \sim 2.3$, the corresponding gas-phase superadiabaticity is ~11%. An example of the distribution of total energy (chemical and sensible) in the gas is illustrated in Fig. 3.5, pertaining to a $\varphi = 5$ H_2/air mixture in a planar catalytic channel with a full-height of 1.2 mm. Computed 2D distributions of the local equilibrium temperature $T_{eq}(y) = T(y) + Y_{H2}(y)\,Q/c_p$ (i.e., based on the local temperature and hydrogen composition) and the normalized $\widetilde{T}_{eq} = \left(T_{eq} - T_{IN}\right)/\left(T_{ad} - T_{IN}\right) =$ are given in Fig. 3.5 for infinitely fast catalytic hydrogen conversion and adiabatic channel operation without heat recirculation in the solid (no heat conduction in the wall). It is seen that the peak normalized energy excess $(\widetilde{T}_{eq} - 1)$ occurs at the channel center ($y = 0.6$ mm) and reaches the value of 10.8%.

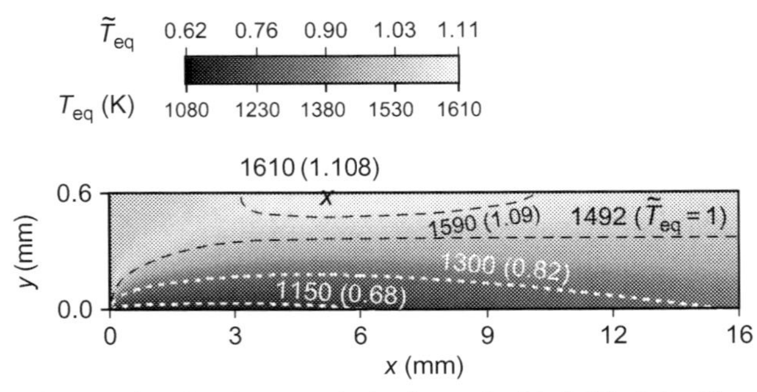

Figure 3.5 H_2/air combustion in a catalytic channel with full-height $2b = 1.2$ mm (the catalytic wall is located at $y = 0$ and the center at $y = 0.6$ mm), $\varphi = 5.0$, $T_{IN} = 400$ K, $U_{IN} = 1$ m/s, $p = 10$ bar. Distribution of local equilibrium temperature $T_{eq}(y) = T(y) + Y_{H_2}(y)Q/c_p$ and nondimensional temperature excess (or equivalently energy excess) $\tilde{T}_{eq} = (T_{eq} - T_{IN})/(T_{ad} - T_{IN})$. *Adapted from Zheng and Mantzaras (2014) (with permission).*

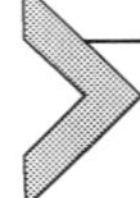

5. VALIDATION OF HETERO-/HOMOGENEOUS HYDROGEN KINETICS

5.1. Heterogeneous Kinetics of Hydrogen on Noble Metals

As mentioned in Section 2.1, the development of surface reaction mechanisms proceeds with optimization of theoretical reaction parameters through comparisons with measurements obtained in a variety of reactors. Catalytic ignition of hydrogen over noble metals has been extensively investigated in stagnation point flow geometries.

Experiments and kinetic model predictions for hydrogen catalytic ignition (light-off) over platinum foils are firstly presented. In Fig. 3.6A, light-off temperatures are plotted as a function of the partial pressure ratio $\alpha = p_{H_2}/(p_{H_2} + p_{O_2})$ for a nitrogen dilution parameter $(p_{H_2} + p_{O_2})/p_{total} = 0.059$ (Behrendt et al., 1996). Therein, an electric coil heated the platinum foil until ignition was attained. The symbols in Fig. 3.6A are measurements and the lines are stagnation point flow simulations (numerical model of Section 3.2) using an optimized H_2/O_2 surface mechanism (Deutschmann et al., 1996). The predictions indicate a bistable behavior for hydrogen concentrations $\alpha < 0.3$; the line marked "1" in Fig. 3.6A refers to predictions with a surface initially covered with H(s), while the line marked "2" to a surface initially covered with Pt(s) or O(s). For the investigated fuel-lean range ($\alpha < 0.667$), hydrogen self-inhibits its ignition (higher ignition temperatures

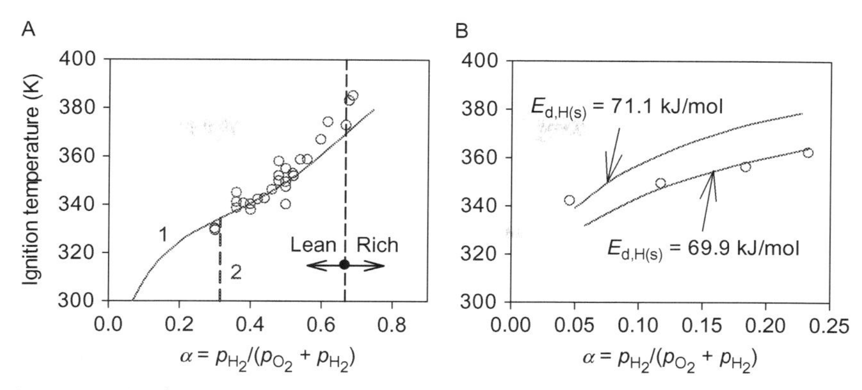

Figure 3.6 Catalytic ignition temperatures in a stagnation point flow over a platinum foil. (A) Measurements (symbols) and predictions (lines) for $H_2/O_2/N_2$ mixtures with $(p_{H_2}+p_{O_2})/p_{total}=0.059$). Dashed- and solid-line predictions for $\alpha<0.3$ indicate bistability. (B) Measurements (symbols) and predictions (lines, for two values of the hydrogen desorption activation energy) for H_2/air mixtures. *Panel (A) is adapted from Behrendt et al. (1996) (with permission) and panel (B) is adapted from Vlachos and Bui (1996) (with permission).*

for larger α), which is a result of the higher sticking coefficient of H_2 than that of O_2 (compare in Table 3.1 reaction S3 for H_2 with the duplicate reactions S1 and S2 for O_2 and further consider that S3 has order one with respect to Pt). This leads to increased blockage of the surface by H(s) with rising α, which in turn inhibits O_2 adsorption. Additional stagnation-flow hydrogen catalytic ignition studies over Pt are shown in Fig. 3.6B (Vlachos and Bui, 1996), for fuel-lean H_2/air mixtures with $\alpha<0.25$. The self-inhibition of hydrogen is again evident. Kinetic optimization is also illustrated in Fig. 3.6B with predictions using a detailed surface reaction mechanism (Vlachos and Bui, 1996) with two different activation energies for the hydrogen desorption reaction.

Self-inhibition of hydrogen catalytic ignition has also been observed over palladium (Behrendt et al., 1996; Deutschmann et al., 1996), albeit with a higher reactivity than that of platinum (i.e., lower ignition temperatures). This is because the sticking coefficients of both hydrogen and oxygen are more than an order of magnitude larger in Pd than in Pt. On the other hand, rhodium catalysts have an opposite ignition behavior (no H_2 self-inhibition) compared to Pt or Pd. This is a result of the higher oxygen-to-hydrogen sticking coefficient ratio in rhodium. Catalytic ignition temperatures of hydrogen over Rh are shown in Fig. 3.7. In Fig. 3.7A, experimental ignition temperatures are depicted in a stagnation point flow reactor for lean and

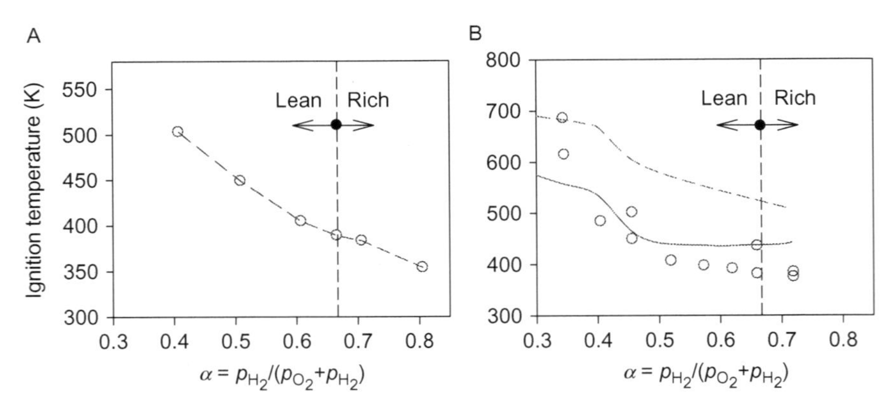

Figure 3.7 Catalytic ignition temperatures over rhodium. (A) Stagnation-flow experiments and (B) experiments (symbols) and stagnation-flow simulations for two surface mechanisms (solid-line: screening mechanism, dashed-line: optimized mechanism). *Panel (A) is adapted from Bär et al. (2013) (with permission) and panel (B) is adapted from Mhadeshwar and Vlachos (2005) (with permission).*

modestly rich stoichiometries (Bär et al., 2013), while in Fig. 3.7B, experiments and stagnation point flow simulations are shown for a screening and an optimized detailed surface reaction mechanism (Mhadeshwar and Vlachos, 2005). Both literature experiments in Fig. 3.7 clearly indicate promotion of catalytic ignition with increasing ratio α. It is finally noted that the H_2 catalytic self-inhibition over Pt and Pd is only chemical. It is pointed out that in practical Pt or Pd reactors the thermal effects (higher exothermicity for mixtures with larger ratio α) can overtake the chemical self-inhibition and thus lead to easier catalytic ignition at larger hydrogen concentrations.

5.2. Gas-Phase Kinetics of Hydrogen in Hetero-/Homogeneous Combustion

Homogeneous combustion of fuel-lean hydrogen/air mixtures over Pt was investigated in the reactor of Fig. 3.2 by Appel et al. (2002a) at atmospheric pressure, nonpreheated reactants, and surface temperatures $950\ \mathrm{K} \leq T_w \leq 1220\ \mathrm{K}$. The water-cooling arrangement (see Fig. 3.2B) at the channel entry was pivotal in suppressing the superadiabatic surface temperatures that would otherwise appear in fuel-lean H_2/air catalytic combustion, as discussed in Fig. 3.4. LIF-measured and predicted 2D maps of the OH radical are depicted in Fig. 3.8A. Homogeneous ignition positions (x_{ig}) are marked with arrows and are defined as the far-upstream positions whereby OH rose to 5% of its maximum value in the reactor. Simulations were performed with

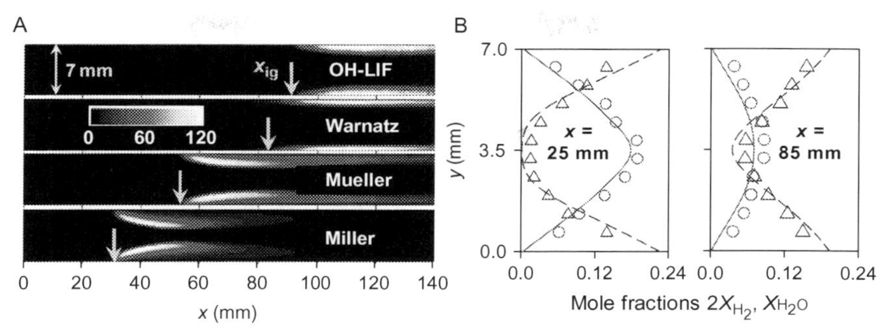

Figure 3.8 Combustion of H_2/air over Pt in the reactor of Fig. 3.2, $p=1$ bar, $\varphi=0.28$, $T_{IN}=312$ K, $U_{IN}=2$ m/s. (A) LIF-measured and numerically predicted distributions of the OH radical. Predictions refer to the catalytic scheme of Deutschmann et al. (2000) and the gas-phase schemes of Warnatz et al. (1996), Mueller et al. (1999), and Miller and Bowman (1989). Arrows denote homogeneous ignition locations (x_{ig}) and the color bar OH levels in ppmv. (B) Measured (symbols) and predicted (lines) transverse mole fraction profiles of H_2 (circles, solid lines) and H_2O (triangles, dashed lines) at two selected axial locations preceding the measured onset of homogeneous ignition. *Adapted from Appel et al. (2002a) (with permission).*

the model of Section 3.3 (steady-state version, 2D Cartesian coordinates, boundary condition of Eq. 3.31) using the catalytic scheme of Deutschmann et al. (2000) and three different gas-phase schemes: Warnatz et al. (1996), Mueller et al. (1999), and Miller and Bowman (1989). The catalyst coating in the reactor of Fig. 3.2 was achieved via plasma vapor deposition of a nonporous 1.5-μm thick Al_2O_3 layer on the Si[SiC] ceramic plates, followed by a 2.2-μm-thick Pt layer. Such a thick Pt layer on the top of a nonporous Al_2O_3 layer closely resembled a polycrystalline Pt surface; this was further verified by BET and depth X-ray Photoelectron Spectroscopy (Reinke et al., 2004). Hence, the parameter B in Eqs. (3.28) and (3.30) was unity.

Raman measurements and computed transverse profiles of major species are shown in Fig. 3.8B at two axial locations preceding the measured position of gas-phase ignition ($x < x_{ig}$). Transport-limited conversion of hydrogen is attested in Fig. 3.8B (manifested by the nearly zero concentration of the limiting hydrogen reactant at both channel walls, $y=0$ and 7 mm). Predictions with the employed catalytic scheme reproduced the measurements in Fig. 3.8B. On the other hand, full validation of hydrogen catalytic kinetics could not be obviously performed under transport-limited operation: the surface temperatures required to achieve homogeneous ignition were too high (in excess of 1200 K as will be elaborated in Section 6) for kinetically controlled hydrogen catalytic conversion (see also magnitude of catalytic ignition temperatures in Fig. 3.6). It is nonetheless emphasized that an incorrect prediction of the catalytic hydrogen conversion over the gaseous induction length could greatly affect the location of homogeneous ignition in Fig. 3.8A and could thus falsify the gaseous kinetics (Mantzaras and Appel, 2002). In this respect, the Raman measurements and predictions in Fig. 3.8B removed any uncertainty originating from the catalytic pathway that could affect the validation of gaseous chemistry in Fig. 3.8A.

The gas-phase mechanism from Warnatz et al. (1996) mildly underpredicted the measured homogenous ignition distance, while the other two gaseous schemes led to considerably shorter ignition distances (Fig. 3.8A). The appreciable differences between the three gaseous schemes, were primarily attributed to the catalytically produced water (Appel et al., 2002a), which was a very efficient third body in the chain terminating step $H+O_2+M \Leftrightarrow HO_2+M$ (see reaction R9 in Table 3.2 and its associated H_2O third body efficiency in the table footnote). The differences in effective ignition delay times among the gaseous schemes in Fig. 3.8A (evaluated using the ignition distances and the flow velocity) were as large as 20 ms, particularly long times for practical systems (i.e., gas turbines using the catalytically stabilized combustion approach). It is finally noted that the differences in Fig. 3.8A were germane to the presence of catalytic reactions (due to the catalytically produced water) and to the particular operating conditions of catalytic systems (low temperatures and very lean equivalence ratios) as discussed in Appel et al. (2002a).

The foregoing fuel-lean atmospheric pressure H_2/air homogeneous ignition studies were extended to pressures up to 15 bar, mixture preheats up to 780 K, and surface temperatures $910\ K \leq T_w \leq 1300\ K$ (Ghermay et al., 2010, 2011; Mantzaras et al., 2009). The catalytic scheme of Deutschmann et al. (2000) and the recent gas-phase mechanism from Princeton (Li et al., 2004) were used in these simulations (see Tables 3.1 and 3.2). Measured and predicted OH distributions are provided in Fig. 3.9 at three different pressures with preheats in the range 653–679 K. The agreement between measured and predicted ignition distances and flame shapes in Fig. 3.9 was particularly good. Moreover, Raman measurements over the gaseous induction zones were in good agreement with the predictions (Ghermay et al., 2011), demonstrating a transport-limited catalytic conversion of H_2, similar to the previous atmospheric pressure studies in Fig. 3.8b. Of great interest was the observed suppression of gaseous combustion at 15 bar in Fig. 3.9 (suppression was reported for

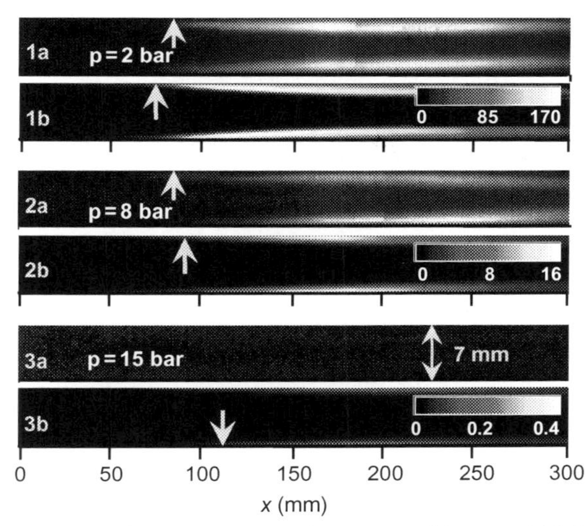

Figure 3.9 Combustion of H_2/air over Pt in the reactor of Fig. 3.2. (a) LIF-measured and (b) -predicted distributions of the OH radical, for three cases at elevated pressures and preheats: (1) $p=2$ bar, $\varphi=0.11$, $T_{IN}=653$ K, $U_{IN}=6.5$ m/s, (2) $p=8$ bar, $\varphi=0.11$, $T_{IN}=676$ K, $U_{IN}=2.1$ m/s, and (3) $p=15$ bar, $\varphi=0.11$, $T_{IN}=679$ K, $U_{IN}=0.9$ m/s. Gas-phase combustion is suppressed in (3). Predictions with the catalytic scheme of Deutschmann et al. (2000) and the gas-phase mechanism of Li et al. (2004). Arrows denote the onset of homogeneous ignition and color bars the predicted OH levels (ppmv). *Adapted from Ghermay et al. (2011) (with permission).*

pressures higher than 12 bar in Ghermay et al. (2011)). The reason for this behavior and the implication for hetero-/homogeneous hydrogen combustion will be discussed in Section 6.

Fuel-rich H_2/air hetero-/homogeneous combustion was less studied than its fuel-lean counterpart. Maestri et al. (2007, 2008) investigated fuel-rich H_2/air combustion on Rh-based catalysts at atmospheric pressure in a nearly isothermal annular reactor with channel gap of 2.1 mm. Fundamental homogeneous ignition studies were reported only recently (Schultze et al., 2013) in the Pt-coated reactor of Fig. 3.2, at pressures up to 5 bar, nonpreheated fuel-rich H_2/air mixtures, and surface temperatures $760\ \mathrm{K} \leq T_w \leq 1200\ \mathrm{K}$. Fig. 3.10 provides comparisons between LIF-measured and -predicted OH distributions at two different pressures. Simulations were performed with the hetero-/homogeneous reactions schemes from Deutschmann et al. (2000) and Li et al. (2004). Raman measurements further attested a nearly transport-limited conversion of the deficient O_2 reactant, which was captured by the numerical model (Schultze et al.,

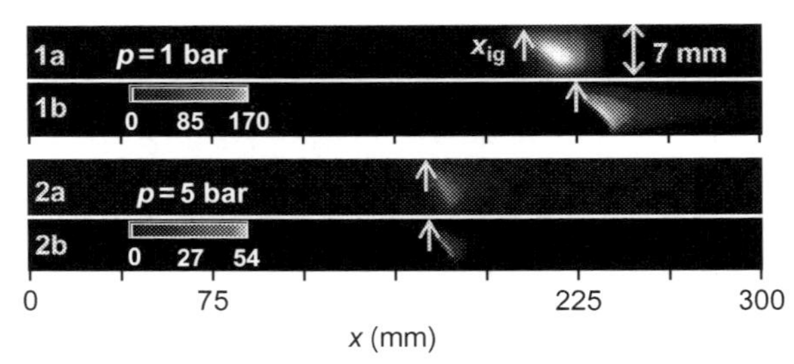

Figure 3.10 Combustion of H_2/air over Pt in the reactor of Fig. 3.2 at fuel-rich stoichiometries. (a) LIF-measured and (b) -predicted distributions of the OH radical, for two cases with practically nonpreheated reactants: (1) $p=1$ bar, $\varphi=2.0$, $T_{IN}=307$ K, $U_{IN}=3.1$ m/s (2) $p=5$ bar, $\varphi=4.0$, $T_{IN}=310$ K, $U_{IN}=0.8$ m/s. Predictions with the catalytic scheme of Deutschmann et al. (2000) and the gas-phase mechanism of Li et al. (2004). Arrows denote the onset of homogeneous ignition (x_{ig}) and color bars the predicted OH levels (ppmv). *Adapted from Schultze et al. (2013) (with permission).*

2013). The agreement between measurements and predictions in Fig. 3.10 was again very good. Slight asymmetries in the flames shapes were due to small differences between the upper and lower wall temperatures, an effect fully accounted for in the model via the boundary conditions in Eq. (3.31). A notable difference between the fuel-rich and fuel-lean flames is that the former established a connected V-shaped flame, while the latter comprised two separate flame branches near both walls (compare Figs. 3.8 and 3.9 with Fig. 3.10). Flame shapes were dictated by the Lewis number of the deficient reactant ($Le_{H_2} < 1$ in fuel-lean, and $Le_{O_2} > 1$ in fuel-rich combustion). Implications of this behavior will be discussed in Section 7.

The lower pressure range (1–5 bar) used for the fuel-rich flames in Fig. 3.10 compared to the higher pressures (up to 15 bar) for the fuel-lean flames in Fig. 3.9 was due to strong quenching of the OH-LIF at elevated pressures and rich stoichiometries. Very recently, a combination of OH-LIF (1–5 bar) with O_2-LIF (5–14 bar) was established for the investigation of fuel-rich H_2/CO homogeneous combustion over platinum (Schultze et al., 2014). This technique can also be applied to fuel-rich H_2/air mixtures.

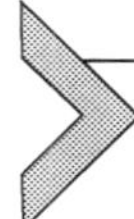

6. COUPLING OF HYDROGEN HETERO-/HOMOGENEOUS CHEMISTRY AND TRANSPORT

The foregoing results in Fig. 3.9 indicated suppression of gaseous combustion at 15 bar for preheated hydrogen/air mixtures, and this was also the case for pressures higher than 12 bar (Schultze et al., 2013). To

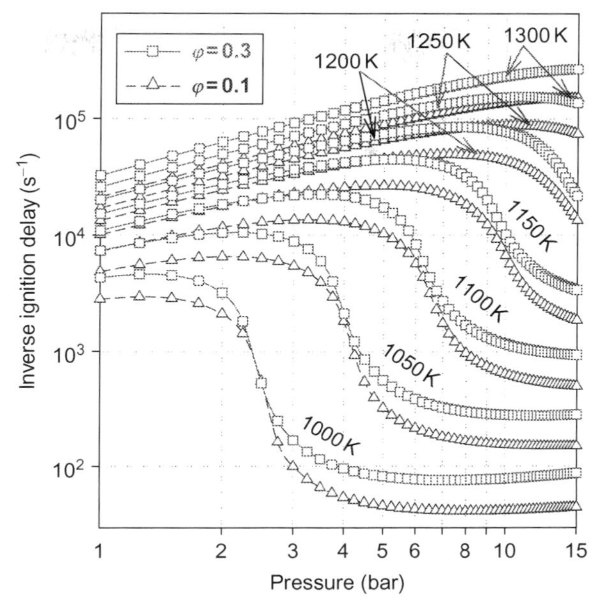

Figure 3.11 Computed inverse ignition delays of fuel-lean H_2/air mixtures ($\varphi = 0.30$ and 0.10) in a constant pressure batch reactor at different pressures and initial temperatures. *Adapted from Ghermay et al. (2011) (with permission).*

understand the origins of this suppression, homogeneous ignition characteristics of hydrogen are firstly investigated. Ignition delays of fuel-lean H_2/air mixtures ($\varphi = 0.10$ and 0.30) are calculated in a constant pressure batch reactor using the mechanism from Li et al. (2004) and their inverses (which are quantities proportional to the gaseous reactivity) are plotted in Fig. 3.11 as a function of pressure with parameter the initial temperature. For a moderately high temperature of 1000 K (which, for the spatially inhomogeneous reactor in Fig. 3.2, corresponds to an average between the inlet temperature and the wall temperature, with an added weight on the latter), the gaseous reactivity firstly decreases rapidly with rising pressure and then changes modestly above 3 bar. For temperatures $T > 1000$ K, Fig. 3.11 shows the reactivity initially increasing with rising pressure and then dropping, with the turning point shifted to higher pressures for higher temperatures.

The complicated hydrogen gaseous ignition characteristics in Fig. 3.11 are caused by the competition between the chain branching step $H + O_2 \Leftrightarrow O + OH$ (R1 in Table 3.2), the chain terminating step $H + O_2 + M \Leftrightarrow HO_2 + M$ (R9), and the chain branching sequence $HO_2 + HO_2 \Leftrightarrow H_2O_2 + O_2$ (R14 and R15) and $H_2O_2 + M \Leftrightarrow 2OH + M$

(R16) that overtakes the stability of HO_2 in the termination step (Glassman, 1996). Implication of Fig. 3.11 for a hetero-/homogeneous reactor operating at moderate temperatures (ca. 1000 K) is that, at sufficiently high pressures, the catalytic pathway has the opportunity to consume significant amounts of H_2 during the elongated gas-phase induction zones, thus depriving H_2 from the gaseous pathway and inhibiting homogeneous ignition. It is herein noted that the catalytic pathway is very efficient in converting hydrogen due to the large molecular diffusivity of this species and its very high reactivity on platinum even at low-to-moderate temperatures (Fig. 3.6). Thus, the suppression of gaseous combustion at modest reactor temperatures and elevated pressures is a result of intrinsic gas-phase hydrogen kinetics and competition between the heterogeneous and homogeneous reaction pathways for hydrogen consumption. This behavior can further explain the hetero-/homogeneous combustion behavior for nonpreheated fuel-lean H_2/air mixtures ($T_{IN} \approx 300$ K) in Mantzaras et al. (2009) and the associated suppression of gaseous combustion even at pressures as low as 4 bar.

For the preheated reactant experiments in Fig. 3.9, gas-phase combustion could not be sustained at $p > 12$ bar even at preheats as high as 780 K and wall temperatures in excess of 1200 K (Ghermay et al., 2011). This appeared counterintuitive since, for such high preheats and wall temperatures, the corresponding behavior in Fig. 3.11 (temperatures above 1150 K) dictated for $p > 12$ bar either a modest drop or a small increase in reactivity with rising pressure. It turns out that for this high preheat/pressure regime the controlling factor is not the ignition chemistry but the flame propagation characteristics. Laminar burning rates for 1D freely propagating flames are plotted as a function of pressure in Fig. 3.12 for two fuel-lean H_2/air equivalence ratios, $\varphi = 0.30$ and 0.20, and two mixture preheats, $T_o = 673$ K and 773 K. For $T_o = 673$ K and $\varphi = 0.30$, the burning rate at 15 bar was 34% lower than the corresponding peak value at ~5 bar, while for $T_o = 673$ K and $\varphi = 0.20$, the burning rate at 15 bar was 73% lower than the peak value at 1 bar; similar trends are also seen for $T_o = 773$ K. The reduction of laminar burning rates at higher pressures led to a push of the gaseous reaction zone closer to the wall, then to an increased leakage of hydrogen through the gaseous reaction zone (resulting in catalytic conversion of the escaped fuel at the wall) and finally to flame extinction (Ghermay et al., 2011). Therefore, the observed significant suppression of gaseous combustion at $p > 12$ bar and high mixture preheats bore the combined effects of the reduced laminar burning rates and also of the high molecular transport (low Lewis number) of hydrogen that decreased the availability of fuel in the near-wall hot ignitable region.

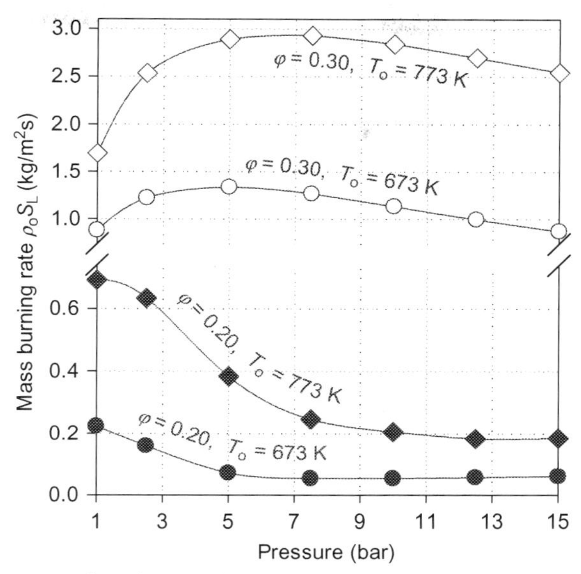

Figure 3.12 Computed 1D laminar burning rates versus pressure for H_2/air mixtures with $\varphi = 0.30$ and 0.20 and fresh reactant temperatures $T_o = 673$ and 773 K. *Adapted from Ghermay et al. (2011) (with permission).*

Similar to the fuel-lean results in Fig. 3.11, homogeneous ignition analysis was also carried out for fuel-rich H_2/air mixtures (Schultze et al., 2013). It was shown that, although below a critical pressure (that depended on temperature) the intrinsic gas-phase kinetics of hydrogen dictated higher reactivity for fuel-lean stoichiometries than for fuel-rich ones, homogeneous ignition was actually achieved easier in the latter. This was due to the lower molecular transport of the limiting O_2 reactant that in turn negated its appreciable catalytic consumption over the gaseous induction zone. Above the aforementioned critical pressures, the intrinsic gaseous hydrogen kinetics yielded higher reactivity for the fuel-rich mixtures, which in conjunction with the slower catalytic depletion of oxygen, resulted in vigorous gaseous combustion at elevated pressures (even above the pressures where gas-phase combustion was suppressed in the fuel-lean mixtures).

In terms of hetero-/homogeneous hydrogen chemistry coupling, it was shown that the formation of the major product H_2O from the catalytic pathway inhibited chemically the homogeneous ignition of both fuel-lean (Appel et al., 2002a) (as also discussed in Section 5.2 within the context of Fig. 3.8) and fuel-rich mixtures (Schultze et al., 2013). Radical adsorption/desorption reactions had a more modest (inhibitive) impact on steady homogeneous combustion. The most significant radical adsorption/desorption reactions affecting steady homogeneous combustion were those of OH radical at fuel-lean stoichiometries (Appel et al., 2002a; Bui et al., 1996; Ghermay et al., 2011) and of H radical at fuel-rich stoichiometries (Schultze et al., 2013). Nonetheless, very recent 2D channel simulations (using the full transient model in Section 3.3) at fuel-lean H_2/air stoichiometries have shown that the onset of homogeneous ignition was very sensitive to the OH desorptive fluxes (Brambilla et al., 2014). Thus, while the startup of homogeneous combustion was particularly sensitive to catalytically produced radicals, once the flame finally stabilized over the catalytic surface the impact of radical adsorption/desorption fluxes dropped significantly.

The lack of appreciable gas-phase hydrogen combustion in certain pressure regimes and reactant preheats is of concern in reactor design. It has been shown (Appel et al., 2002a; Mantzaras, 2008) that, contrary to many premises, the presence of a flame inside a hydrogen catalytic reactor is beneficial. This is because the flame shields the catalytic surfaces from the hydrogen-rich channel core, thus reducing the heterogeneous conversion which is responsible for the superadiabatic surface temperatures. This was also shown in Fig. 3.4, where the wall temperatures for $\varphi = 0.3$ in Case 3 (predictions with catalytic and gas-phase chemistry) were lower than those for Case 2 (only catalytic chemistry). The absence of this surface temperature moderation can be of concern, particularly for reactor thermal management at elevated pressures, thus exemplifying the need for proper gaseous chemistry description when dealing with either hydrogen or hydrogen-rich syngas fuels. The importance of gaseous chemistry in practical catalytic channels depends on the pressure, inlet temperature, and geometrical confinement (channel hydraulic diameter). Extensive simulations in Fig. 3.13 have delineated the minimum channel wall temperatures above which gas-phase combustion could not be neglected (it accounted for at least 5% of the total hydrogen conversion). The simulations in Fig. 3.13 were carried out using a steady-state, 2D axisymmetric version of the model in Section 3.3 with prescribed constant wall temperatures and the validated hetero-/homogeneous reaction schemes (Deutschmann et al., 2000; Li et al., 2004). Fig. 3.13 illustrates that appreciable homogeneous combustion of hydrogen in practical catalytic reactors requires surface temperatures in excess of 1200 K.

Another intricate coupling between transport and hetero-/homogeneous chemistry occurs in turbulent catalytic combustion. In catalytic reactors of large turbines operating at full load, the inlet Reynolds numbers in each individual honeycomb channel can exceed 30,000. In entry channel flows with heat transfer from the hot wall to the flowing gas, which

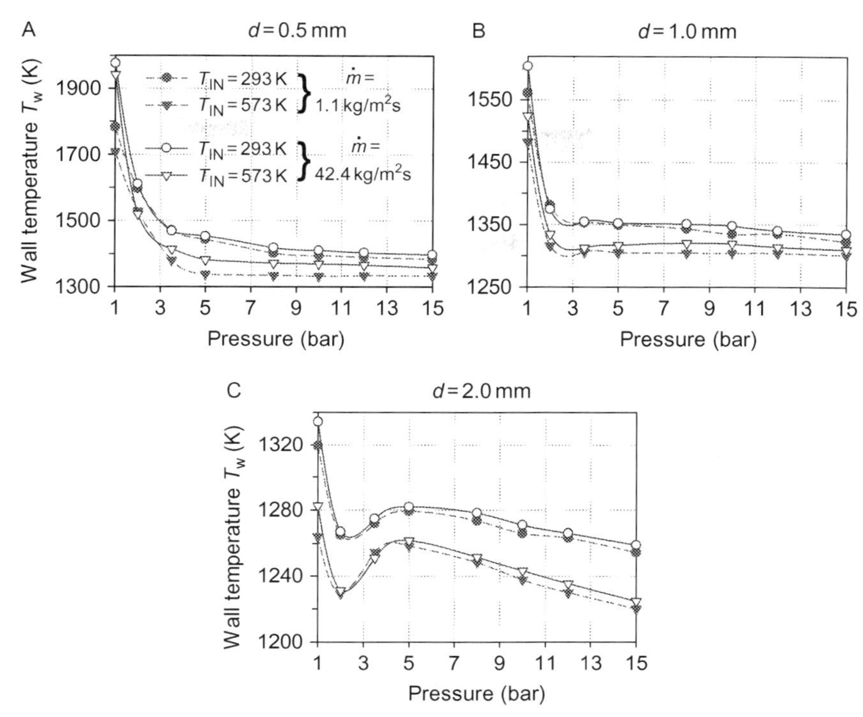

Figure 3.13 Computations in tubular catalytic (Pt-coated) channels with diameters (A) $d=0.5$ mm, (B) $d=1.0$ mm, and (C) $d=2.0$ mm. Calculated threshold channel wall temperatures leading to 5% gaseous hydrogen conversion, as a function of pressure for a $\varphi=0.30$ H_2/air mixture at two different preheats $T_{IN}=293$ and 573 K and two mass fluxes: $\dot{m}=42.4\,kg/m^2s$ (open symbols) and $\dot{m}=1.1\,kg/m^2s$(filled symbols). *Adapted from Ghermay et al. (2011) (with permission).*

is relevant to catalytic combustion systems, the controlling parameters are the magnitude of the incoming turbulence and the flow laminarization due to the increase in kinematic viscosity with rising gas temperature. Experiments in the catalytic (Pt-coated) reactor of Fig. 3.2 have been carried out at inlet Reynolds numbers Re_{IN} up to 30,000 and involved Raman, OH-LIF, and particle image velocimetry along the x–y symmetry plane (Appel et al., 2002b, 2005). Simulations in the framework of Reynolds-Averaged Navier-Stokes 2D turbulence modeling with appropriate near-wall low-Reynolds number heat transfer turbulence submodels (Appel et al., 2005; Mantzaras et al., 2000b) and also very recent full 3D direct numerical simulations (Lucci et al., 2013) have attested the importance of laminarization in turbulent catalytic combustion.

Fig. 3.14 provides comparison between measured and predicted OH maps for two cases with $\varphi = 0.24$, $T_{IN} = 300$ K, $p = 1$ bar, with $Re_{IN} = 15{,}080$ (Fig. 3.14(1c,d)) and 30,150 (Fig. 3.14(2c,d)). The increased transport under turbulent flow conditions had a profound impact on wall-bounded flames, by reducing the available residence times across the gaseous combustion zone. Early laminar stagnation-flow catalytic combustion studies (Law and Sivashinsky, 1982) have shown that for deficient reactants with Lewis numbers less than unity, a rise in strain rate pushes the flame against the catalytic wall, leading to incomplete combustion through the gaseous reaction zone and to a subsequent catalytic conversion of the leaked fuel. A further rise in the strain rate extinguishes the flame.

Turbulent transport in channel flows played a role analogous to that of the strain rate in stagnation point flows. This is illustrated in Fig. 3.14(1a,b and 2a,b) providing Raman-measured and predicted transverse profiles of H_2 and H_2O mean mole fractions at two selected axial positions ($x = 85$ and 205 mm) for the two investigated cases. At $x = 85$ mm (position upstream of homogeneous ignition), Fig. 3.14(1a and 2a) indicated mass-transport-limited catalytic conversion of hydrogen. For the higher Reynolds number case, the predicted transverse gradient of hydrogen mole fraction at the wall was nonzero at the postignition location $x = 205$ mm (shown with the thick solid line passing through $y = 0$ in Fig. 3.14(2b)), clearly indicating hydrogen leakage through the flame zone. This result was also supported by the measurements, despite the near-wall limitations of the Raman data. Therefore, even well-downstream the homogeneous ignition position, the catalytic pathway converted hydrogen in parallel to the gaseous pathway. On the other hand, for the lower Reynolds number case, the corresponding hydrogen gradient at the wall was practically zero (Fig. 3.14(1b)), indicating that catalytic conversion ceased after the onset of homogeneous ignition. The reduction (and eventually complete suppression) of gaseous hydrogen conversion in favor of catalytic conversion with increasing Re_{IN} was detrimental to the catalyst integrity since, as discussed previously, the presence of a flame moderated the surface temperatures of catalytic reactors.

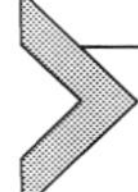

7. METHODOLOGIES FOR HYDROGEN HETERO-/HOMOGENEOUS COMBUSTION

As mentioned in Section 1, the fuel-lean (CST) and fuel-rich catalytic/fuel-lean gaseous combustion concepts in Fig. 3.1 can also be used for hydrogen or hydrogen-rich fuels. The fuel-lean concept in Fig. 3.1A is

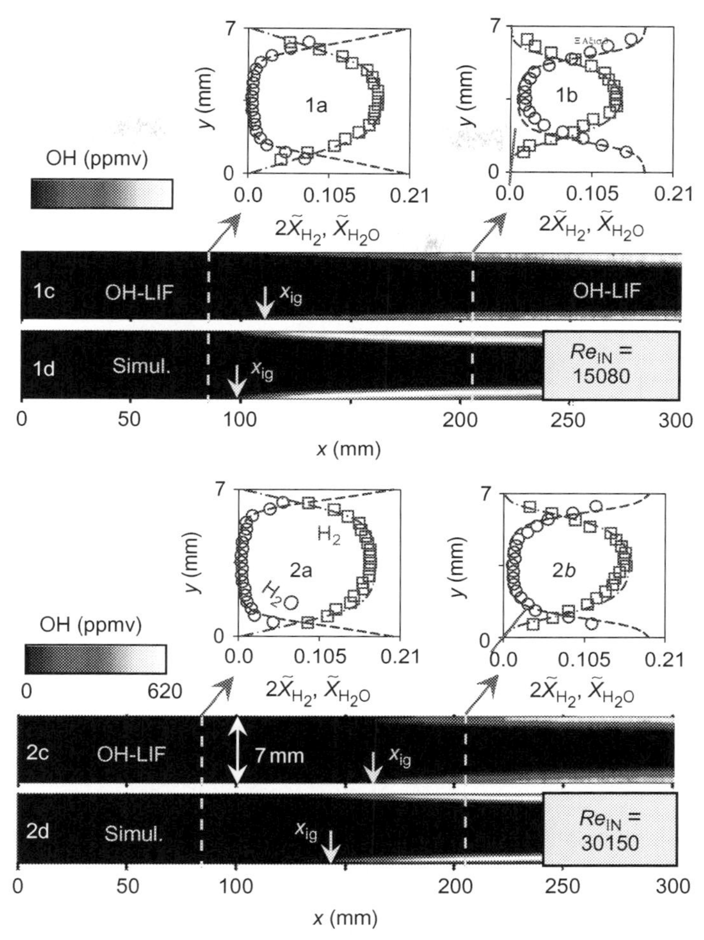

Figure 3.14 Measurements and simulations in turbulent H_2/air catalytic (Pt-coated) combustion in the reactor of Fig. 3.2. Case 1 (plates 1a-d): $\varphi = 0.24$, $T_{IN} = 300$ K, $Re_{IN} = 15080$; Case 2 (plates 2a-d): $\varphi = 0.24$, $T_{IN} = 300$ K, $Re_{IN} = 30150$. Raman-measured (symbols) and -predicted (lines) transverse profiles of mean mole fractions at two axial positions: (a) $x = 85$ mm and (b) $x = 205$ mm (H_2: squares and dashed-dotted lines; H_2O: circles and dashed lines). The thick solid lines through $y = 0$ in (b) are the predicted transverse gradients of the mean hydrogen mole fraction at the wall. (c) LIF-measured and (d) -predicted 2D OH maps. The color bars provide predicted OH in ppmv and the vertical arrows in (c) and (d) define the location of homogeneous ignition (x_{ig}). *Adapted from Appel et al. (2005) (with permission).*

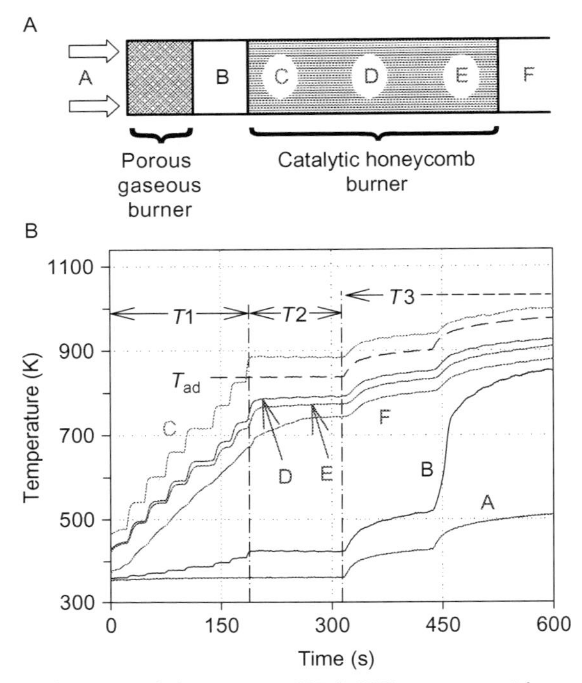

Figure 3.15 (a) Schematic of the inverse-CST (i-CST) concept with a porous gaseous burner preceding the catalytic burner. (b) Measured temperatures for the porous burner and catalytic reactor configuration. Inlet volumetric composition: 6.1% H_2, 9.1% O_2, 77.6% N_2, and 7.2% CO_2, $p=5$ bar, $U_{IN}=2.3$ m/s. The time intervals $T1$, $T2$, and $T3$ denote the fuel-ramping to its final composition, quasisteady operation, and ramping of the mixture preheat (thermocouple A), respectively. *Adapted from Ghermay et al. (2010) (with permission).*

compounded by the superadiabatic flame temperatures elaborated in Section 4, while the fuel-rich catalytic concept in Fig. 3.1B is considerably more complex due to the bypass air arrangement. To this direction, a new fuel-lean concept for hydrogen hybrid combustion has been proposed in Ghermay et al. (2010), which is termed inverse catalytically stabilized thermal combustion or "i-CST." Therein, gaseous combustion precedes catalytic combustion (Fig. 3.15a). Homogeneous combustion is initiated inside an inert porous burner, which is in turn ignited via heat conduction and radiation heat transfer from a downstream catalytic honeycomb reactor. At steady operation, the catalytic module is exposed to the hot combustion products of the homogeneous combustion zone. As long as complete fuel conversion is achieved homogeneously in the porous burner, the maximum temperature encountered by the catalytic reactor is limited by the adiabatic equilibrium temperature of the reactive mixture. Superadiabatic surface temperatures can still be attained in the i-CST concept due to heat recirculation in the solid structure of the gaseous burner. Nevertheless, this type of superadiabaticity is generally smaller than the catalytic combustion-induced one.

The start-up in i-CST entails the flow of fuel-lean H_2/air mixtures through the porous/catalytic burners, achieving initially light-off of the catalytic honeycomb reactor. Afterward, the hot honeycomb structure transfers heat upstream to the porous structure and initiates gaseous combustion within it. The experimentally assessed start-up process for the porous/catalyst modules is illustrated in Fig. 3.15b, providing the time history of measured temperatures at various reactor locations (Ghermay et al., 2010). The pressure is 5 bar, the volumetric inlet composition (location A) 6.1% H_2, 9.1% O_2, 77.6% N_2, and 7.2% CO_2 ($\varphi=0.33$), while the inlet velocity (referring to an inlet temperature 360 K at A) is 2.30 m/s. The zone marked T1 in Fig. 3.15b denotes the ramping of hydrogen to its final content (6.1% vol.) while maintaining the inlet temperature (thermocouple A) at 360 K; zone T2 designates quasisteady operation with the established final mixture composition and inlet temperature still at 360 K; finally, T3 indicates the ramping of inlet temperature (thermocouple A) until gaseous ignition is achieved in the porous burner. The adiabatic equilibrium temperatures based on the inlet temperatures (A) are also shown in Fig. 3.15b. Strong gas-phase combustion is achieved inside the porous structure for inlet temperatures above ~430 K (see time record of thermocouple A), as indicated by the porous burner exit temperature (thermocouple B). When the honeycomb reactor is replaced by an identical inert (noncatalytic) structure, no gas-phase ignition could be achieved in the porous burner even for inlet temperatures (A) as high as 720 K.

A comparative modeling study of the three concepts (the two in Fig. 3.1 and the aforementioned i-CST in Fig. 3.15a) is presented next for H_2/air combustion over platinum. The numerical model for the CST concept of Fig. 3.1A is shown in Fig. 3.3 and refers to a single channel with FeCr-alloy walls. For the fuel-rich concept in Fig. 3.1B, the single-channel model with bypass air in Fig. 3.16A is adopted, whereas for the i-CST concept the model in Fig. 3.16B is employed (due to symmetry, only half the domains are shown in Fig. 3.16). In Fig. 3.16B, the initial channel length $L=15$ mm was chemically inert (noncatalytic), while the remaining 60 mm was coated with platinum. This approach allowed for the establishment of an upstream

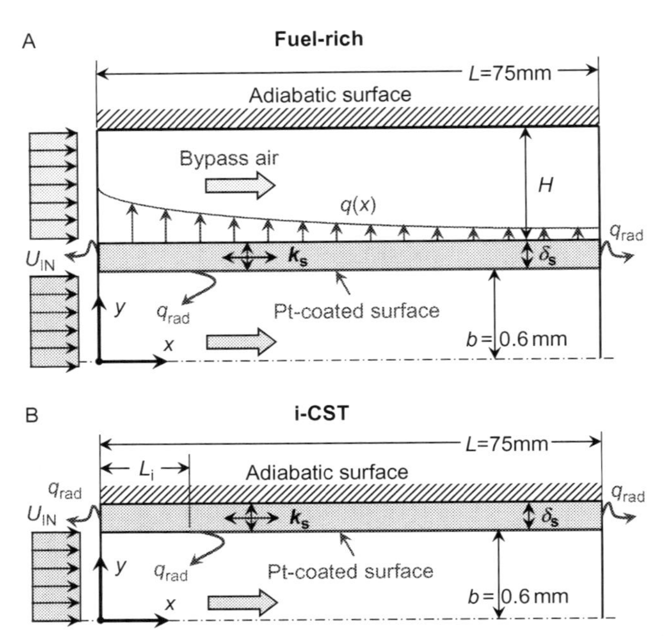

Figure 3.16 Simulated domains for (A) the fuel-rich concept in Fig. 3.1B and the (B) i-CST concept in Fig. 3.15a. In both concepts, the catalytic channel has a length $L=75$ mm and a half-height $b=0.6$ mm. In (A), bypass air flows in adjacent channels with height $H=2b=1.2$ mm.

homogeneous combustion zone in the inert channel section. Thus, the same combustion concept as that in Fig. 3.15a could be obtained, while avoiding the complex modeling of homogeneous combustion in a porous medium and its thermal coupling to the follow-up gas-phase combustion zone. The postcatalyst gas-phase zone for the CST and fuel-rich concepts were not included in the simulations, as the focus was on the thermal management of the catalytic reactors. Simulations were performed using the steady 2D model described in Section 3.3, with detailed hetero-/homogeneous chemistry (Deutschmann et al., 2000; Li et al., 2004).

The numerical parameters for each concept are summarized in Table 3.3. The inlet pressure was $p=1$ bar, the preheat temperature ranged from $T_{IN}=300$–800 K and the total equivalence ratio (φ_{TOT}, based on the total amount of air) was fuel-lean and varied from 0.1 to 0.5. For the fuel-rich concept in Fig. 3.16A, the equivalence ratios inside the catalytic channel ranged from 1.0 to 5.0, while T_{IN} was the same for both reacting and bypass air flows. Two total inflow velocities U_{IN} were considered, 10 and 20 m/s. In the CST and i-CST concepts (Figs. 3.3 and 3.16B), the total U_{IN} was the

Table 3.3 Operating Conditions for the Three Concepts in Figs. 3.3 and 3.16[a]

	φ_{TOT}	U_{IN} (m/s)	T_{IN} (K)	φ_{CAT}	Bypass Air (%)
CST (Fig. 3.3)	0.1–0.3	10, 20	300–800	0.1–0.3	0
(fuel-rich) (Fig. 3.16A)	0.2–0.5	(2.2–7.8), (4.4–15.6)	300–800	1.0–5.0	32.7–91.3
i-CST (Fig. 3.16B)	0.1–0.3	10, 20	300–800	0.1–0.3	0

[a]Total equivalence ratio (φ_{TOT}), inlet velocity in the catalytic channel (U_{IN}), inlet temperature (T_{IN}), equivalence ratio in the catalytic channel (φ_{CAT}, different than φ_{TOT} for the fuel-rich concept), and bypass air (% of the total mass flux).

same with the inflow velocity in the catalytic channel. For the fuel-rich concept in Fig. 3.16A, however, the inflow velocities in the catalytic channel and the bypass air were adjusted to yield a total mass flux equal to that of the CST and i-CST cases that shared the same φ_{TOT}, T_{IN}, and total U_{IN}. This led to bypass air accounting for 32.7–91.3% of the total incoming air flow (see Table 3.3).

Results of the CST simulations are summarized in Fig. 3.17. For the three investigated equivalence ratios, plots are provided for the computed maximum wall temperatures $T_{w,max}$ (symbols) as a function of the inlet temperature, for inlet velocities $U_{IN}=10$ and 20 m/s. The attained wall temperatures in the presence of only catalytic chemistry (no gas-phase chemistry included) are further shown for three selected T_{IN} at $\varphi=0.2$ and 0.3 and for $U_{IN}=10$ m/s (filled triangles). Also plotted in Fig. 3.17 are the corresponding adiabatic equilibrium temperatures, T_{ad}, the theoretically maximum wall temperatures, $T_{w,max-theory}$, from Eq. (3.32) and an upper limit for the tolerable wall temperature $T_{w,limit}=1500$ K that warranted thermal stability of the FeCr-alloy solid structure. The attained maximum surface temperatures $T_{w,max}$ for $\varphi=0.1$ never exceeded the limiting temperature $T_{w,limit}=1500$ K (Fig. 3.17A). Moreover, the attained $T_{w,max}$ at $\varphi=0.3$ were always higher than $T_{w,limit}$ (Fig. 3.17C), and this was of course the case for equivalence ratios up to 0.5. Therefore, the standard CST concept in Fig. 3.3 with hydrogen fuel was primarily limited to sufficiently lean stoichiometries $\varphi\leq 0.2$.

Computed surface temperatures for the i-CST concept in Fig. 3.16B are presented in Fig. 3.18. For a given equivalence ratio, there existed a T_{IN} above which the maximum wall temperature approached the adiabatic equilibrium temperature T_{ad}. This behavior was a result of flames anchored close to $x\sim 0$ when T_{IN} increased above a certain value. Hence, for practical

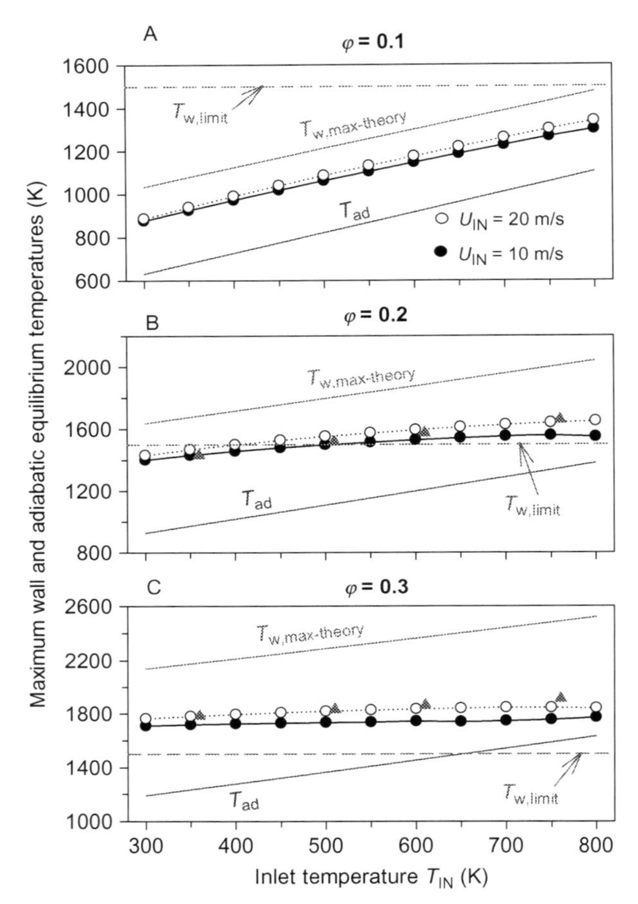

Figure 3.17 Predicted maximum wall temperatures for $U_{IN} = 10$ m/s (filled circles) and $U_{IN} = 20$ m/s (open circles) versus inlet temperature (T_{IN}) for the CST concept (model in Fig. 3.3) at three different equivalence ratios: (A) $\varphi = 0.1$, (B) $\varphi = 0.2$, and (C) $\varphi = 0.3$. Solid triangles provide the computed maximum wall temperatures in the presence of only catalytic reactions at selected cases for $\varphi = 0.2$ and 0.3. The adiabatic equilibrium temperatures, T_{ad}, the maximum theoretical wall temperatures, $T_{w,max-theory}$, from Eq. (3.32) and a limiting wall temperature for reactor thermal stability $T_{w,limit} = 1500$ K are also shown. *Adapted from Schultze and Mantzaras (2013) (with permission).*

preheats of large power generation systems (600 K $< T_{IN} <$ 800 K), the i-CST concept was limited by the value of T_{ad} and did not yield super-adiabatic surface temperatures. The dependence of the wall temperature on the inlet velocity was stronger in the i-CST than the CST concept (Figs. 3.17 and 3.18), the reason being that the magnitude of U_{IN} dictated

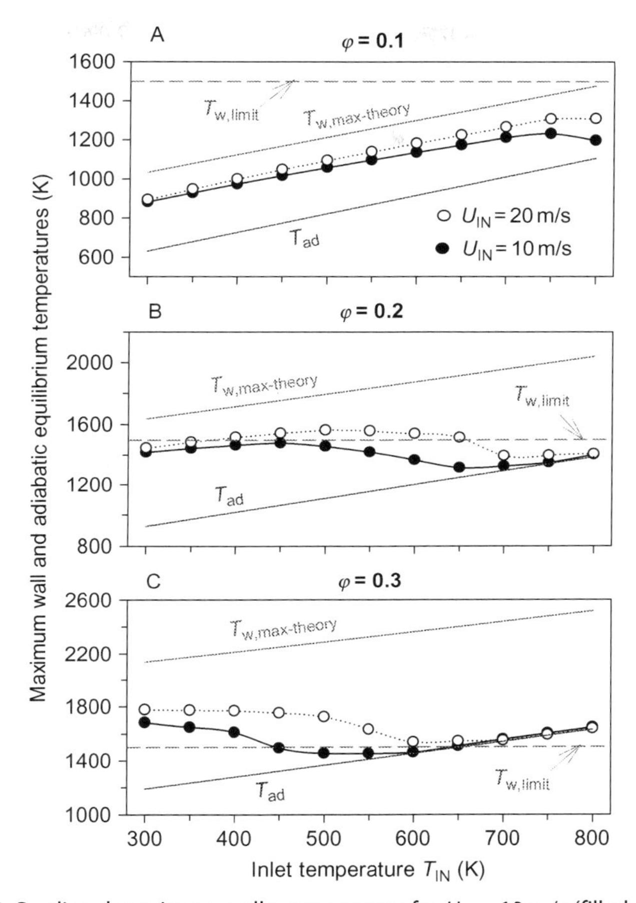

Figure 3.18 Predicted maximum wall temperatures for $U_{IN} = 10$ m/s (filled circles) and $U_{IN} = 20$ m/s (open circles) versus inlet temperature (T_{IN}) for the i-CST concept (model in Fig. 3.16b) at three different equivalence ratios: (a) $\varphi = 0.1$, (b) $\varphi = 0.2$, and (c) $\varphi = 0.3$. The adiabatic equilibrium temperatures, T_{ad}, the maximum theoretical wall temperatures, $T_{w,max-theory}$, from Eq. (3.32) and a limiting wall temperature for reactor thermal stability $T_{w,limit} = 1500$ K are also shown. *Adapted from Schultze and Mantzaras (2013) (with permission).*

how close the flame anchored to $x \sim 0$ and hence the degree of suppression of the heat-recirculation-induced superadiabaticity. When compared to the CST results in Fig. 3.17, the i-CST concept had a wider safe operational envelope ($T_{w,max} < T_{w,limit}$), which for $U_{IN} = 10$ m/s encompassed the entire examined range of T_{IN} (Table 3.3) at $\varphi = 0.2$ and a considerable part of the T_{IN} range at $\varphi = 0.3$. Design of more complex upstream gaseous combustion zones (such as porous burners with increased residence times) would undoubtedly further improve the higher U_{IN} characteristics in Fig. 3.18 by reducing heat recirculation in the solid matrix.

Simulations for the fuel-rich combustion concept (Fig. 3.16A) are depicted in Fig. 3.19 for a total $U_{IN} = 10$ m/s, three different total fuel-lean equivalence ratios φ_{TOT} and five fuel-rich catalytic equivalence ratios φ_{CAT}. Additional results for a total $U_{IN} = 20$ m/s are shown for $\varphi_{CAT} = 3.0$.

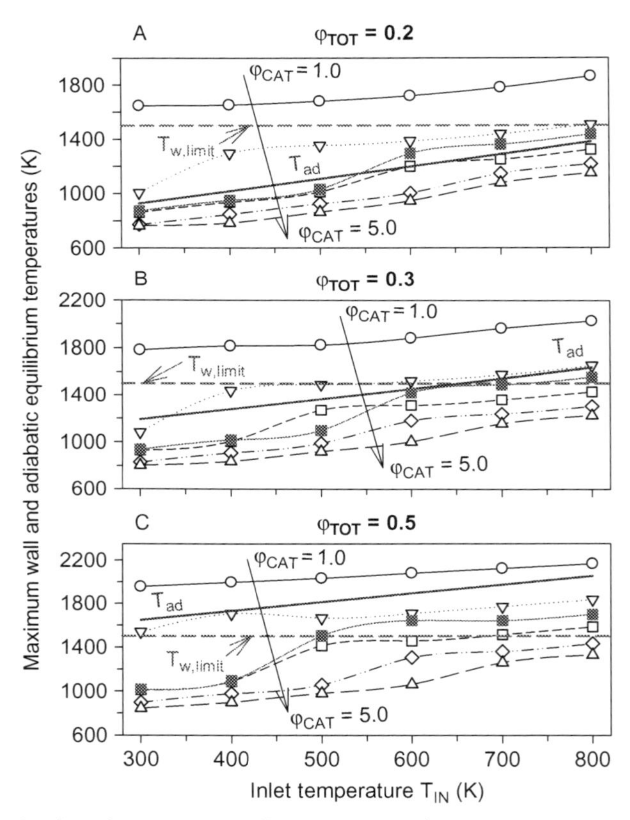

Figure 3.19 Predicted maximum wall temperatures for $U_{IN} = 10$ m/s (open symbols) versus inlet temperature (T_{IN}) for the fuel-rich mode (model in Fig. 3.16A) at three total equivalence ratios: (a) $\varphi_{TOT} = 0.2$, (b) $\varphi_{TOT} = 0.3$, (d) $\varphi_{TOT} = 0.5$. Equivalence ratios in the catalytic channel are $\varphi_{CAT} = 1.0$ (open circles), $\varphi_{CAT} = 2.0$ (lower open triangles), $\varphi_{CAT} = 3.0$ (open squares), $\varphi_{CAT} = 4.0$ (open diamonds), and $\varphi_{CAT} = 5.0$ (upper open triangles). Filled squares give the maximum wall temperatures for $\varphi_{CAT} = 3.0$ and $U_{IN} = 20$ m/s. The adiabatic equilibrium temperatures (T_{ad} based on φ_{TOT}) and a limiting wall temperature for reactor thermal stability $T_{w,limit} = 1500$ K are also shown. *Adapted from Schultze and Mantzaras (2013) (with permission).*

Comparison of Fig. 3.19 with Figs. 3.17 and 3.18 clearly shows that the fuel-rich concept had a much wider range of applicability ($T_{w,max} < T_{w,limit}$). In particular, the fuel-rich mode could safely operate at gas-turbine relevant conditions (i.e., total equivalence ratios $\varphi_{TOT} = 0.5$ and inlet temperatures $T_{IN} > 650$ K) as long as $\varphi_{CAT} > 3.0$. The wall temperatures could be maintained below the limit wall temperature $T_{w,limit} = 1500$ K, at least for sufficiently large φ_{CAT}. This was due to the cooling of the catalytic channel by the bypass air and also due to the catalytic process itself that led to under-adiabatic surface temperatures for the $Le_{O_2} > 1$ deficient reactant, as discussed in Section 4 and the accompanying Fig. 3.4. Despite these factors that led to reduced wall temperatures, flames were established at high T_{IN}. Computations have shown (Schultze and Mantzaras, 2013) that for $\varphi_{TOT} \geq 0.3$ and $T_{IN} \geq 700$ K homogeneous combustion was always present for all examined φ_{CAT}. Moreover, the flames established in the fuel-rich catalytic stage did not approach the channel walls (see also Fig. 3.10 and further discussion in Schultze et al. (2013)) and hence did not pose a reactor thermal management concern. Reason for this behavior was that in fuel-rich stoichiometries the flames were preferentially established at the channel core, where the total energy (thermal and chemical) of the reactive mixture was highest (Fig. 3.5).

The three investigated concepts are finally compared in Fig. 3.20 for $U_{IN} = 10$ m/s. Lines demarcate operating envelopes in the total equivalence

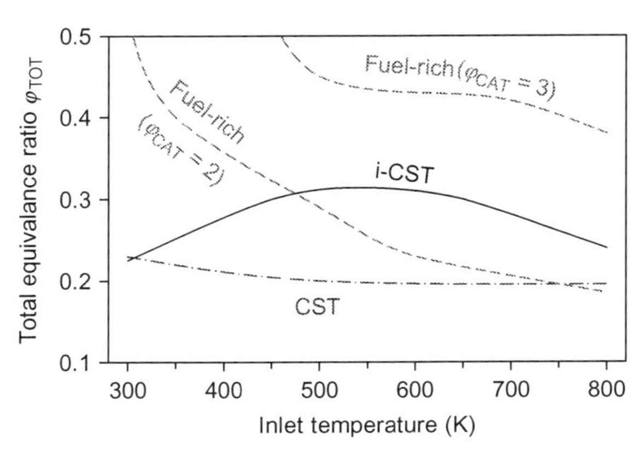

Figure 3.20 Safe operating regimes for three combustion concepts (CST in Fig. 3.3, i-CST in Fig. 3.16B, and fuel-rich in Fig. 3.16A) as a function of total equivalence ratio (φ_{TOT}) and mixture preheat (T_{IN}) for $U_{IN} = 10$ m/s. Areas below the lines define safe operating envelopes for each concept. *Adapted from Schultze and Mantzaras (2013) (with permission).*

ratio versus inlet temperature parameter space. Areas below the lines signified safe reactor operation ($T_{w,max} < 1500\,K$). CST had the narrower operating envelope, but it offered a simple reactor design solution for applications with $\varphi_{TOT} < 0.2$. The operating envelope in i-CST was broader than that of CST. The highest acceptable φ_{TOT} in i-CST exceeded 0.3 in the intermediate temperature range 450 K < T_{IN} < 650 K. For the fuel-rich concept, results are shown for $\varphi_{CAT} = 2$ and 3. For $\varphi_{CAT} = 3$, operation with $\varphi_{TOT} > 0.4$ and $T_{IN} < 750$ K was possible. Moreover, all of the $\varphi_{TOT} - T_{IN}$ parametric space in Fig. 3.20 was accessible for $\varphi_{CAT} = 4$ and 5 . Therefore, this concept was suitable for hydrogen-fueled gas turbines (φ_{TOT} up to 0.5 and T_{IN} up to 800 K) and justified its application despite the more complex design requirements. It is further noted that although $\varphi_{CAT} = 6$ was feasible (experiments in a subscale gas-turbine reactor with hydrogen have shown safe and stable combustion up to $\varphi_{CAT} = 6$, Bolaños et al., 2013), it was appropriate to limit φ_{CAT} below 4. This was because the resulting NO_x emissions at the end of the postcatalyst gaseous combustion zone in Fig. 3.1B dropped appreciably with rising hydrogen catalytic conversion and hence with decreasing φ_{CAT} (Bolaños et al., 2013).

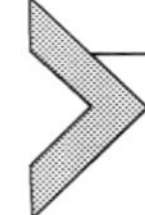

8. CATALYTIC COMBUSTION OF HYDROGEN WITH OTHER FUELS

Pure hydrogen combustion is a long-term goal for power generation, while hydrogen-rich syngas fuels (mixtures comprising mostly of H_2 and CO) are of immediate concern in IGCC plants (Daniele et al., 2013). Moreover, the addition of hydrogen to natural gas can improve the combustion stability in conventional gas-fired power plants (Winkler et al., 2009); therein, hydrogen can be produced by reforming part of the natural gas. While hydrogen catalytic combustion requires careful consideration of reactor thermal management issues (Section 7), it has the advantage of high catalytic reactivity on noble metals that in turn ensures a fast light-off and good combustion stability. Transient 2D simulations of fuel-lean H_2/air in a Pt-coated catalytic planar channel with 10-mm length and 2-mm height were recently performed in Brambilla et al. (2014) using the full transient model in Section 3.3 and the hetero-/homogeneous reaction schemes from Deutschmann et al. (2000) and Li et al. (2004). The pressure was atmospheric, with inlet temperatures 450–650 K (the lower inlet temperature range encompassing gas turbine burners at idling or partial load operation) and an inlet velocity of 10 m/s. Very short light-off times (less than 0.3 ms) were obtained, which were highly desirable as they ensured combustion stability at extreme operating conditions. On the other hand, these simulations were computationally very intensive, as more affordable quasisteady-state models could not be used to investigate light-off. This was due to the much shorter light-off times compared to the characteristic time for heat conduction in the thin solid wall. Nonetheless, the high catalytic reactivity of hydrogen is appealing for enhancing the combustion stability of other fuels.

The impact of hydrogen addition on methane catalytic combustion has been investigated in recent works (Deutschmann et al., 2000; Karagiannidis and Mantzaras, 2012). Hydrogen primarily has a thermal impact on the light-off of catalytic reactors fueled with CH_4/H_2 blends, by providing significant exothermicity already from the first time instances of reactor heat-up. Simulations with CH_4 and CH_4/H_2 fuels were performed in Karagiannidis and Mantzaras (2012), using a quasisteady-state transient 2D model (see Section 3.3) for a microscale Pt-coated channel reactor with 10-mm length, 1-mm height, wall thickness $\delta_s = 50\,\mu m$ and cordierite wall material. The equivalence ratio of the CH_4/air mixture was $\varphi = 0.40$. For the CH_4/H_2/air mixture the total equivalence ratio was $\varphi = 0.37$, obtained by substituting volumetrically 10% CH_4 by H_2. The inlet temperature was set to $T_{IN} = 850$ K, a value practically achievable in recuperated micro-reactor thermal cycles. Results in Fig. 3.21A provide the computed ignition

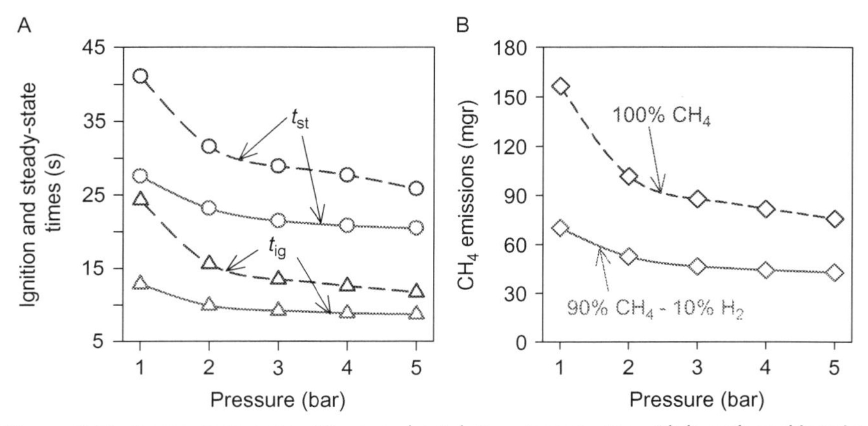

Figure 3.21 Computations in a Pt-coated catalytic microreactor with length and height of 10 mm and 1 mm, respectively. (A) Ignition (t_{ig}) and steady-state (t_{st}) times versus inlet pressure. Triangles: ignition times; circles: steady-state times. Solid lines: 90% CH_4/10% H_2 fuel blend ($\varphi = 0.37$). Dashed lines: 100% CH_4 fuel ($\varphi = 0.40$). (B) Cumulative emissions of unburned methane. *Adapted from Karagiannidis and Mantzaras (2012) (with permission).*

times (t_{ig}, defined as the elapsed time required to reach 50% fuel conversion at the channel outlet), and the steady-state times (t_{st}, defined as the elapsed time for the outlet gas temperature to vary by less than 10^{-3} K) as a function of reactor pressure. Hydrogen addition considerably shortened the ignition and steady-state times. As the catalytic reactivity of methane on platinum had a $p^{0.47}$ pressure dependence (Reinke et al., 2004), the beneficial impact of hydrogen addition was reduced with increasing pressure. During the start-up of microreactors, pollutants such as unburned hydrocarbons or CO were also formed. Cumulative emissions (computed until steady state is reached) of unburned methane are compared for the CH_4 and CH_4/H_2 fuels in Fig. 3.21B. Since most of the emissions occurred during the pre-ignition phase, the 90% CH_4/10% H_2 fuel blend showed reduced cumulative emissions of unburned methane due to its appreciably shorter ignition times.

While hydrogen affected mainly thermally the catalytic combustion of methane, the catalytic kinetic interactions of hydrogen and carbon monoxide were more involved. Although it is well known that CO inhibits the H_2 catalytic oxidation (a result simply attributed to competitive adsorption of the two fuel components), there is no consensus on the effect of H_2 on CO oxidation. Salomons et al. (2006) measured the exhaust gas response in a monolithic Pt-coated reactor to changes in the inlet CO and H_2 compositions. They reported that addition of H_2 promoted CO oxidation in a nonlinear fashion: small quantities (500 ppmv) of H_2 reduced the CO light-off temperature by 23 K, whereas larger H_2 amounts (up to 2000 ppmv) resulted in nearly the same reduction of the light-off temperature (by 26 K). The behavior could not be solely attributed to the hydrogen exothermicity as the additional 1500 ppm H_2 ought to have further reduced the light-off temperature of CO. Neutral or even inhibitive effects of H_2 addition on CO oxidation have also been reported. Experiments in a Pt-coated porous reactor by Federici and Vlachos (2011) revealed neutral H_2 effects for sufficiently large H_2 contents (H_2:CO volumetric ratios larger than unity), while a promoting effect of H_2 addition was observed at lower H_2 contents. Simulations in Mantzaras (2008) reported inhibition of CO oxidation with H_2 addition below a critical temperature of about 580 K, which was nonetheless strongly depended on the employed H_2/CO surface reaction mechanism.

Raman experiments in the reactor of Fig. 3.2 along with kinetic simulations have recently shown that, below certain transition temperatures, H_2 inhibited the catalytic oxidation of CO (Zheng et al., 2013b, 2014). This is illustrated in Fig. 3.22, providing comparisons between Raman-measured and simulated H_2 and CO transverse profiles (simulations were carried out using the 2D steady code of Section 3.3, with the boundary condition of Eq. 3.31). Contrary to the transpor-limited catalytic conversion of hydrogen (Fig. 3.8B), the catalytic conversion of both H_2 and CO in Fig. 3.22 was kinetically controlled at the lower temperature upper wall, as manifested by the Raman data. Two simulations are provided in Fig. 3.22, one with the full H_2/CO catalytic scheme from Zheng et al. (2014) and another with only the CO kinetic subset of this scheme. Comparison between the CO Raman data and the two types of CO predictions at the upper catalytic wall indicated that H_2 inhibited the oxidation of CO at the first two axial positions in Fig. 3.22B and C and promoted CO oxidation at the last two positions in Fig. 3.22D

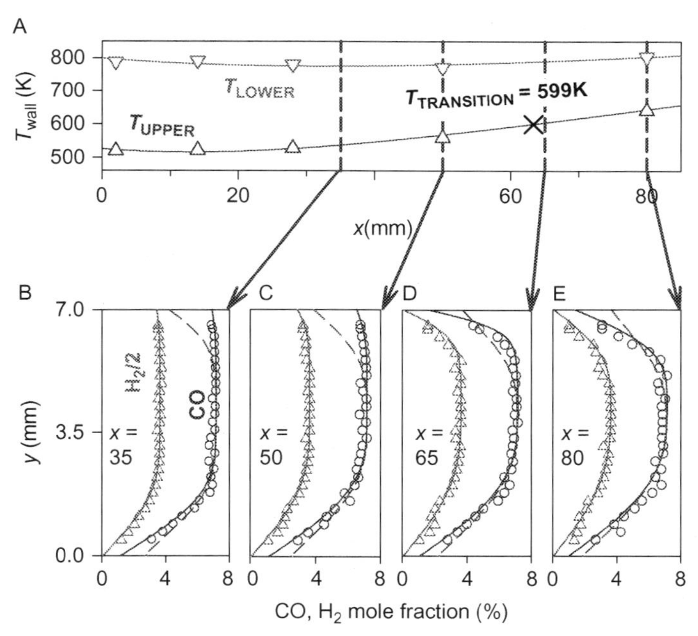

Figure 3.22 Experiments with H_2/CO fuel blends over platinum in the reactor of Fig. 3.2 for H_2:CO volumetric ratio 1:1, total equivalence ratio $\varphi=0.26$, $T_{IN}=317$ K, $U_{IN}=0.74$ m/s, and $p=5$ bar. (A) Temperature measurements (upper triangles: upper wall, lower triangles: lower wall) and fitted temperature profiles (lines). (B–E) Raman-measured transverse profiles of H_2 (triangles) and CO (circles) and predicted transverse species profiles at four axial positions (solid lines: H_2/CO catalytic chemistry from Zheng et al. (2014), dashed line for CO: only CO chemistry). For clarity, the H_2 mole fractions have been divided by two. *Adapted from Zheng et al. (2014) (with permission).*

and E. For the operating conditions in Fig. 3.22, the transition temperature below which H_2 inhibited CO oxidation was 599 K. This inhibition was attributed from the one side to the competition of H_2 with CO and O_2 for adsorption on the platinum sites and from the other side to the competition for surface-deficient O(s) species between the adsorbed H(s) and CO(s) that in turn decelerated CO(s) oxidation. The second mechanism was stronger in causing the inhibiting H_2 impact on CO oxidation (Zheng et al., 2014). For reactor residence times relevant to power generation systems, the transition temperatures demarcating H_2 inhibition depended primarily on the total equivalence ratio (550 $\pm$ 5 K at $\varphi = 0.13$ and 600 $\pm$ 5 K at $\varphi = 0.26$) and weakly on the H_2:CO volumetric ratio (for the investigated range 0.2–3.0).

Practical example of H_2 inhibition on CO oxidation is illustrated with the simulations in Fig. 3.23, pertaining to the channel geometry in Fig. 3.3. The quasisteady state 2D transient version of the model in Section 3.3 has been used, along with the heterogeneous scheme from Zheng et al. (2014) and the H_2/CO gas-phase mechanism from Li et al. (2007). For an inlet temperature of 500 K, hydrogen inhibited the oxidation of CO as manifested by the longer ignition (t_{ig}) and steady state (t_{st}) times of the H_2/CO fuel mixture compared to the corresponding times of the only CO fuel.

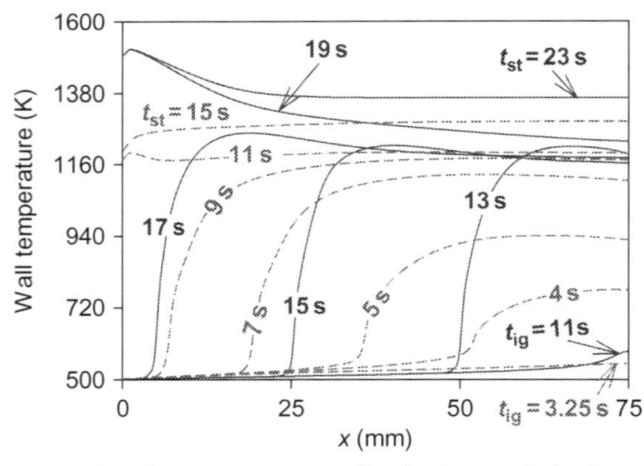

Figure 3.23 Computed wall temperature profiles in the catalytic channel of Fig. 3.3 at selected times for a H_2/CO/air mixture (solid lines) with $\varphi = 0.26$, H_2:CO vol. ratio 1:1, $T_{IN} = 500$ K, $U_{IN} = 4$ m/s, and $p = 5$ bar. Dashed lines are predictions with only CO fuel, having the same exothermicity as the H_2/CO mixture. The times marked t_{ig} and t_{st} denote catalytic ignition and steady state, respectively. *Adapted from Zheng et al. (2014) (with permission).*

9. CONCLUSIONS

Catalytic combustion of hydrogen and high-hydrogen content syngases has attracted attention in IGCC power generation, as it provides a direct way for mitigating flame flashback of these highly reactive mixtures. Multidimensional numerical modeling and *in situ* spatially resolved measurements of gas-phase thermoscalars over the catalyst boundary layer have recently advanced the fundamental investigation of hydrogen heterogeneous and homogeneous chemical reaction schemes and their coupling at industrially relevant operating conditions. Moreover, combination of advanced numerical modeling and near-wall measurements of thermoscalars and flow velocity has clarified the intricate interplay of interphase fluid transport (laminar or turbulent) and hetero-/homogeneous kinetics. Homogeneous ignition chemistry, flame propagation characteristics, competition between the two reaction pathways for hydrogen consumption, diffusional imbalance of the limiting reactant, flow laminarization due to heat transfer from the catalytic walls and fuel leakage through the gaseous combustion zone are the controlling processes.

Fuel-lean and fuel-rich hybrid (heterogeneous and homogeneous) combustion methodologies for hydrogen were assessed. The former were compounded by the attainment of superadiabatic surface temperatures due to the low Lewis number of the deficient hydrogen reactant while the latter provided a viable combustion approach for power generation applications. A new fuel-lean combustion concept was further demonstrated, based on an inverse catalytically stabilized combustion methodology, whereby gaseous combustion preceded catalytic combustion. The impact of hydrogen addition on fuels of major interest to power generation (methane and carbon monoxide) was finally outlined.

ACKNOWLEDGMENTS

Support by the European Union Project HRC-Power is gratefully acknowledged.

REFERENCES

Aghalayam P, Park YK, Vlachos DG: Construction and optimization of complex surface-reaction mechanisms, *AIChE J* 46:2017–2029, 2000.

Ahn T, Pinczewski WV, Trimm DL: Transient performance of catalytic combustors for gas-turbine applications, *Chem Eng Sci* 41:55–64, 1986.

Alavandi SK, Etemad S, Baird BD: Low single digit NOx emissions catalytic combustor for advanced hydrogen turbines for clean coal power systems, ASME GT2012-68128.

In Proceedings of the ASME. New York, NY, 2012, American Society of Mechanical Engineers, pp 53–62.
Appel C, Mantzaras J, Schaeren R, et al. An experimental and numerical investigation of homogeneous ignition in catalytically stabilized combustion of hydrogen/air mixtures over platinum, *Combust Flame* 128:340–368, 2002a.
Appel C, Mantzaras J, Schaeren R, Bombach R, Kaeppeli B, Inauen A: An experimental and numerical investigation of turbulent catalytically stabilized channel flow combustion of hydrogen/air mixtures over platinum, *Proc Combust Inst* 29:1031–1038, 2002b.
Appel C, Mantzaras J, Schaeren R, Bombach R, Inauen A: Turbulent catalytically stabilized combustion of hydrogen/air mixtures in entry channel flows, *Combust Flame* 140:70–92, 2005.
Bär JN, Karakaya C, Deutschmann O: Catalytic ignition of light hydrocarbons over Rh/Al_2O_3 studied in a stagnation-point flow reactor, *Proc Combust Inst* 34:2313–2320, 2013.
Beebe KW, Cairns KD, Pareek VK, Nickolas SG, Schlatter JC, Tsuchiya T: Development of catalytic combustion technology for single-digit emissions from industrial gas turbines, *Catal Today* 59:95–115, 2000.
Behrendt F, Deutschmann O, Schmidt R, Warnatz J. In: Warren B, editor: *Heterogeneous hydrocarbon oxidation*, vol 638, Washington, DC, 1996, American Chemical Society, ACS Symposium Series.
Boehman AL: Radiation heat transfer in catalytic monoliths, *AIChE J* 44:2745–2755, 1998.
Bolaños F, Winkler D, Piringer F, Griffin T, Bombach R, Mantzaras J: Study of a rich/lean staged combustion concept for hydrogen at gas turbine relevant conditions, ASME GT2013-94420. In Proceedings of the ASME. New York, NY, 2013, American Society of Mechanical Engineers, pp V01AT04A031.
Brambilla A, Frouzakis CE, Mantzaras J, Tomboulides A, Kerkemeier S, Boulouchos K: Detailed transient numerical simulation of H_2/air hetero-/homogeneous combustion in platinum-coated channels with conjugate heat transfer, *Combust Flame*, 2014. http://dx.doi.org/10.1016/j.combustflame.2014.04.003 (in press).
Bui PA, Vlachos DG, Westmoreland PR: Homogeneous ignition of hydrogen/air mixtures over platinum, *Proc Combust Inst* 26:1763–1770, 1996.
Bui PA, Wilder EA, Vlachos DG, Westmoreland PR: Hierarchical reduced models for catalytic combustion: H_2/air mixtures near platinum surfaces, *Combust Sci Technol* 129:243–275, 1997.
Carroni R, Griffin T: Catalytic hybrid lean combustion for gas turbines, *Catal Today* 155:2–12, 2010.
Carroni R, Griffin T, Mantzaras J, Reinke M: High-pressure experiments and modeling of methane/air catalytic combustion for power generation applications, *Catal Today* 83:157–170, 2003.
Cerkanowicz AE, Cole RB, Stevens JG: Catalytic combustion modeling—comparisons with experimental data, *J Eng Gas Turbines Power* 99:593–600, 1977.
Ciambelli P, Palma V, Tikhov SF, Sadykov VA, Isupova LA, Lisi L: Catalytic activity of powder and monolith perovskites in methane combustion, *Catal Today* 47:199–207, 1999.
Coltrin ME, Kee RJ, Evans GH, Meeks E, Rupley FM, Grcar JF: *SPIN (version 3.83): a FORTRAN program for modeling one-dimensional rotating-disk/stagnation-flow chemical vapor deposition reactors: Report No. SAND91-8003,* 1991, Sandia National Laboratories.
Coltrin ME, Moffatt HK, Kee RJ, Rupley FM: *CRESLAF (version 4.0): a Fortran program for modeling laminar, chemically reacting, boundary-layer flow in cylindrical or planar channels: Report SAND93-0478,* 1996, Sandia National Laboratories.
Cormos CC: Evaluation of power generation schemes based on hydrogen-fuelled combined cycle with carbon capture and storage (CCS), *Int J Hydrogen Energ* 36:3726–3738, 2011.
Czekaj I, Kacprzak KA, Mantzaras J: CH4 combustion cycles at Pd/Al2O3 - important role of support and oxygen access, *Phys Chem Chem Phys* 15:11368–11374, 2013.

Daniele S, Jansohn P, Mantzaras J, Boulouchos K: Turbulent flame speed for syngas at gas turbine relevant conditions, *Proc Combust Inst* 33:2937–2944, 2011.

Daniele S, Mantzaras J, Jansohn P, Denisov A, Boulouchos K: Flame front/turbulence interaction for syngas fuels in the thin reaction zones regime: turbulent and stretched laminar flame speeds at elevated pressures and temperatures, *J Fluid Mech* 724:36–68, 2013.

Deutschmann O, Schmidt R, Behrendt F, Warnatz J: Numerical modeling of catalytic ignition, *Proc Combust Inst* 26:1747–1754, 1996.

Deutschmann O, Maier LI, Riedel U, Stroemman AH, Dibble RW: Hydrogen assisted catalytic combustion of methane on platinum, *Catal Today* 59:141–150, 2000.

Dogwiler U, Mantzaras J, Benz P, Kaeppeli B, Bombach R, Arnold A: Homogeneous ignition of methane/air mixtures over platinum: comparison of measurements and detailed numerical predictions, *Proc Combust Inst* 27:2275–2282, 1998.

Dogwiler U, Benz P, Mantzaras J: Two-dimensional modelling for catalytically stabilized combustion of a lean methane-air mixture with elementary homogeneous and heterogeneous chemical reactions, *Combust Flame* 116:243–258, 1999.

Egolfopoulos FN: Dynamics and structure of unsteady, strained, laminar premixed flames, *Proc Combust Inst* 25:1365–1373, 1994.

Eriksson S, Wolf M, Schneider A, et al. Fuel rich catalytic combustion of methane in zero emissions power generation processes, *Catal Today* 117:447–453, 2006.

Evans CJ, Kyritsis DC: Operational regimes of rich methane and propane/oxygen flames in mesoscale non-adiabatic ducts, *Proc Combust Inst* 32:3107–3114, 2009.

Federici JA, Vlachos DG: Experimental studies on syngas catalytic combustion on Pt/Al_2O_3 in a microreactor, *Combust Flame* 158:2540–2543, 2011.

Forsth M, Gudmundson F, Persson JL, Rosen A: The influence of a catalytic surface on the gas-phase combustion of H_2+O_2, *Combust Flame* 119:144–153, 1999.

Fridell E, Rosen A, Kasemo B: A laser-induced fluorescence study of OH desorption from Pt in H_2O/O_2 and H_2O/H_2 mixtures, *Langmuir* 10:699–708, 1994.

Ghermay Y, Mantzaras J, Bombach R: Effects of hydrogen preconversion on the homogeneous ignition of fuel-lean $H_2/O_2/N_2/CO_2$ mixtures over platinum at moderate pressures, *Combust Flame* 157:1942–1958, 2010.

Ghermay Y, Mantzaras J, Bombach R, Boulouchos K: Homogeneous combustion of fuel lean $H_2/O_2/N_2$ mixtures over platinum at elevated pressures and preheats, *Combust Flame* 158:1491–1506, 2011.

Glassman I: *Combustion,* ed 3, London, 1996, Academic Press, p 67.

Griffin T, Winkler D, Wolf M, Appel C, Mantzaras J: Staged catalytic combustion method for the advanced zero emissions gas turbine power plant, ASME GT2004-54101. In Proceedings of the ASME. New York, NY, 2004, American Society of Mechanical Engineers, pp 705–711.

Groppi G, Belloli A, Tronconi E, Forzatti P: A Comparison of lumped and distributed models of monolith catalytic combustors, *Chem Eng Sci* 50:2705–2715, 1995.

Groppi G, Ibashi W, Tronconi E, Forzatti P: Structured reactors for kinetic measurements in catalytic combustion, *Chem Eng J* 82:57–71, 2001.

Hayes RE, Kolaczkowski ST: Mass and heat transfer effects in catalytic monolith reactors, *Chem Eng Sci* 49:3587–3599, 1994.

Hegedus LL: Temperature excursions in catalytic monoliths, *AIChE J* 21:849–853, 1975.

Hellsing B, Kasemo B, Zhdanov VP: Kinetics of the hydrogen oxygen reaction on platinum, *J Catal* 132:210–228, 1991.

Hougen OA, Watson KM: Solid catalysts and reaction rates - general principles, *J Ind Eng Chem* 35:529–541, 1943.

Kacprzak KA, Czekaj I, Mantzaras J: DFT studies of oxidation routes for Pd_9 clusters supported on γ-alumina, *Phys Chem Chem Phys* 14:10243–10247, 2012.

Karagiannidis S, Mantzaras J: Numerical investigation on the start-up of methane-fueled catalytic microreactors, *Combust Flame* 157:1400–1413, 2010.

Karagiannidis S, Mantzaras J: Numerical investigation on the hydrogen-assisted start-up of methane-fueled, catalytic microreactors, *Flow Turbul Combust* 89:215–230, 2012.

Karagiannidis S, Mantzaras J, Bombach R, Schenker S, Boulouchos K: Experimental and numerical investigation of the hetero-/homogeneous combustion of lean propane/air mixtures over platinum, *Proc Combust Inst* 32:1947–1955, 2009.

Karagiannidis S, Mantzaras J, Boulouchos K: Stability of hetero-/homogeneous combustion in propane and methane fueled catalytic microreactors: channel confinement and molecular transport effects, *Proc Combust Inst* 33:3241–3249, 2011.

Kee RJ, Miller JA, Evans GH, Dixon-Lewis G: A computational model of the structure and extinction of strained, opposed flow, premixed methane-air flames, *Proc Combust Inst* 22:1479–1494, 1988.

Kee RJ, Dixon-Lewis G, Warnatz J, Coltrin ME, Miller JA: *A Fortran computer code package for the evaluation of gas-phase multicomponent transport properties: Report No. SAND86-8246,* 1996, Sandia National Laboratories.

Kee RJ, Coltrin ME, Glaborg P: *Chemically reacting flow, theory and practice,* Hoboken, NJ, 2003, Wiley.

Kissel-Osterrieder R, Behrendt F, Warnatz J, Metka U, Volpp HR, Wolfrum J: Experimental and theoretical investigation of the CO-oxidation on platinum: bridging the pressure and the materials gap, *Proc Combust Inst* 28:1341–1348, 2000.

Kramer JF, Reihani SAS, Jackson GS: Low temperature combustion of hydrogen on supported Pd catalysts, *Proc Combust Inst* 29:989–996, 2002.

Kurdyumov VN, Pizza G, Frouzakis CE, Mantzaras J: Dynamics of premixed flames in a narrow channel with a step-wise wall temperature, *Combust Flame* 156:2190–2200, 2009.

Law CK, Sivashinsky GI: Catalytic extension of extinction limits of stretched premixed flames, *Combust Sci Technol* 29:277–286, 1982.

Li J, Zhao Z, Kazakov A, Dryer FL: An updated comprehensive kinetic model of hydrogen combustion, *Int J Chem Kinet* 36:566–575, 2004.

Li J, Zhao ZW, Kazakov A, Chaos M, Dryer FL, Scire JJ: A comprehensive kinetic mechanism for CO, CH_2O, and CH_3OH combustion, *Int J Chem Kinet* 39:109–136, 2007.

Lucci F, Frouzakis CE, Mantzaras J: Three-dimensional direct numerical simulation of turbulent channel flow catalytic combustion of hydrogen over platinum, *Proc Combust Inst* 34:2295–2302, 2013.

Ludwig J, Vlachos DG: Molecular dynamics of hydrogen dissociation on an oxygen covered Pt(111) surface, *J Chem Phys* 128:154708, 2008.

Maestri M, Reuter K: Semiempirical rate constants for complex chemical kinetics: first-principles assessment and rational refinement, *Angew Chem Int Ed* 50:1194–1197, 2011.

Maestri M, Beretta A, Faravelli T, Groppi G, Tronconi E: Role of gas-phase chemistry in the rich combustion of H_2 and CO over a Rh/Al_2O_3 catalyst in annular reactor, *Chem Eng Sci* 62:4992–4997, 2007.

Maestri M, Beretta A, Faravelli T, Groppi G, Tronconi E, Vlachos DG: Two-dimensional detailed modeling of fuel-rich H_2 combustion over Rh/Al_2O_3 catalyst, *Chem Eng Sci* 63:2657–2669, 2008.

Mantzaras J: Understanding and modeling of thermofluidic processes in catalytic combustion, *Catal Today* 117:394–406, 2006.

Mantzaras J: Catalytic combustion of syngas, *Combust Sci Technol* 180:1137–1168, 2008.

Mantzaras J, Appel C: Effects of finite rate heterogeneous kinetics on homogeneous ignition in catalytically stabilized channel-flow combustion, *Combust Flame* 130:336–351, 2002.

Mantzaras J, Benz P: An asymptotic and numerical investigation of homogeneous ignition in catalytically stabilized channel flow combustion, *Combust Flame* 119:455–472, 1999.

Mantzaras J, Appel C, Benz P: Catalytic combustion of methane/air mixtures over platinum: homogeneous ignition distances in channel flow configurations, *Proc Combust Inst* 28:1349–1357, 2000a.

Mantzaras J, Appel C, Benz P, Dogwiler U: Numerical modelling of turbulent catalytically stabilized channel flow combustion, *Catal Today* 53:3–17, 2000b.

Mantzaras J, Bombach R, Schaeren R: Hetero-/homogeneous combustion of hydrogen/air mixtures over platinum at pressures up to 10 bar, *Proc Combust Inst* 32:1937–1945, 2009.

Markatou P, Pfefferle LD, Smooke MD: A computational study of methane-air combustion over heated catalytic and non-catalytic surfaces, *Combust Flame* 93:185–201, 1993.

Marks CM, Schmidt LD: Hydroxyl radical desorption in catalytic combustion, *Chem Phys Lett* 178:358–362, 1991.

Maruta K: Micro and mesoscale combustion, *Proc Combust Inst* 33:125–150, 2010.

Mhadeshwar AB, Vlachos DG: Hierarchical, multiscale surface reaction mechanism development: CO and H_2 oxidation, water-gas shift, and preferential oxidation of CO on Rh, *J Catal* 234:48–63, 2005.

Miller JA, Bowman CT: Mechanism and Modeling of Nitrogen Chemistry in Combustion, *Prog Energy Combust Sci* 15:287–338, 1989.

Mladenov N, Koop J, Tischer S, Deutschmann O: Modeling of transport and chemistry in channel flows of automotive catalytic converters, *Chem Eng Sci* 65:812–826, 2010.

Motz H, Wise H: Diffusion and heterogeneous reaction. III. Atom recombination at a catalytic boundary, *J Chem Phys* 32:1893–1894, 1960.

Mueller MA, Kim TJ, Yetter RA, Dryer FL: Flow reactor studies and kinetic modeling of the H_2/O_2 reaction, *Int J Chem Kinet* 31:113–125, 1999.

Nord LO, Anantharaman R, Bolland O: Design and off-design analyses of a pre-combustion CO_2 capture process in a natural gas combined cycle power plant, *Int J Greenh Gas Con* 3:385–392, 2009.

Norton DG, Wetzel ED, Vlachos DG: Fabrication of single-channel catalytic microburners: effect of confinement on the oxidation of hydrogen/air mixtures, *Ind Eng Chem Res* 43:4833–4840, 2004.

Park YK, Aghalayam P, Vlachos DG: A generalized approach for predicting coverage-dependent reaction parameters of complex surface reactions: application to H_2 oxidation over platinum, *J Phys Chem A* 103:8101–8107, 1999.

Pfefferle WC, Pfefferle LD: Catalytically stabilized combustion, *Prog Energy Combust Sci* 12:25–41, 1986.

Pfefferle LD, Griffin TA, Winter M, Crosley DR, Dyer MJ: The influence of catalytic activity on the ignition of boundary layer flows part I: hydroxyl radical measurements, *Combust Flame* 76:325–338, 1989.

Pizza G, Frouzakis CE, Mantzaras J, Tomboulides AG, Boulouchos K: Dynamics of premixed hydrogen/air flames in microchannels, *Combust Flame* 152:433–450, 2008a.

Pizza G, Frouzakis CE, Mantzaras J, Tomboulides AG, Boulouchos K: Dynamics of premixed hydrogen/air flames in mesoscale channels, *Combust Flame* 155:2–20, 2008b.

Pizza G, Mantzaras J, Frouzakis CE, Tomboulides AG, Boulouchos K: Suppression of combustion instabilities of premixed hydrogen/air flames in microchannels using heterogeneous reactions, *Proc Combust Inst* 32:3051–3058, 2009.

Pizza G, Frouzakis CE, Mantzaras J, Tomboulides AG, Boulouchos K: Three-dimensional simulations of premixed hydrogen/air flames in microtubes, *J Fluid Mech* 658:463–491, 2010a.

Pizza G, Mantzaras J, Frouzakis CE: Flame dynamics in catalytic and non-catalytic mesoscale microreactors, *Catal Today* 155:123–130, 2010b.

Pizza G, Frouzakis CE, Mantzaras J: Chaotic dynamics in premixed hydrogen/air channel flow combustion, *Combust Theor Model* 16:275–299, 2012.

Raja LL, Kee RJ, Petzold LR: Simulation of the transient, compressible, gas-dynamic behavior of catalytic combustion ignition in stagnation flows, *Proc Combust Inst* 27:2249–2257, 1998.
Raja LL, Kee RJ, Deutschmann O, Warnatz J, Schmidt LD: A critical evaluation of Navier-Stokes, boundary-layer, and plug-flow models of the flow and chemistry in a catalytic-combustion monolith, *Catal Today* 59:47–60, 2000.
Reinke M, Mantzaras J, Schaeren R, Bombach R, Kreutner W, Inauen A: Homogeneous ignition in high-pressure combustion of methane/air over platinum: comparison of measurements and detailed numerical predictions, *Proc Combust Inst* 29:1021–1029, 2002.
Reinke M, Mantzaras J, Schaeren R, Bombach R, Inauen A, Schenker S: High-pressure catalytic combustion of methane over platinum: in situ experiments and detailed numerical predictions, *Combust Flame* 136:217–240, 2004.
Reinke M, Mantzaras J, Bombach R, Schenker S, Inauen A: Gas phase chemistry in catalytic combustion of methane/air mixtures over platinum at pressures of 1 bar to 16 bar, *Combust Flame* 141:448–468, 2005a.
Reinke M, Mantzaras J, Schaeren R, Bombach R, Inauen A, Schenker S: Homogeneous ignition of CH_4/air and H_2O- and CO_2-diluted CH_4/O_2 mixtures over platinum; an experimental and numerical investigation at pressures up to 16 bar, *Proc Combust Inst* 30:2519–2527, 2005b.
Reinke M, Mantzaras J, Bombach R, Schenker S, Tylli N, Boulouchos K: Effects of H_2O and CO_2 dilution on the catalytic and gas phase combustion of methane over platinum at elevated pressures, *Combust Sci Technol* 179:553–600, 2007.
Robbins FA, Zhu HY, Jackson GS: Transient modeling of combined catalytic combustion/ CH_4 steam reforming, *Catal Today* 83:141–156, 2003.
Salomons S, Votsmeier M, Hayes R, Drochner A, Vogel H, Gieshof J: CO and H_2 oxidation on a platinum monolith diesel oxidation catalyst, *Catal Today* 117:491–497, 2006.
Satterfield CN, Resnick H, Wentworth RI: Simultaneous heat and mass transfer in a diffusion-controlled chemical reaction: part I: studies in a tubular reactor, *Chem Eng Prog* 50:460–466, 1954.
Schefer RW: Catalyzed combustion of H_2/air mixtures in a flat boundary layer: II. Numerical model, *Combust Flame* 45:171–190, 1982.
Schlegel A, Benz P, Griffin T, Weisenstein W, Bockhorn H: Catalytic stabilization of lean premixed combustion: method for improving NOx emissions, *Combust Flame* 105:332–340, 1996.
Schneider A, Mantzaras J, Bombach R, Schenker S, Tylli N, Jansohn P: Laser induced fluorescence of formaldehyde and Raman measurements of major species during partial catalytic oxidation of methane with large H_2O and CO_2 dilution at pressures up to 10 bar, *Proc Combust Inst* 31:1973–1981, 2007.
Schneider A, Mantzaras J, Eriksson S: Ignition and extinction in catalytic partial oxidation of methane-oxygen mixtures with large H_2O and CO_2 dilution, *Combust Sci Technol* 180:89–126, 2008.
Schultze M, Mantzaras J: Hetero-/homogeneous combustion of hydrogen/air mixtures over platinum: fuel-lean versus fuel-rich combustion modes, *Int J Hydrogen Energ* 38:10654–10670, 2013.
Schultze M, Mantzaras J, Bombach R, Boulouchos K: An experimental and numerical investigation of the hetero-/homogeneous combustion of fuel-rich hydrogen/air mixtures over platinum, *Proc Combust Inst* 34:2269–2277, 2013.
Schultze M, Mantzaras J, Grygier F, Bombach R: Hetero-/homogeneous combustion of syngas over platinum at fuel-rich stoichiometries and pressures up to 14 bar, *Proc Combust Inst* 35, 2014. http://dx.doi.org/10.1016/j.proci.2014.05.018 (in press).

Smith LL, Karim H, Castaldi MJ, Etemad S, Pfefferle WC: Rich-catalytic lean-burn combustion for fuel-flexible operation with ultra-low emissions, *Catal Today* 117:438–446, 2006.

Song X, Williams WR, Schmidt LD, Aris R: Ignition and extinction of homogeneous-heterogeneous combustion: CH4 and C3H8 oxidation on Pt, *Proc Combust Inst* 23:1129–1137, 1990.

Stefanidis GD, Vlachos DG, Kaisare NS, Maestri M: Methane steam reforming at microscales: operation strategies for variable power output at millisecond contact times, *AIChE J* 55:180–191, 2009.

Tien JS: Transient catalytic combustor model, *Combust Sci Technol* 26:65–75, 1981.

Tock L, Marechal F: H_2 processes with CO_2 mitigation: thermo-economic modeling and process integration, *Int J Hydrogen Energ* 37:11785–11795, 2012.

Vlachos DG, Bui PA: Catalytic ignition and extinction of hydrogen: comparison of simulations and experiments, *Surf Sci* 364:1625–1630, 1996.

Warnatz J, Dibble RW, Maas U: *Combustion, physical and chemical fundamentals, modeling and simulation,* New York, 1996, Springer-Verlag.

Williams FA: *Combustion theory,* ed 2, Menlo Park, CA, 1985, Benjamin/Cummings, p164.

Williams WR, Marks CM, Schmidt LD: Steps in the reaction $H_2+O_2=H_2O$ on Pt - OH desorption at high-temperatures, *J Phys Chem* 96:5922–5931, 1992.

Winkler D, Mueller P, Reimer S, et al. Improvement of gas turbine combustion reactivity under flue gas recirculation condition with in situ hydrogen addition, ASME GT2009-59182. In Proceedings of the ASME. New York, NY, 2009, American Society of Mechanical Engineers, pp 137–145.

Yan X, Maas U: Intrinsic low-dimensional manifolds of heterogeneous combustion processes, *Proc Combust Inst* 28:1615–1621, 2000.

Zheng X, Mantzaras J: An analytical and numerical investigation of hetero-/homogeneous combustion with deficient reactants having larger than unity Lewis numbers, *Combust Flame* 161:1911–1922, 2014. http://dx.doi.org/10.1016/j.combustflame.2013.12.018.

Zheng X, Mantzaras J, Bombach R: Hetero-/homogeneous combustion of ethane/air mixtures over platinum at pressures up to 14 bar, *Proc Combust Inst* 34:2279–2287, 2013a.

Zheng X, Schultze M, Mantzaras J, Bombach R, Boulouchos K: Effects of hydrogen addition on the catalytic oxidation of carbon monoxide over platinum at power generation relevant temperatures, *Proc Combust Inst* 34:3343–3350, 2013b.

Zheng X, Mantzaras J, Bombach R: Kinetic interactions between hydrogen and carbon monoxide oxidation over platinum, *Combust Flame* 161:332–346, 2014.

CHAPTER FOUR

Novel Developments in Fluidized Bed Membrane Reactor Technology

Ivo Roghair, Fausto Gallucci, Martin van Sint Annaland
Chemical Process Intensification, Eindhoven University of Technology, Eindhoven, The Netherlands

Contents

Abstract

This chapter outlines recent and ongoing investigations on the effect of using membranes in a fluidized bed reactor (FBR), using numerical simulations and experiments.

Fluidized bed membrane reactors are a novel, integrated type of reactors where heterogeneously catalyzed reactions can be performed with simultaneous reactant feeding or product extraction in a single unit operation. While this operating technique is beneficial for various reasons (e.g., shift of chemical equilibrium, very good mass and heat transfer), addition or extraction of components can significantly change the behavior of the fluidized bed compared to traditional FBRs, hydrodynamics (bubble and emulsion phase behavior), and mass and heat transfer may be severely affected by the presence of the membranes.

Advances in Chemical Engineering, Volume 45
ISSN 0065-2377
http://dx.doi.org/10.1016/B978-0-12-800422-7.00004-2

A number of experimental measurement techniques are discussed, with a focus on noninvasive optical techniques such as particle image velocimetry and digital image analysis, as well as a number of academic numerical modeling tools such as discrete particle model and two-fluid model. Not only hydrodynamic aspects, such as the emergence of defluidized zones and solids circulation profile inversion, but also the effect on the bubble size distributions are discussed for wall-mounted membranes and horizontally immersed membranes.

The development of two novel experimental techniques, which may be used for studying concentration profiles in the gas phase, and for studying the fluidized bed at reaction conditions, are outlined in Section 5.

NOMENCLATURE

Bold face indicates tensor quantity, overline arrow indicates vector quantity

$\vec{C}$ particle fluctuating velocity (m/s)
δ overlap (m)
$\mathbf{K_{be}}$ bubble-to-emulsion phase mass transfer coefficient (s^{-1})
$\vec{T}$ torque (N m)
d_b bubble diameter (m)
d_p particle diameter (m)
g_0 radial distribution function
p_c critical solids pressure (Pa)
q_s pseudo-Fourier fluctuating kinetic energy flux (kg/m/s)
λ_s bulk viscosity (Pa s)
$\vec{\omega}$ angular velocity (rad/s)
ℓ target length
A area (m^2 or px^2)
C concentration (mol/l)
E extent of densified zones
F force (N)
I moment of inertia (kg m^2)
$I(x,y,z)$ **or** $I(z)$ intensity function (gray value)
N number
R radius (m)
Re Reynolds number
S source term (kg/m^2/s^2)
T transmittance
V volume (m^3)
e restitution coefficient
g gravitational acceleration (m/s^2)
k spring stiffness (N/m)
m mass (kg)

$\boldsymbol{p}$ pressure (Pa)
$\boldsymbol{t}$ time (s)
$\boldsymbol{u}$ (superficial) fluid velocity (m/s)
$\boldsymbol{v}$ particle velocity (m/s)
$\boldsymbol{x, y, z}$ coordinates (m)
$\boldsymbol{D}$ rate of strain tensor
$\boldsymbol{I}$ unit tensor
$\boldsymbol{\beta}$ interphase momentum transfer coefficient (kg/m^3/s)
$\boldsymbol{\gamma}$ dissipation of granular energy due to inelastic particle–particle collisions (kg/m/s^3)
$\boldsymbol{\varepsilon}$ phase fraction
$\boldsymbol{\varepsilon}$ molar absorbance (mol^{-1} cm^{-2}) Eq. 4.40
$\boldsymbol{\eta}$ damping coefficient
$\boldsymbol{\theta}$ granular temperature (m^2/s^2)
$\boldsymbol{\kappa}$ pseudothermal conductivity (kg/m/s)
$\boldsymbol{\mu}$ viscosity (Pa s)
$\boldsymbol{\rho}$ density (kg/m^3)
$\boldsymbol{\tau}$ stress tensor (Pa)
$\vec{\boldsymbol{u}}$, $\vec{\boldsymbol{v}}$ velocity, particle velocity (m/s)

ABBREVIATIONS

CARPT computer-automated radioactive particle tracking
CFD computational fluid dynamics
CSTR Continuously stirred tank reactor
DBM discrete bubble model
DIA digital image analysis
DNS direct numerical simulations
DPM discrete particle model
ePIV endoscopic PIV
FBMR fluidized bed membrane reactor
FBR fluidized bed reactor
IBM immersed boundary method
IR infra-red
KTGF kinetic theory of granular flow
MAFBR Membrane-assisted fluidized bed reactor
MFC mass flow controller
MRI magnetic resonance imaging
PBMR packed bed membrane reactor
PEPT positron emission particle tracking
PIV particle image velocimetry
RHS right-hand side
TFM two-fluid model
VIS visual spectrum

SUBSCRIPTS

0 at inlet position
b bubble
f fluid
g gas
i index
mf minimum fluidization conditions
n normal
p particle
s solids
t tangential

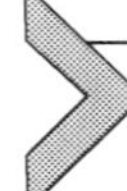

1. INTRODUCTION

1.1. From Packed Bed to Fluidized Bed Membrane Reactors

Fluidized bed reactors (FBRs) are chemical reactors in which (catalytic) particles interact with a gas stream that is fed from the bottom, such that the mixture (emulsion phase) behaves as a fluid. This type of reactors is often used in the chemical and process industries, where they have gained their popularity due to their excellent heat and mass transfer characteristics. FBRs are used for instance for gas-phase polymerization reactions for polyolefin production (polyethylene, polypropylene), chemical looping combustion or reforming processes, and gas-phase Fischer–Tropsch synthesis.

Membrane reactors (often multiphase reactors) integrate a catalytic reaction and a separation through a membrane in a single unit. These reactors are often used to circumvent the equilibrium constraints which limit the conversion in conventional reactor systems. The success of membrane reactors is basically associated with: (i) the advances in membrane production methods; (ii) the design of innovative reactor concepts, which allow the integration of separation and energy exchange, (iii) the reduction of mass and heat transfer resistances and simplification of the housing of the membranes (Gallucci et al., 2011).

Important advantages of processes in membrane reactors, as compared to conventional reactors, include the shift of the reaction equilibrium to the product side, enhanced conversion, prevention of side reactions, distributive feeding of reactants to allow for more control on reaction selectivity and integration of the separation step in one unit. Packed bed membrane reactors have been very popular as a research topic (see, e.g., Andrés et al., 2011;

Coroneo et al., 2010; Godini et al., 2013; Mahecha-Botero et al., 2006; Rakib et al., 2011), mainly due to the possibility for a straightforward comparison between packed bed reactors with and without a membrane, hereby avoiding the complication of complex fluid dynamics that prevail in fluidized beds. The main drawbacks of packed bed membrane reactors (PBMRs), however, are related to the temperature profiles occurring in these reactors (such as possible hotspot formation which are detrimental for the membrane stability), the bed-to-wall mass transfer limitations and, to some extent, also the intraparticle mass transfer limitations because relatively large particle sizes are often applied to prevent large pressure drops. As membrane technology advances, the extent of bed-to-wall mass transfer limitations becomes even more prominent due to the anticipated availability of state-of-the-art high-flux membranes.

All these detrimental phenomena can be circumvented by using a fluidized bed membrane reactor (FBMR). In this case, the membranes are immersed in a fluidized bed of small catalytic particles. The fluidization regime results in a virtually uniform temperature throughout the reactor even in case of highly exothermic reactions. At the same time, the bed-to-wall mass transfer limitations are strongly reduced while the small particle size also results in negligible intraparticle mass transfer limitations. Possible bubble-to-emulsion mass transfer limitations can be reduced by optimizing the positions of the membranes in the fluidized bed. When the combination of (conventional) FBRs with membranes is more beneficial as a whole than it is as the sum of its parts, a form of process intensification (PI) has been employed. PI can be broadly described as "any chemical engineering development that leads to a substantially smaller, cleaner, safer, and more energy efficient technology." PI is a broad topic, and is not limited to the field of FBMRs, or indeed to separation integration. Still, much research is performed on the integration of membranes (see, for example, Brunetti et al., 2012; Drioli and Curcio, 2007; Drioli et al., 2003, 2012).

The use of membranes in FBRs comes at a price; the membranes influence the hydrodynamics, mass and heat transfer in the reactor and it is important to understand the implications of this effect. For instance, the membranes may incur steep concentration gradients, also known as concentration polarization, due to the extraction of a single chemical species, or induce stagnant zones due to particles being drawn to the membranes due to large extraction flows. Additionally, many actual processes are operated at elevated temperatures and under reactive (viz., gas-producing) conditions, increasing the complexity of these reactors further. Finally, there are a number of practical and economic considerations; the construction of membranes into a reactor is challenging, because a good, leak-free sealing of the membranes may be difficult to attain, and since membranes are created using costly materials, the initial investment in membrane reactors on industrial scale will be higher than with conventional reactors. This is partly compensated by reduced capital costs (due to the reduction in the number of process units), and reduced downstream separation costs (the separation is integrated) can also be foreseen. The advantages and disadvantages described earlier have been summarized in Table 4.1 for conventional reactors, PBMRs, and FBMRs.

While conventional FBRs have been the topic of scientific research for decades, the incorporation of membranes as internals or in the reactor walls has only recently received more attention. This introduction continues with an outline of several processes that have been anticipated to benefit specifically from operation in an FBMR, followed by the current state of modeling and experiments that has been performed on conventional fluidized beds to serve as a starting point.

1.2. Fluidised Bed Membrane Reactor Concepts

This section serves as an illustrating example of the benefits of FBMRs as opposed to conventional and PBMRs.

1.2.1 Product Extraction

One area of interest of membrane reactors is product extraction via membranes, for instance in the production of hydrogen. The interest toward the use of membranes and membrane reactors for hydrogen production has increased especially because mechanically and chemically stable membranes with high permselectivity toward hydrogen are available and are continuously further improved in terms of stability and hydrogen flux (Gallucci et al., 2013). Most of the previous work on hydrogen production in membrane reactors has been focused on PBMRs, e.g., via reforming of methane, reforming of alcohols, autothermal reforming, partial oxidation of methane, water gas shift, etc. (Gallucci et al., 2011). However, it has been shown that the extent of bed-to-wall mass transfer limitations (also known as concentration polarization) is the limiting factor for the permeation of hydrogen through the membrane and thus it determines the amount of membrane area required for a given hydrogen recovery. These limitations are not seen in FBMRs.

Table 4.1 Advantages and Disadvantages of Conventional Reactors Versus Packed Bed Membrane Reactors Versus Fluidized Bed Membrane Reactors

	Conventional Reactor	Packed Bed Membrane Reactor	Fluidized Bed Membrane Reactor
Advantages	• Simple construction • Proved technology	• Integrated separation • Distributive feeding of reactants • No mechanical stress at membrane • Simple construction	• Integrated separation • Excellent heat and mass transfer (isothermal operation) • Negligible pressure drop • Flexibility in size and placement of membrane surface area • Submerged membranes decrease bubble size, hence increase bubble-to-emulsion phase mass transfer • Compartmentalization, i.e., reduced axial gas back-mixing.
Disadvantages	Additional separation steps needed Limited by reaction equilibrium	• Large pressure drop • Mass transfer limitations • Possible hotspot formation • More complex construction • Membrane costs	• Complex construction • Submerged membranes influence hydrodynamics, require detailed studies • Possible formation of stagnant zones • Membranes experience mechanical stresses • Membrane costs

Adris et al. (1991, 1994, 1997), who patented a reactor concept for selectively removing hydrogen from a fluidized bed for methane steam reforming, were among the first to use an FBMR for the production of ultrapure hydrogen utilizing palladium membranes. This is still a very actively researched topic. Among many others, Grace et al. (2005), Patil et al. (2007), Gallucci et al. (2008), and Chen et al. (2008) continued to explore the subject of hydrogen production with FBMRs in the bubbling or fast fluidization regimes, employing mainly phenomenological models in combination with experiments. It is worth mentioning here that some groups (Chen et al., 2008; Rakib et al., 2010) are investigating potential concepts for circulating FBMRs, not only for steam reforming of methane but also for steam reforming of higher hydrocarbons.

A typical fluidized membrane reactor (or membrane-assisted fluidized bed reactor, MAFBR) for hydrogen production consists of a bundle of hydrogen-selective membranes immersed in a catalytic bed operated in the bubbling or turbulent regime. The use of FBMRs not only reduces the bed-to-wall mass transfer limitations but also allows operating the reactor at a virtually isothermal condition (due to the movement of catalyst). This possibility can be used for operating the autothermal reforming of hydrocarbons inside the membrane reactor. In fact, as indicated by Tiemersma et al. (2006), the autothermal reforming of methane in a PBMR is quite difficult due to the hotspot at the reactor inlet which can melt down the membrane. This problem is completely circumvented in FBMRs. In this case, both autothermal reforming and hydrogen recovery can be performed in a single reactor.

Two FBMR concepts for the production of ultrapure hydrogen with integrated CO_2 capture have been proposed by our group (see Patil et al. (2005), Van Sint Annaland et al. (2006)) and patented in collaboration with Shell Global Solutions BV. The two concepts are depicted in Fig. 4.1.

In Concept 1, both H_2 permselective (dense Pd- or Pd/Ag-based membranes) and O_2 permselective membranes (perovskite) are integrated in two different sections, because of the differences in required operating conditions for the membranes: H_2 permselective membranes should be operated at temperatures below 600–700 °C because of membrane stability, while acceptable oxygen fluxes through available perovskite type O_2 permselective membranes can only be realized above 900–1000 °C. The perovskite membranes are integrated in an oxidation section at the bottom, and the hydrogen permselective membranes are integrated in a reforming/shift section at the top. In the oxidation section, methane or light hydrocarbons are partially oxidized in order to achieve the high temperatures required for oxygen permeation through the perovskite membranes and to simultaneously preheat part of the hydrocarbon/steam feed. The preheated feed is mixed with additional hydrocarbons and steam streams and fed to the reforming/shift section, where the hydrocarbons are completely reformed to carbon dioxide and hydrogen due to the selective hydrogen extraction via the membranes. Autothermal operation is achieved by tuning the overall hydrocarbon and steam feeds to the permeated oxygen fluxes. By selecting the proper ratio of the hydrocarbon and steam fed at the oxidation and reforming/shift sections, the temperatures in both sections

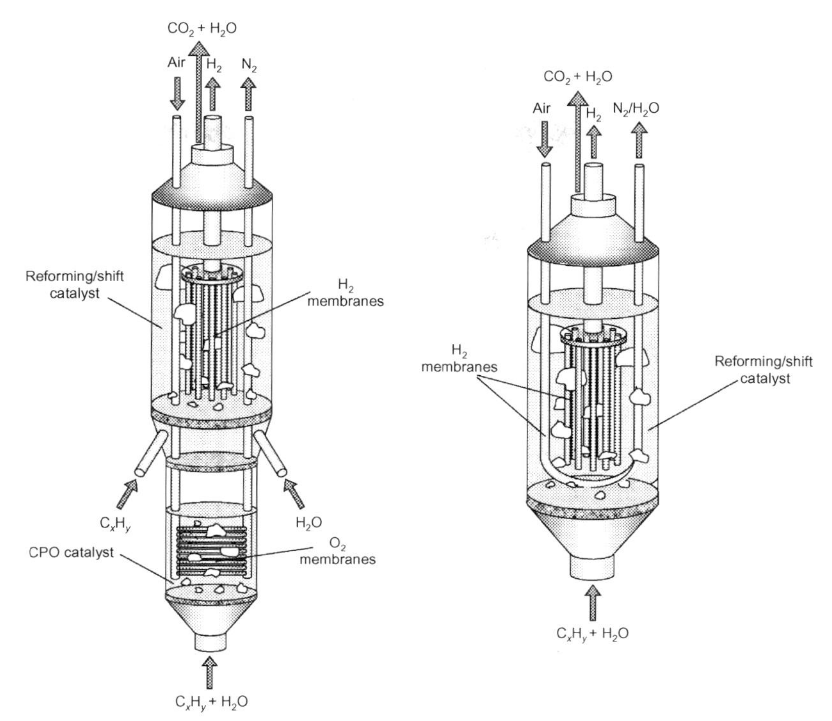

Figure 4.1 Schematic representation of two novel multifunctional fluidized bed membrane reactors for the autothermal coproduction of ultrapure H_2 and pure CO_2 (Kuipers et al., 1992; Van Sint Annaland et al., 2006). *Reprinted from Gallucci et al. (2008).*

can be controlled, while maintaining overall autothermal operation. Finally, also recuperative heat exchange for the air flow is incorporated by preheating air and cooling nitrogen inside the reforming section. The distinct advantage of the proposed concept is its flexibility in adjusting the hydrogen capacity by tuning the hydrocarbon and steam flows to the bottom and top sections and in maintaining very high overall energy efficiencies by recuperative heat exchange of air within the reactor.

In Concept 2, both the O_2 and H_2 permselective membranes are used in a single section. The reactor resembles largely the top section of Concept 1, but now the reaction energy for the steam reforming is supplied by hydrogen combustion via an oxidative sweep with air through a U-tube H_2 permselective membrane.

The advantage over Concept 1 is that the use of perovskite membranes is circumvented (and the associated problems with mechanical and chemical stability and sealing of these ceramic membranes in the reactor) and that the entire process is carried out in a single unit. The disadvantage is the higher required steam feed ratio. Nevertheless, the overall energy efficiencies of both concepts are very comparable and much higher than that can be obtained with the conventional process, especially when scaled down and particularly when including CO_2 capture.

1.2.2 Reactant Dosing

Another interesting application of FBMRs is selective dosing of a reactant—mostly oxygen—via membranes. The partial oxidation of methane to syngas has been extensively studied by Mleczko et al. (1996). They stressed that the interaction between the membrane and the fluidized bed hydrodynamics is of primary importance. A related area of research is the oxidative dehydrogenation of hydrocarbons. Al-Sherehy et al. (2005) studied the conversion of ethane into ethylene, and Deshmukh et al. (2005a,b) studied the conversion of methanol to formaldehyde. Using two-phase models and comparing it to experiments, both concluded that the oxygen distribution is beneficial in expanding the range of reactant feed compositions beyond those normally allowed by safety constraints, while the selectivity of undesired by-products was reduced. Distributed oxygen dosing was investigated for the production of hydrogen (Abashar et al., 2008) and for the oxidative dehydrogenation of ethane (Ahchieva et al., 2005). They found that especially in the case of moderate oxygen excess, an FBMR outperforms the FBR without membranes. Beside a higher productivity, a broader favorable operation range with respect to the oxygen–hydrocarbon ratio was reported.

Other application areas of FBMRs include the Fischer–Tropsch synthesis in gas-to-liquid technology and methanol synthesis from syngas (Rahimpour and Elekaei, 2009; Rahimpour et al., 2010), the reaction of butane to maleic anhydride (Marin et al., 2010), and the dehydrogenation of propane (Gimeno et al., 2009). These examples clearly illustrate the growing interest in FBMRs.

Deshmukh et al. (2003, 2005a, 2007a,b) investigated the effect of the presence of—and permeation through—the membranes on the extent of gas-phase back-mixing and the bubble-to-solids phase mass transfer rate. With ultrasound gas tracer experiments, they showed that due to the presence of membranes, but particularly due to gas permeation through the membranes, the macroscale solids circulation was strongly reduced, resulting in a near plug-flow behavior for the gas phase, while the heat transfer coefficients (although somewhat reduced in magnitude due to compartmentalization of the bed) were still sufficiently high to assure near isothermal conditions even for exothermic reactions. They also found smaller average bubble diameters for higher permeation ratios relative to the total gas flow. Christensen et al. (2008a–c) went further into the fundamental research of distributed feed, confirming that such systems indeed lead to a decrease in bubble size and bubble holdup, and therefore to an increase in the total number of bubbles. This effect increases the bubble-to-emulsion phase mass transfer.

1.2.3 Microstructured FBMRs

As FBMRs circumvent the external mass transfer limitations that adversely affect the performance of PBMRs to a large extent, the volumetric production capacity in FBMRs is limited by the relatively low permeation rate through the membranes (provided that the catalytic activity is sufficiently high).

There are two ways to improve the membrane permeation, namely (i) decreasing the thickness of the membranes and thereby increasing the membrane permeability and (ii) increasing the number of membranes installed per unit volume of the reactor.

Decreasing the thickness of the membranes to increase the permeation flux has a clear limitation. For instance, for Pd-based hydrogen permselective membranes, the lower limit of the thickness for stable membranes has apparently been reached as membranes as thin as 0.1–1 μm are nowadays available on the market (Gallucci et al., 2013). It is not foreseen that thinner membranes can be produced without compromising the membrane permselectivity or stability (lifetime).

Therefore, installing more membranes per unit of volume seems to be the only reasonable way to improve the membrane fluxes. Installing more membrane area (thus more membranes) will drastically reduce the space between membranes where the catalyst is suspended in fluidization. For instance, using planar membranes (Mahecha-Botero et al., 2008) close to each other would result in a small compartment that can be seen as "microstructured" FBMR as theoretically studied by Wang et al. (2011). Their simulation study has also elucidated that in these small confinements the turbulent fluidization regime (with anticipated improved mass transfer characteristics) can be achieved at lower superficial gas velocities.

It can be foreseen that microstructured FBMRs will have different design paradigms compared to conventional fluidized bed (membrane) reactors. It is for this reason that we also highlight several design aspects of these microreactors in this chapter.

1.3. An Overview of this Chapter

This chapter deals with the recent research efforts on the benefits and pitfalls of FBMRs. Recent studies on the topic, using an experimental or numerical approach, will be covered. In Section 2, we cover several techniques to investigate the hydrodynamics of membrane FBRs, including several experimental techniques and numerical models, followed by a number of recent results that have been obtained with these techniques.

Subsequently, several very recent novel modeling and state-of-the-art experimental techniques that have been (and are being) developed specifically for studying FBMRs are discussed.

2. HYDRODYNAMICS IN FBMRs

The hydrodynamics in FBMRs consist of the behavior of the emulsion phase, i.e., the mixture of the solid particles and the interstitial gas, along with the behavior of the gas bubbles. A complete overview of the characteristics and related models for fluidized beds is reported in the excellent book of Kunii and Levenspiel (1991). It is quite accepted that the most difficult FBR to be simulated is a bubbling fluidized bed, where the description of bubble behavior should be taken into account along with the description of solid movement and reactions occurring on the solid surface.

For conventional FBRs, computational fluid dynamics (CFD) modeling studies complemented with nonintrusive measurements have already paved the road toward a good understanding of the hydrodynamics in conventional FBRs, which is the topic of the following section. While this provides a good starting point in the study on FBMRs, the influence of submerged membranes, additive/extractive gas fluxes and chemical reactions, and heat effects (viz., the influence of reaction heat on the gas-phase density and viscosity) are only a few of the many intricating factors that influence the hydrodynamics in FBMRs.

2.1. Experimental Techniques

Measuring the hydrodynamics in fluidized bed (membrane) reactors is not straightforward, due to the opaqueness of the system. A number of nonintrusive techniques (as opposed to intrusive, flow-disturbing techniques such as using reflective fibers) have been developed to shed more light in the movement of the solids phase and the bubbles without disturbing the flow, which can be classified into three general principles.

2.1.1 General Principles

2.1.1.1 Optical Measurements of Pseudo-2D Setup

By confining the fluidized bed in one direction and using a translucent wall, visual access is restored so that the bed behavior can be studied fully and non-intrusively using optical techniques, such as particle image velocimetry (PIV) or digital image analysis (DIA), which are discussed in detail below. With these techniques, it is possible to obtain information on the instantaneous flow fields, but it remains difficult to translate the 2D results quantitatively to 3D. As a learning tool that allows to see and verify different aspects of the bed behavior (e.g., bubble size distribution, instantaneous particle fluxes) however, such techniques are unrivaled. The main focus of this chapter therefore lies on these optical techniques.

A typical pseudo-2D experimental setup for optical measurements is shown below in Fig. 4.2. A glass plate makes up the front of the reactor, which is filled with inert particles. Fluidization gas, possibly treated with a small amount of steam to prevent static charging of the particles, is fed from below through a porous plate. To prevent blurring of the particles on the images, a fast shutter time is required (in the order of milliseconds), and therefore additional lighting is usually necessary.

This type of setup allows for further modifications (see Fig. 4.3) to investigate the effect of immersed or wall-mounted membranes. De Jong et al.

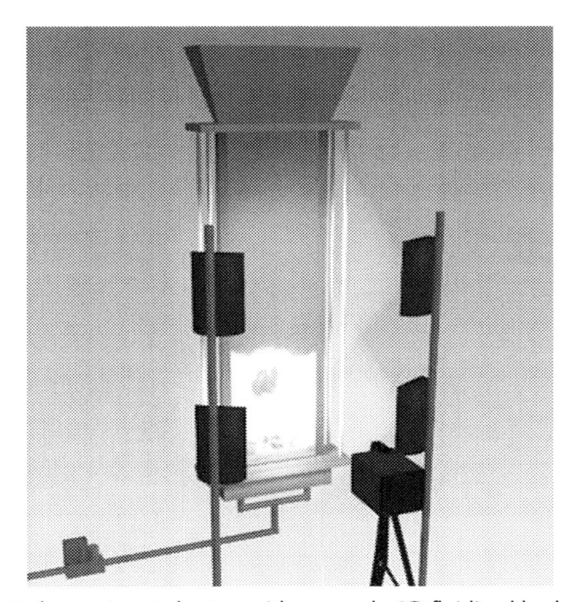

Figure 4.2 Typical experimental setup with a pseudo-2D fluidized bed used for optical measurement techniques PIV and DIA.

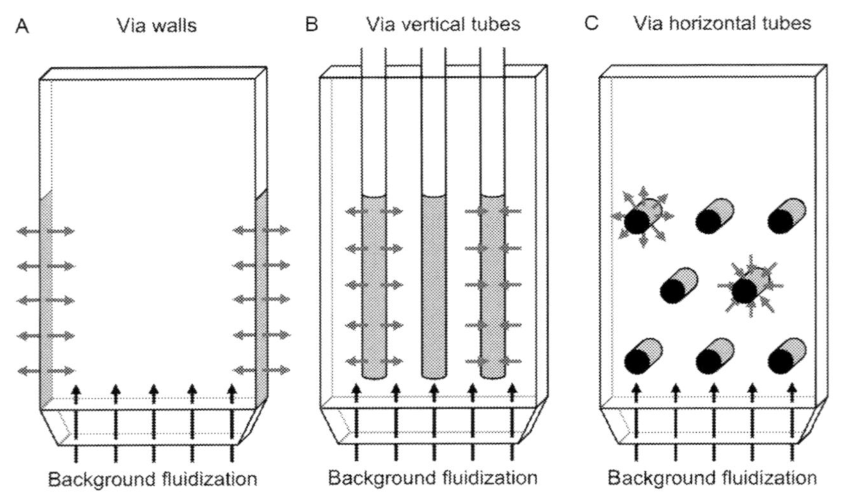

Figure 4.3 Possible membrane configurations inside fluidized beds, with wall-mounted membranes (A), vertically immersed membranes (B) and horizontally immersed membranes (C). The arrows indicate extraction or addition of gas via the membranes. *Reprinted from De Jong et al. (2012c) with permission from Elsevier.*

(2011) have described the construction of an experimental setup that allows to add or extract gas via the reactor walls in a detailed fashion. The setup has been constructed to investigate the effects of gas permeation via the walls on the fluidized bed hydrodynamics in detail. For the front wall of the bed, a glass plate is used. The rear wall of the bed is constructed from anodized aluminum to provide good contrast between the solids phase and the background. The distributor is a porous plate with a mean pore size of 40 μm. At both sides of the fluidized bed, along a height of 30 cm from the distributor, gas can be added to or extracted from the fluidized bed through a 10-μm porous plate.

In another work, De Jong et al. (2013) explain the modification of this setup to include up to 121 horizontally immersed membranes. Via holes in the rear wall, membrane tubes consisting of a porous cylinder with a mean pore size of 10 μm can be inserted.

In these setups, the background fluidization gas has been controlled with a 1000 L/min mass flow controller (MFC) from Brooks Instruments. A 50-ml/min membrane water pump, connected to an evaporator (operated at 200 °C), was used to feed steam to the background fluidization gas. This prevents static charging of the particles, which may strongly obfuscate the measured solids movements.

As an example, the P&ID of the setup with wall-mounted permeable membranes is shown in Fig. 4.4.

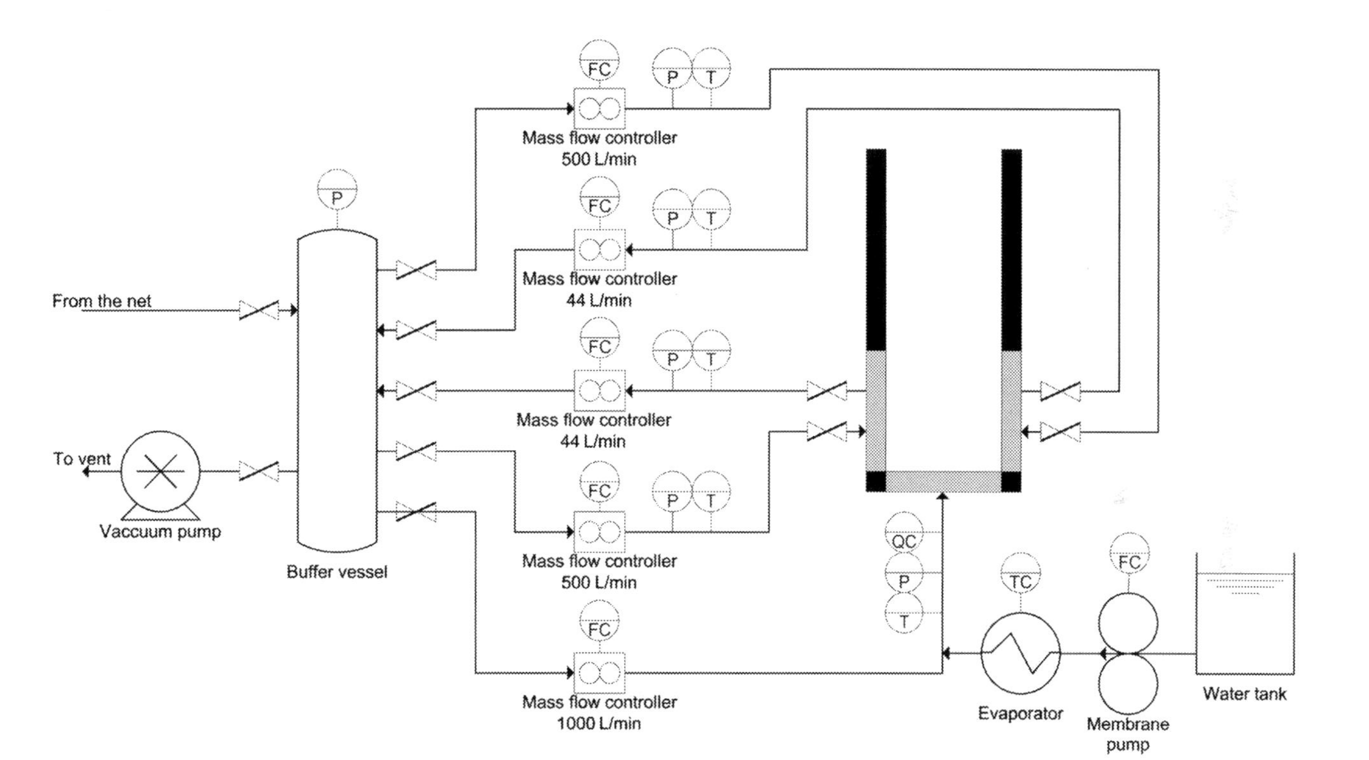

Figure 4.4 Process flow diagram of an experimental pseudo-2D membrane-assisted fluidized bed reactor for optical measurements. *Reprinted from De Jong et al. (2011) with permission from Elsevier.*

2.1.1.2 Particle Tracking Techniques

Particle tracking techniques allow the remote detection of the location of a single particle in a 3D FBR. Performing an experiment for an extended amount of time allows the construction of a time-averaged particle flow field. These techniques do not give information on the bubble phase behavior. Techniques that are used often include magnetic particle tracking (Patterson et al., 2010), computer-automated radioactive particle tracking (CARPT), and positron emission particle tracking (PEPT). CARPT (Chen et al, 1999) involves the placement of a gamma-emitter in the fluidized bed, which is tracked in 3D space via an array of detectors. PEPT requires a small source of radioactivity, emitting positrons, which upon annihilation with an electron releases a back-to-back gamma ray. Via triangulation, the location of the particle can then be reconstructed (Seville, 2010) (Fig. 4.5).

However, apart from relatively high costs and radiation hazards, the major drawback of these methods is the poor spatial and temporal resolution, and that often either information on the solids holdup or particle velocity is obtained.

2.1.1.3 Tomographic Techniques

Tomographic techniques allow to reconstruct the instantaneous particle holdup (porosity) of a 3D fluidized bed, thus allows to extract information especially on the bubble phase. Well-known techniques include electrical capacitance tomography, X-ray tomography, and magnetic resonance imaging (MRI).

The experimental techniques described in Sections 2.1.2–2.1.5 have been developed initially for FBRs with a focus on the large-scale solid

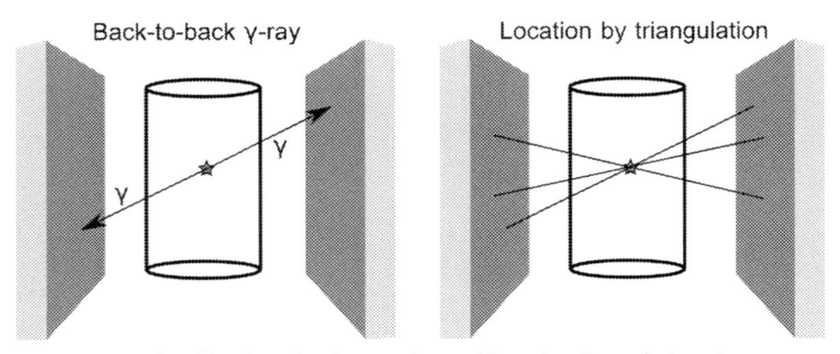

Figure 4.5 Principle of back-to-back particle tracking via triangulation. *Image recreated from Seville (2010).*

circulation patterns and bubble size distributions. These techniques are used as a basis to construct novel nonintrusive measurement techniques that are very well suited for FBMRs. In this respect, they have been extended to measure concentration profiles and to allow measurements at realistic (e.g., elevated temperatures) conditions.

2.1.2 Particle Image Velocimetry

PIV is a noninvasive measuring technique developed originally to investigate liquid or gas–liquid systems, but recently extended to gas–solid dispersed flows. The basic principle of PIV is to divide the recorded images into small interrogation areas and use a spatial cross-correlation on two consecutive images so that the instantaneous velocity of each interrogation zone can be obtained. PIV was first applied to dense phase gas-fluidized beds by Bokkers et al. (2004), who measured the emulsion phase circulation patterns in freely bubbling gas–solid fluidized beds, in order to validate the extent of particle drift induced by rising gas bubbles predicted by Euler–Lagrange and Euler–Euler models. Link et al. (2004) used PIV to establish fluidization regime maps in spouted fluidized beds and found excellent agreement with their discrete particle simulations. Dijkhuizen et al. (2007) extended the PIV technique to enable the measurement of the granular temperature distribution simultaneously in the fluidized bed.

The advantage of PIV is that the entire, instantaneous flow field can be obtained from two high-speed images of the bed. It is a well-known technique and several software packages exist that allow to postprocess the images. PIV, however, does not take the particle fraction in each interrogation zone into account, and therefore requires a correction step if one is interested in the particle flux rather than the average particle velocity. This correction step is discussed in Section 2.1.5.

2.1.3 Digital Image Analysis

The principle of DIA is to use the pixel intensity to discriminate between the bubble and the emulsion phase. If the pixel intensity is below a certain threshold value, the pixel area is assigned to the bubble phase, and otherwise to the emulsion phase. Commonly, DIA algorithms import the raw images and start by performing a number of preprocessing steps in order to remove influences of inhomogeneous lighting and other irregularities. The next step is to evaluate for every pixel whether it belongs to the emulsion phase, the gas phase, or whether it contains a mixture of the two. Several different algorithms have been proposed (see Section 2.1.5.2), which translate the measured image intensity to a particle porosity value. The last step is to extract the pixels that belong to a bubble, and to store the information regarding the bubbles, e.g., positions, size, and velocity, which can be used later on for statistical analysis. For instance, the equivalent bubble diameter can be obtained from the area of adjacent pixels A_b that represent a bubble via:

$$d_b = \sqrt{\frac{4A_b}{\pi}} \tag{4.1}$$

Bubbles can be tracked in time (based upon their displacement and size) to yield their velocity.

When necessary, this DIA algorithm can be quickly adapted to account for a specific situation. In Fig. 4.6, a number of postprocessing steps are shown on images created from a fluidized bed with horizontally submerged membranes, where obviously one has to account for the membranes to get the right bubble sizes.

2.1.4 Time Averaging of PIV and DIA Results

Although PIV delivers instantaneous velocity fields, mostly the focus lies on the time-averaged flow field of the solids phase, which can be acquired by

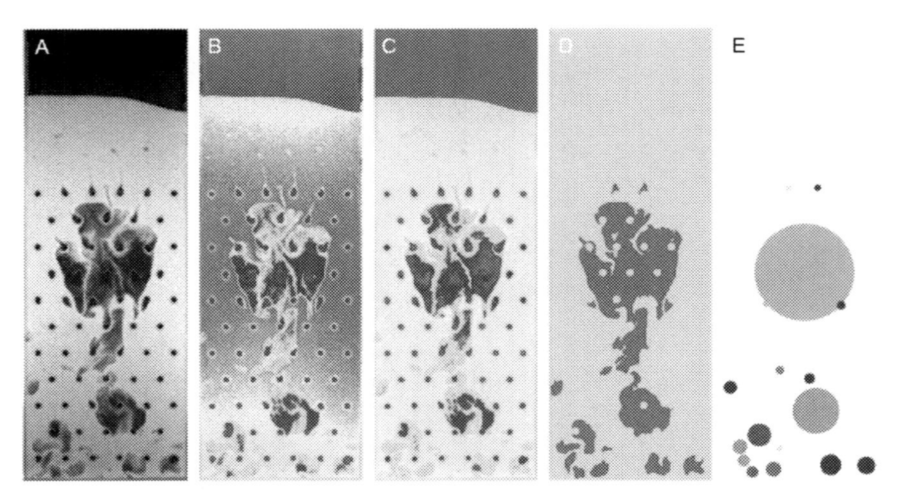

Figure 4.6 Selection of processing steps of the digital image analysis (DIA) script; (A) original image, (B) imported image by DIA, (C) corrected and smoothed image with freeboard removed, (D) bubble detection, and (E) representation of the equivalent bubble diameter of all bubbles found. *Reprinted from De Jong et al. (2013) with permission from Elsevier.*

recording many sets of images, each delivering an instantaneous velocity field and averaging them. Similarly, DIA delivers instantaneous positions and sizes of bubbles in a fluidized bed, but in a time-averaged fashion they give meaning to the measurements.

For reliable and reproducible results, it is essential to investigate the sensitivity of the data acquired. Therefore, it was investigated if sufficient data points have been used for time-averaging of the PIV and DIA results (De Jong et al., 2011). Subsequently, an estimation of the error in the presented results is given based on this analysis.

For a reference series (containing 4320 images, thus 2160 image pairs), PIV images were time-averaged over time intervals of 100, 200, 350, 500, 700, 900, and 1080 time steps of 0.25 s, respectively. Each time, these independent series were compared to each other, and the relative velocity deviation between these two series was calculated using

$$\Delta u = \frac{100}{N}\sum_{N}\frac{\left|\vec{u}_a - \vec{u}_b\right|}{\left|\vec{u}_a\right|} + \frac{100}{N}\sum_{N}\frac{\left|\vec{u}_a - \vec{u}_b\right|}{\left|\vec{u}_b\right|} \tag{4.2}$$

Because of its characteristics, a power trend line was used to estimate the error for a larger number of image pairs, resulting in an estimated error of 5.1% for the 2160 pairs of the vector plots presented in this chapter (see Fig. 4.7).

Second, DIA data files were compared for the reference series in terms of equivalent bubble diameter. However, contrary to the PIV data files (which always have the same amount of data), the deviation in the bubble diameter

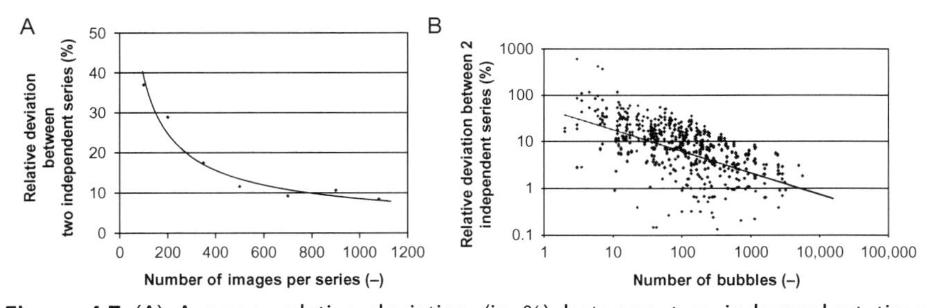

Figure 4.7 (A) Average relative deviation (in %) between two independent time-averaged vector images over a certain number of data files as a function of that number and (B) average relative deviation (in %) between two independent time-averaged bubble size data files as a function of the number of bubbles encountered in that series on a double logarithmic scale. *Reprinted from De Jong et al. (2011) with permission from Elsevier.*

strongly depends on how many bubbles were detected at that specific height in the fluidized bed. Therefore, in Fig. 4.7B, the average relative deviation between two independent series is plotted on a log–log scale as a function of the number of bubbles that were present in the fluidized bed in those series. Again, a power trend line is used to provide an estimate of the experimental error as a function of the number of bubbles. It has been found that when over 100 bubbles were encountered, the statistical error reduces to below 5%.

As can be seen in Fig. 4.8A and B, the accuracy of the equivalent bubble diameter varies over the height of the fluidized bed, because in the bottom section of the bed more bubbles are present compared to the top section. In the top section, the bubbles typically have coalesced to very big, but very few, large bubbles and therefore a lack of good statistics plays a role in this region. If one discards the bubble information gathered for the very top of the bed, error margins less than 5% can be obtained using 1000–1500 images of the bed.

2.1.5 PIV/DIA

Although PIV measures the instantaneous, average particle velocity in every interrogation zone, the measurement technique does not account for the different number of particles in different interrogation zones. A combination of the two techniques discussed above is essential for a good

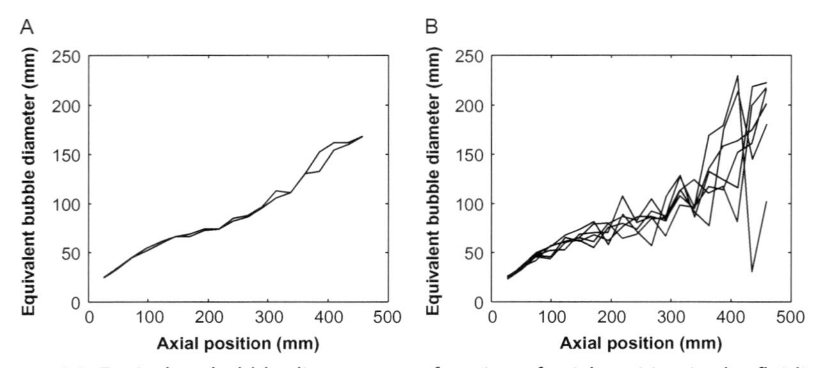

Figure 4.8 Equivalent bubble diameter as a function of axial position in the fluidized bed for (A) two series based on 1350 independent images each and (B) six series based on 50 independent images each. *Reprinted from De Jong et al. (2011) with permission from Elsevier.*

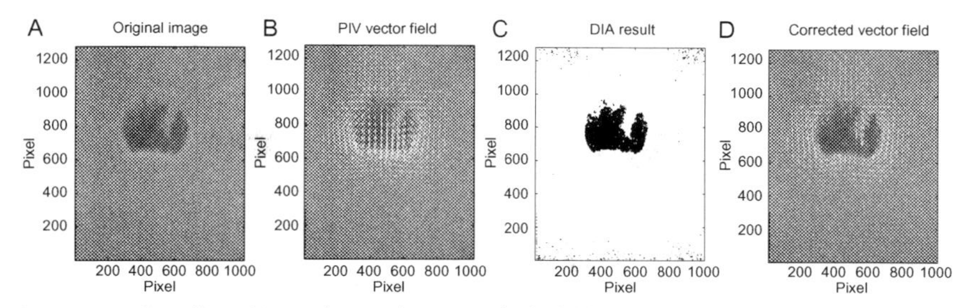

Figure 4.9 Coupling PIV with DIA for a single bubble injected into a fluidized bed at incipient fluidization conditions: (A) original recorded digital image, (B) PIV velocity field without correction, (C) phase separation by DIA, and (D) instantaneous flow field after PIV–DIA correction. *Reprinted from Laverman et al. (2008) with permission from John Wiley and Sons.*

interpretation of the emulsion phase mass fluxes. The DIA algorithm provides information that is used to correct for differences in particle number density, which is needed especially because of particle raining through the bubbles, where a small number of particles have a very high velocity, while the particle mass flux is small. The influence of particle raining is demonstrated in Fig. 4.9, for the case of a single bubble injected into a fluidized bed at incipient fluidization conditions. To correct for particle raining, information on the local particle number density (for every PIV interrogation area) is required. Although PIV is carried out such that all the particles are individually distinguishable (using 2–3 pixels), the exact number of particles in an interrogation area is difficult to determine automatically for systems with relatively small particles. One could use the average intensity of an interrogation zone as an estimate for the number of particles in that particular zone. However, in this case, a very homogeneous illumination is required, and a relation between the intensity and the packing degree has to be determined.

2.1.5.1 Porosity Corrections Using Conventional DIA

The conventional DIA algorithm applied to bubbly fluidized beds (Laverman et al, 2008) discriminates between the bubble and solid phases on the basis of the pixel intensity, employing a prescribed threshold value. The algorithm corrects the instantaneous flow field produced by PIV for particles raining through the roof of larger bubbles (see Fig. 4.9).

The large influence of filtering out of particle velocities of particles inside bubbles on the time-averaged emulsion phase velocity profiles can

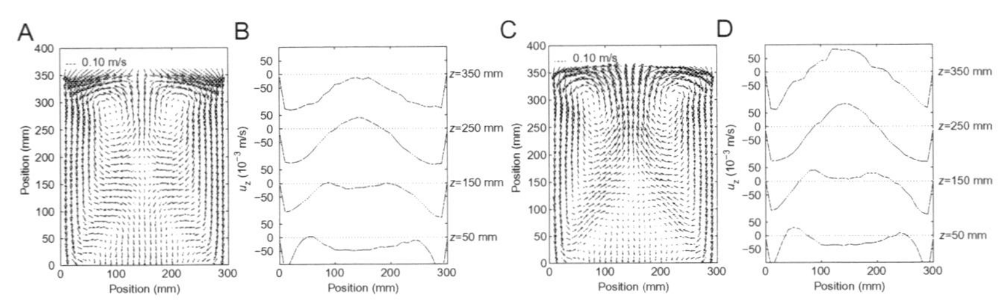

Figure 4.10 Time-averaged emulsion phase velocity profiles and vertical velocity component at different heights for $u/u_{mf}=2.5$ in a 0.30-m pseudo-2D fluidized bed filled with 0.30 m of glass beads: (A and B) before filtering using DIA and (C and D) after filtering using DIA. *Reprinted from Laverman et al. (2008) with permission from John Wiley and Sons.*

be discerned from Fig. 4.10, showing the difference of the time-averaged emulsion phase velocity profiles before (A and B) and after (C and D) filtering making use of DIA. Note that for clarity of presentation not all obtained vectors are plotted in the figure. The figure clearly shows that without filtering the upflow of the emulsion phase in the center of the fluidized bed is strongly underestimated. Since most bubbles move through the fluidized bed at the center of the bed, the effect of the filtering procedure is most pronounced at the center, while the extent of downflow is hardly affected by the filtering. This is also very clear from the (B) and (D) plots, showing the axial emulsion phase velocity profile at different heights in the bed. Note that the time-averaged emulsion phase velocity profiles obtained directly from the PIV results without filtering wrongly indicates the absence of upflow of the emulsion phase at lower positions, while it underestimates the maximum longitudinal emulsion phase velocity at higher positions in the bed by a factor of 2!

It has thus been established that it is essential to measure the particle fluxes taking into account the porosity information as well. However, using merely the pixel intensity of the acquired images as a measure of the local solids fraction introduces an error, because the pixel intensity does not contain any depth information of the solids holdup, since this is not visible from the frontal projection. Moreover, the conventional DIA algorithm does not account for local variations in the solids fractions, but instead employs a "binary" approach, setting every pixel either to "0" (for the bubble phase) or to "1" (for the emulsion phase), thereby limiting the application of this technique to cases where the interface between the phases can be clearly discerned. These shortcomings have led to the development of a new DIA algorithm with improved accuracy and wider applicability.

2.1.5.2 Improved Image Intensity to Porosity Algorithm

A reliable correlation between the 2D image intensity and the true 3D solids fraction in fluidized beds is required to obtain more accurate information on the solid particle flux profiles in granular systems. Using only experimental data, it is difficult—if not impossible—to obtain such a correlation, but by using data from discrete particle model (DPM) simulations (which are explained in detail in Section 2.2.1) to create artificial 2D images, the 2D to 3D correlation for the solids fraction could be reconstructed.

From a series of simulation snapshots, artificial images, similar to those created in actual experiments, have been created. The script used for the generation of these images, overlays a Gaussian intensity distribution function $I(x,y,z)$ as function of the particle diameter d_p for every particle, whose exact position (x_i,y_i,z_i) is known from the DPM simulation:

$$I(x, y, z) = \frac{8I(z)}{\pi d_p^2} e^{-\frac{8}{d_p^2}\left[(x-x_i)^2 + (y-y_i)^2\right]} \tag{4.3}$$

Thus, each particle obtains the highest intensity at the center of the particle. Particles which lie behind other particles will not appear in the resulting artificial image, thus creating a realistic frontal snapshot of that situation. Except for the factor $I(z)$, this equation is a 2D intensity representation.

However, the intensity should depend on the depth position of the particle; particles that are located more toward the back of the bed are (partially) shielded from the light by particles in the front and have a lower pixel intensity on the recorded image. Unfortunately, the actual dependency $I(z)$ is unknown. Different intensity profiles (see Eqs. 4.4–4.8 and Fig. 4.11) have been tested by De Jong et al. (2012a–d), with I_{max} and Δz being the maximum intensity and bed depth, respectively. Because the majority of the image consists of dense emulsion phase with particles immediately at the front of the image, I_{max}—and not an average value—has been kept the same in these profiles in order to avoid large variations in pixel intensity.

$$I_{linear}(z) = I_{max}\left(1 - 0.5\frac{z}{\Delta z}\right) \tag{4.4}$$

$$I_{high-linear}(z) = I_{max}\left(1 - 0.3\frac{z}{\Delta z}\right) \tag{4.5}$$

$$I_{low-linear}(z) = I_{max}\left(1 - 0.7\frac{z}{\Delta z}\right) \tag{4.6}$$

$$I_{sqrt}(z) = I_{max}\left(1 - 0.5\frac{\sqrt{z}}{\Delta z}\right) \tag{4.7}$$

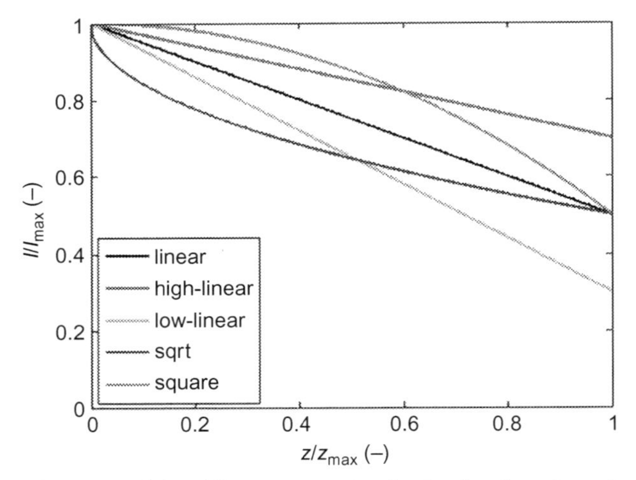

Figure 4.11 Schematic of the different intensity distribution functions. *Reprinted from De Jong et al. (2012a) with permission from Elsevier.*

$$I_{\text{square}}(z) = I_{\max}\left(1 - 0.5\frac{z^2}{\Delta z}\right) \tag{4.8}$$

In dense regions, a strong dependency of the intensity on the bed depth is expected. In dilute regions, on the other hand, this dependency will probably be much less pronounced, because the bed depth is much smaller than the distance to the light source. Therefore, several somewhat arbitrary correlations have been tested to investigate the sensitivity of the chosen correlation on the final result. Two examples of artificial images generated from a DPM simulation with a different pixel intensity depth profile have been given in Fig. 4.12 to highlight the subtle difference (i.e., color intensity) between the different intensity profiles.

Because the PIV software works most accurately when the particle resolution is 2–3 px/d_p (Westerweel, 2000), all artificial images were created with 320 px in width (corresponding to 2.67 px/d_p). Although this resolution is beneficial for a good velocity reconstruction, the quality of the artificial image itself is somewhat decreased, thus sacrificing some accuracy in the determination of the solids fraction profile. By comparing one of these artificial images to a high-resolution artificial image (with 1920 px in width, or 16 px/d_p), it was shown that the error in intensity is less than 5% on average, which was deemed acceptable.

From the DPM simulations, the real 3D porosity is known, and the image intensity is obtained by different DIA algorithms. Fig. 4.13 illustrates

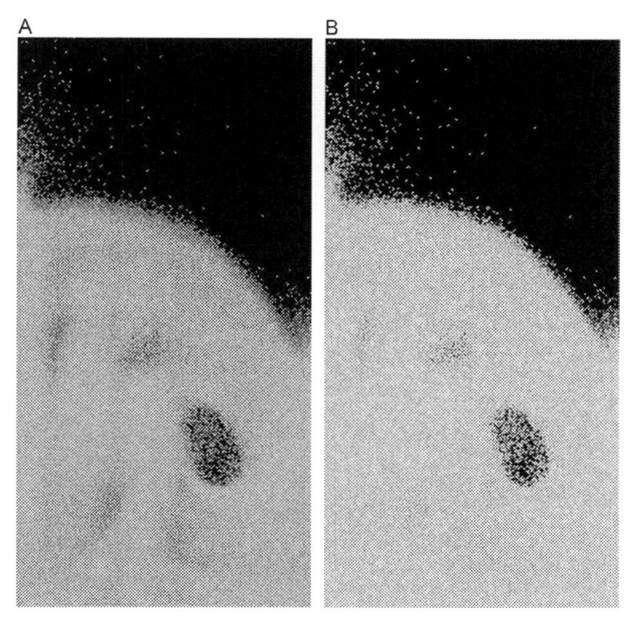

Figure 4.12 Influence of the intensity depth profile on the high-resolution artificial image created from the DPM simulation result: (A) using the sqrt correlation and (B) using the square correlation. *Reprinted from De Jong et al. (2012a) with permission from Elsevier.*

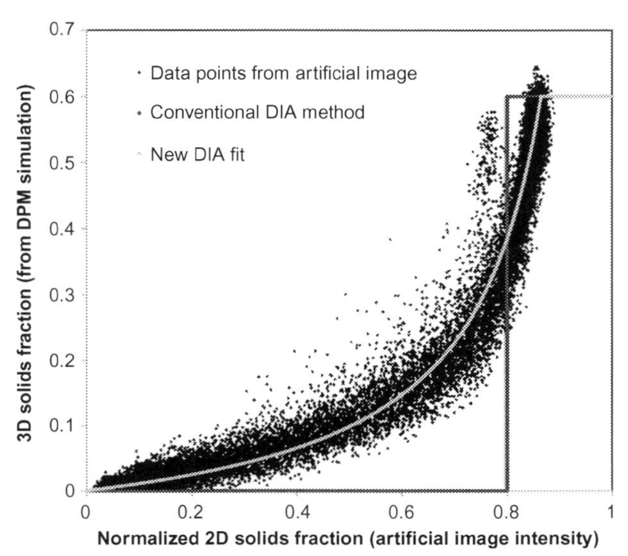

Figure 4.13 Scatter plot obtained from 19,340 nonzero data points from one artificial image for a DPM simulation with 100-μm particles at a fluidization velocity of $u/u_{mf} = 10$, and the developed correlation. In red (dark gray in print version), the conventional approach is schematically included (the exact position of the cut-off value depends on the selected cut-off value). *Reprinted from De Jong et al. (2012a) with permission from Elsevier.*

the image intensity, represented by the normalized 2D solids fraction, plotted against the true 3D solids fraction provided by a DPM simulation, where the "linear" intensity distribution function was used. Fig. 4.14 illustrates how the image intensity of such an artificial image is interpreted using the conventional (threshold) method (Fig. 4.14A) from the image intensity (Fig. 4.14B). Using the new method, the image intensity then translates to a calculated 3D solids fraction (Fig. 4.14C), which can be compared to the true 3D porosity in the DPM simulation. The other simulation cases display a very similar trend; all cases display a linear profile with a small slope for low 2D solids fractions. For solids fractions above 0.8, the 3D solids fraction shows a steep asymptotic increase until a value of approximately 0.6. A suitable function to match this behavior was already proposed by Deen et al. (2006) in the following equation:

$$\varepsilon_{s,3D} = \begin{cases} \varepsilon_{s,3D,\max} & \text{for } \varepsilon_{s,3D} \geq \varepsilon_{s,3D,\max} \\ A \cdot \dfrac{\varepsilon_{s,2D}}{1 - \dfrac{1}{B}\varepsilon_{s,2D}} & \text{for } \varepsilon_{s,3D} < \varepsilon_{s,3D,\max} \end{cases} \tag{4.9}$$

Parameter A determines the slope of the initial linear part, while B determines the location of the asymptote at higher values of the 2D solids fraction $\varepsilon_{s,2D}$. In this research, the maximum particle fraction $\varepsilon_{s,3D,\max}$ was kept fixed at 0.6. Every fit was based on 11 independent random images from the

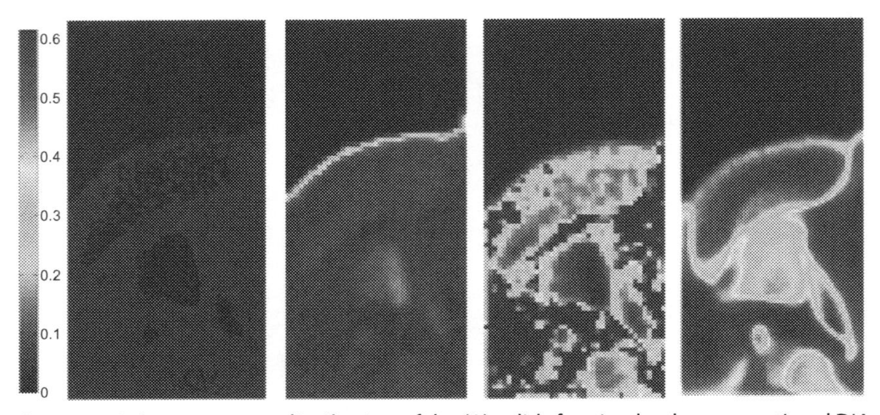

Figure 4.14 Instantaneous distribution of the (A) solids fraction by the conventional DIA algorithm, (B) 2D intensity of the new DIA algorithm, (C) 3D solids fraction of the new DIA algorithm, and (D) solids fraction by the DPM reference. *Reprinted from De Jong et al. (2012a) with permission from Elsevier.*

DPM, each containing approximately 5000 nonzero data points. Parameters A and B were determined by minimization of the root mean square (RMS) error. The fitting results for all series of artificial images have been summarized in Table 4.2.

The work of De Jong et al. (2012a–d) has investigated the influence of particle size, fluidization velocity, bed depth, and intensity distribution function on the 2D–3D correlation, in particular for the application in gas–solid fluidized beds in the bubbling regime. It was found that the particle size and fluidization velocity have no influence on the correlation and the intensity distribution function only slightly affects the results for dilute regions, but the bed depth shows the most prominent effect. A large number of simulations covering all these aspects have been performed, and the artificial images have been generated using the different intensity distribution functions. The average fitted values for all these simulation series have been examined after which the following can be concluded:

- The values found for B do not differ much (maximum difference less than 2%), from which it can be concluded that the type of intensity distribution function used to generate the artificial images from the DPM is not important for high values of $\varepsilon_{s,2D}$ (dense part of the bed). Differences arise for the dilute regions.
- The values found for parameter A vary between 0.011 and 0.025. Basically, for a pixel intensity depth profile with a steeper descent as function of the depth position higher values for A have been found (in case of sqrt or lowlin, for example) and vice versa (square and highlin). The average RMS error in the determined correlation is between 3.0% and 5.1%.

Table 4.2 Fitting Results for A and B Parameters for the Intensity–Porosity Correlation Based on All Series of Simulations

Intensity Distribution Function Used to Generate the Artificial Images	A (–)	B (–)	RMS Error (%)
linear (Eq. 4.4)	0.018	0.874	4.2
highlin (Eq. 4.5)	0.015	0.876	3.4
lowlin (Eq. 4.6)	0.022	0.883	4.8
sqrt (Eq. 4.7)	0.025	0.885	5.1
square (Eq. 4.8)	0.011	0.868	3

To some extent, the intensity distribution function used to generate the artificial images influences fitting parameter A of this correlation. This influence concerns in particular the dilute regions of the fluidized bed. Also, the correlation is influenced by the bed depth; in this case, both fitting parameters (and thus both the dilute and dense parts of the bed) are affected.

The new DIA algorithm has been shown to outperform the conventional algorithm on the representation of the actual porosity (De Jong et al., 2012a–d), decreasing the error in the solids fluxes from 18.7% to 12.3% if the corrected velocity field has been obtained from the DPM model. Using solids velocity reconstructed by the PIV software (hence using a tool stack identical to that used for experimental data), the errors for the conventional and new DIA algorithms were similar to each other: 23.6% and 25.5%, respectively.

2.2. Multiscale Modeling Strategy

To allow a good interpretation of experimental results and to have more flexibility in defining a system, modeling studies are just as essential as their experimental counterparts. There are many aspects to modeling of FBMRs, for instance the efficiency of separation. Boon et al. (2012) have performed a modeling study on palladium-supported membranes, and were able to describe the different contributions to the overall mass transfer resistance using a phenomenological model for the Pd layer, and the dusty gas model for the porous support. After coupling the mass transfer through the membrane to a 2D Navier–Stokes solver with convection/diffusion equation, the mass transfer resistances could accurately be described. Coroneo et al. (2010) have modeled a membrane FBR including membrane separation of the gas mixture, allowing for a geometric optimization of the FBMR. While these examples show that a full FBMR process can be modeled (i.e., incorporating and coupling hydrodynamics, mass transfer and separation), this work will mainly focus on the hydrodynamic aspects of gas extraction or addition in FBMRs, either via wall-mounted membranes or via immersed membrane tubes, not taking separation issues into account at this stage.

A major issue in the modeling of hydrodynamics in FBRs and FBMRs is the large separation of scales that prevails in FBRs (the largest flow structures can be of the order of meters while the particle–gas interactions take place on the scale of millimeters, or even micrometers). To bridge this gap, smaller scale models that take into account the various interactions (gas–particle, particle–particle) in detail are used to develop closure laws which can represent the effective "coarse-grained" interactions in the larger scale models. In Fig. 4.15, a schematic representation of this model hierarchy, developed for FBR studies, is shown. These models are used as a stepping stone for modeling the more complex FBMRs. A general description of the models is given by Van Der Hoef et al. (2006):

1. *Direct numerical simulations (DNS)*: At the most detailed level of description, the gas flow field is modeled at scales smaller than the size of the solid particles. The interaction of the gas phase with the solid phase is incorporated by imposing no-slip boundary conditions at the surface of the solid particles. This model thus allows one to measure the effective momentum exchange between the two phases, which is a key input in all the higher scale models. Many different types of DNS models exist, such as the lattice Boltzmann model (Ladd, 1994; Ladd and Verberg, 2001) or immersed boundary techniques (Peskin (2002), Uhlmann (2005)). The goal of these simulations is to construct drag laws for dense gas–solid systems, which are used in the discrete particle type models.
2. *Discrete particle model*: At one level smaller in detail (and thus larger in scale), the DPM model employs a Lagrangian description of the individual particles on a subgrid scale, while the gas flow field is continuous and solved by the compressible Navier–Stokes equations. The scale at which the gas flow field is described is an order of magnitude larger than the particles, (a CFD grid cell typically contains $O(10^2)$–$O(10^3)$ particles).

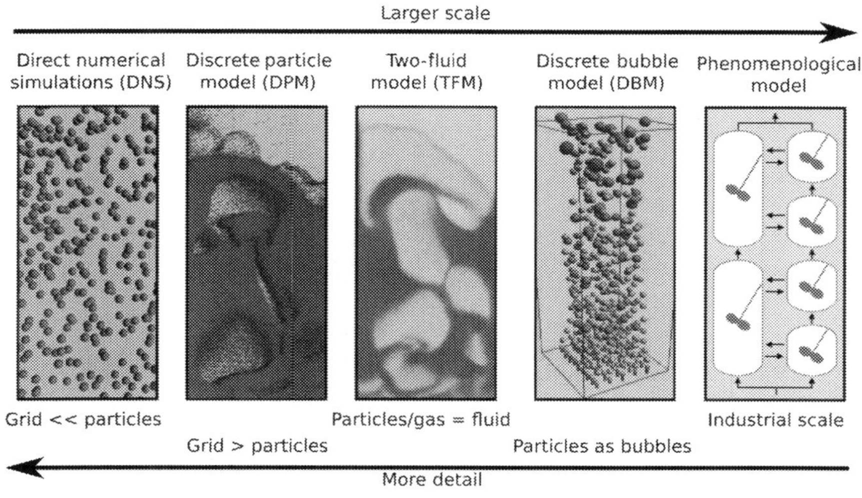

Figure 4.15 Model hierarchy for gas–solid fluidized beds.

The motion of the particles is governed by Newton's law, where the forces on the particles are integrated using standard schemes for ODE's. These forces follow from the interaction with the fluid phase, and from collisions with the other particles. Therefore, both a drag force closure and a collision model have to be specified for this level of modeling. The advantage of this model is that it can account for the particle–wall and particle–particle interactions in a realistic manner. This model allows one to validate (and modify) the viscosity and pressure closures derived from the kinetic theory of granular flow (KTGF), which are used in the two-fluid model (TFM) simulations. Still, a closure law for the effective momentum exchange between the two phases has to specified for this model. The system sizes that can be studied are of the order of $O(10^5)$ particles, which corresponds for millimeter sized particles to systems which have a linear dimension of the order of 0.1 m (i.e., laboratory scale).

3. *Two-fluid model*: At the intermediate scale, a continuum description is employed for both the solid phase and the gas phase, and a CFD-type Eulerian code is used to describe the time evolution of the local mass and momentum density of both phases (see Gidaspow, 1994; Kuipers et al., 1992, among others). In a more sophisticated model, based on the KTGF, also the local granular temperature of the solid phase is a dynamical variable, and thus included in the update. With modern day computers, the TFM model can predict the flow behavior of gas–solid flows of systems with a linear dimension of the order of 1 m, corresponding typically to the size of pilot plants. The TFM relies heavily on closure relations for the effective solid pressure and viscosity, and gas–solid drag. The basic idea of the multiscale modeling is that these relations are obtained from kinetic theory, and from numerical data collected in the more detailed scale models.
4. *Discrete bubble model (DBM)*: In order to study the hydrodynamics in large-scale fluidized beds, the DBM approach is used. In the DBM, the emulsion phase is modeled as a continuum, and the bubbles are regarded as discrete elements (Bokkers et al., 2006). The bubble trajectories are computed by integrating the equations of motion (Newton's second law), accounting for bubble coalescence when two bubbles collide, using closures for the forces acting on the individual bubbles. More detailed and fundamental models could be used in addition to experiments to derive the required closures for the bubble behavior and the emulsion

phase rheology. With continuum models (Euler–Euler) using closures from the KTGF for the solids phase rheology, or DPMs (Euler–Lagrange), where all the individual particles in the fluidized bed are tracked, the bubbles can in principle be completely resolved and the bubble behavior could be computed. However, for the simulation of the macroscale circulation patterns prevailing in large, industrial-scale fluidized beds with these fundamental models, the required number of grid cells or the required number of particles and hence the corresponding calculation times would definitely become prohibitive. Therefore, models that are developed to describe the hydrodynamics of very large systems will have to rely on closures for the bubble behavior. Although, the DBM idealizes the bubbles as perfect spheres, its strong advantage is that no *a priori* assumption is required on the bubble encounter frequency, an important factor determining the bubble coalescence rate.

5. *Phenomenological models*: The most common phenomenological description of the two-phase flow phenomena in FBRs is based on the bubble assemblage model, originally proposed by Kato and Wen. In their model, the fluidized bed is divided in the axial direction into a number of continuously stirred tank reactors (CSTRs) for the bubble phase as well as for the emulsion phase, where the size of the CSTR was related to the local bubble size. The model presented by Deshmukh et al. (2005a) enhances this approach by incorporating an expression for exchanging membranes. Note that this model does not yield information on the hydrodynamics of the system, rather it uses the information derived from smaller scale models to be able to simulate industrial-scale reactors.

Especially, the DPM and TFM models have shown a significant contribution to the understanding of the hydrodynamics in FBMRs. However, these models need to be extended with suitable expressions that account for the addition or extraction of gas via membranes, which can be either submerged into the fluidized bed or incorporated in the walls of the bed. The inner workings of these models and their extension with additive/extractive fluxes will be detailed in the following Sections 2.2.1, 2.2.2 and 2.2.3.

Another aspect is the description of the turbulence in the gas phase due to the movement of the particles. Recent efforts aim to close the Euler–Euler models with appropriate closures for the turbulence and turbulence modulation, or describe the interaction among particles to form clusters which induce turbulence in the continuous phase flow. From here, we refer to the work due to Fox (2012, 2014) and Capecelatro et al. (2014) for further details on this topic. In this chapter, we focus on dense gas–solid flows, where gas-phase turbulence is not very important, but for dilute flows (e.g., risers), this should be incorporated.

2.2.1 Discrete Particle Model

In DPMs, each particle is tracked individually and all collisions are calculated, thus providing a more reliable and detailed representation of the fluidized bed. The model was introduced by Hogue and Newland (1994), Hoomans et al. (1996), and Tsuji et al. (1992), and it employs either a hard-sphere approach for dilute systems or a soft-sphere approach for dense fluidized beds.

The DPM discussed in this section has originally been developed by Hoomans et al. (1996), in which Lagrangian particle tracking is combined with a continuum description of the gas phase. Since the grid size is larger than the particle size, the details of the interaction between gas phase and particles are unresolved and modeled via a drag closure relation. Because the model uses a fixed Cartesian grid, calculation times are small compared to models using unstructured or conformal grids, while arbitrarily shaped objects can easily be added by using an immersed boundary method (IBM). In DPM, translational particle motion is governed by Newton's second law (Eq. 4.10), where $\vec{v}_i$ is the velocity, m_i is the mass, V_i is the volume of particle i, and β is the interphase momentum transfer coefficient (proposed by Beetstra et al., 2007). The forces on the right-hand side (RHS) represent the pressure gradient, drag force, gravity, and collision forces, respectively.

$$m_i \frac{d\vec{u}}{dt} = -V_i \nabla p + \frac{V_i \beta}{\varepsilon_s}\left(\vec{u}_g - \vec{v}_i\right) + m_i g + F_{\text{contact},i} \tag{4.10}$$

The rotational equation of motion is given in Eq. (4.11), where I_i represents the moment of inertia, $\vec{\omega}$ the rotational speed, and $\vec{T}_i$ the torque:

$$I_i \frac{\mathrm{d}\vec{\omega}}{\mathrm{d}t} = \vec{T}_i \tag{4.11}$$

The interphase momentum transfer coefficient β is frequently modeled by a combination of the Ergun equation and the Wen and Yu correlation, but in this model, the improved drag relation by Beetstra et al. (2007), based on a direct numerical simulation study, is implemented to account for the gas–particle interaction:

$$F = \frac{\beta d_p^2}{\mu} = A \cdot \frac{\varepsilon_s^2}{\varepsilon_f} + B \cdot \varepsilon_s \cdot Re \tag{4.12}$$

$$A = 180 + \frac{18\varepsilon_f^4}{\varepsilon_s}\left(1 + 1.5\sqrt{\varepsilon_s}\right) \tag{4.13}$$

$$B = \frac{0.31\left(\varepsilon_f^{-1} + 3\varepsilon_f\varepsilon_s + 8.4Re^{-0.343}\right)}{1 + 10^{3\varepsilon_s} Re^{2\varepsilon_f - 2.5}} \tag{4.14}$$

The interphase momentum transfer coefficient β depends on viscosity, the local voidage, the particle diameter d_p, and the particle Reynolds number. Particle–particle and particle–wall collisions are calculated with a soft-sphere approach, employing a linear spring–dashpot model that distinguishes between sliding and sticking collisions. The gas phase is described with a continuum model. The continuity and volume-averaged Navier–Stokes equations are given in the following equations, where $\boldsymbol{\tau}_f$ is the gas-phase stress tensor (assumed Newtonian) and $\vec{S}_{f\to s} = \beta\vec{u} - \vec{\alpha}$ the source term due to particle–gas interactions (excluding the pressure gradient force):

$$\frac{\partial\left(\varepsilon_g \rho_g\right)}{\partial t} + \left(\nabla \cdot \varepsilon_g \rho_g \vec{u}\right) = 0 \tag{4.15}$$

$$\frac{\partial\left(\varepsilon_g \rho_g \vec{u}\right)}{\partial t} + \left(\nabla \cdot \varepsilon_g \rho_g \vec{u}\,\vec{u}\right) = -\varepsilon_g \nabla p + \varepsilon_g \rho_g g - \left(\nabla \cdot \varepsilon_g \boldsymbol{\tau}_g\right) - \vec{S}_{f\to s} \tag{4.16}$$

For a complete overview of all relevant closures and correlations used in the DPM, the reader is referred to Table 4.3 (De Jong et al., 2012d).

2.2.2 Two-Fluid Model

In contrast to the DPM, the TFM does not longer distinguish single particles. Instead, a continuum description for the solids phase is employed, resulting in a second set of conservation equations for mass and momentum (similar to the gas-phase Navier–Stokes equations of the DPM model, but then for the solids phase):

$$\frac{\partial(\varepsilon_s \rho_s)}{\partial t} + \left(\nabla \cdot \varepsilon_s \rho_s \vec{u}\right) = 0 \tag{4.17}$$

Table 4.3 Equations for the DPM Model

Gas-phase viscous stress tensor	$\boldsymbol{\tau}_g = -\mu_g\left[\nabla\vec{u}_g + (\nabla\vec{u}_g)^T - \frac{2}{3}\boldsymbol{I}(\nabla\vec{u}_g)\right]$
Source term momentum exchange gas–solids	$\vec{S}_{\text{drag}} = \frac{1}{V}\int\sum_{k=0}^{N_p}\frac{V_{p,k}\beta}{1-\varepsilon_g}(\vec{u}-\vec{v}_{p,k})\delta(\vec{r}-\vec{r}_{p,k})\,dV$
Contact force between particles	$\vec{F}_{\text{contact},a} = \sum_b\left(\vec{F}_{ab,n} - \vec{F}_{ab,t}\right)$
Normal contact force	$F_{ab,\text{n}} = -k_\text{n}\delta_\text{n}\vec{n}_{ab} - \eta_\text{n}\vec{v}_{ab,\text{n}}$
Tangential contact force	$F_{ab,\text{t}} = \begin{cases} -k_\text{t}\delta_\text{t} - \eta_\text{t}\vec{v}_{ab,\text{t}} & \text{for}\lvert\vec{F}_{ab,\text{t}}\rvert \le \mu_\text{f}\lvert\vec{F}_{ab,\text{n}}\rvert \\ -\mu_\text{f}\lvert\vec{F}_{ab,\text{n}}\rvert\vec{t}_{ab} & \text{for}\lvert\vec{F}_{ab,\text{t}}\rvert \le \mu_\text{f}\lvert\vec{F}_{ab,\text{n}}\rvert \end{cases}$
Normal displacement	$\delta_\text{n} = (R_a + R_b) - \lvert\vec{r}_a - \vec{r}_b\rvert$
Tangential displacement	$\delta_\text{t}(t) = \int_{t_0}^{t}\vec{v}_{ab,\text{t}}\,dt$
Normal damping coefficient	$\eta_\text{n} = \begin{cases} -2\ln\frac{e_\text{n}\sqrt{m_{ab}k_\text{n}}}{\sqrt{\pi^2+\ln^2 e_\text{n}}} & \text{if } e_\text{n} \neq 0 \\ 2\sqrt{m_{ab}k_\text{n}} & \end{cases}$
Tangential damping coefficient	$\eta_\text{t} = \begin{cases} \frac{-2\ln e_\text{t}\sqrt{\frac{2}{7}m_{ab}k_\text{t}}}{\sqrt{\pi^2+\ln^2 e_\text{t}}} & \text{if } e_\text{t} \neq 0 \\ 2\sqrt{\frac{2}{7}m_{ab}k_\text{t}} & \text{if } e_\text{t} = 0 \end{cases}$
Reduced mass	$m_{ab} = \left(\frac{1}{m_a} + \frac{1}{m_b}\right)^{-1}$
Normal vector unit	$\vec{n}_{ab} = \frac{\vec{r}_b - \vec{r}_a}{\lvert\vec{r}_b - \vec{r}_a\rvert}$
Tangential vector unit	$\vec{t}_{ab} = \frac{\vec{v}_{ab,\text{t}}}{\lvert\vec{v}_{ab,\text{t}}\rvert}$
Relative velocity between particles a and b	$\vec{v}_{ab} = (\vec{v}_a - \vec{v}_b) + \left(R_a\vec{\omega}_a + R_b\vec{\omega}_b\right) \times \vec{n}_{ab}$
Normal relative velocity	$\vec{v}_{ab,\text{n}} = (\vec{v}_{ab}\cdot\vec{n}_{ab})\vec{n}_{ab}$
Tangential relative velocity	$\vec{v}_{ab,\text{t}} = \vec{v}_{ab} - \vec{v}_{ab,\text{n}}$

Table 4.3 Equations for the DPM Model—cont'd

Tangential spring stiffness	$k_t = \dfrac{\frac{2}{7}\left(\pi^2 + (\ln e_t)^2\right)}{\pi^2 + (\ln e_n)^2} k_n$
Moment of inertia	$I = \frac{2}{5} m_a R_a^2$
Torque acting on particle	$\vec{T}_a = \sum_b \left(R_a \vec{n}_{ab} \times \vec{F}_{t,ab}\right)$
Normal contact time	$t_{contact,n} = \sqrt{\dfrac{\pi^2 + (\ln e_n)^2}{B_2 k_n}}$
Tangential contact time	$t_{contact,n} = \sqrt{\dfrac{\pi^2 + (\ln e_t)^2}{B_1 k_t}}$
$B1$, $B2$	$\dfrac{7}{2}\left(\dfrac{1}{m_a} + \dfrac{1}{m_b}\right), \left(\dfrac{1}{m_a} + \dfrac{1}{m_b}\right)$

$$\frac{\partial(\varepsilon_s \rho_s \vec{u})}{\partial t} + \left(\nabla \cdot \varepsilon_s \rho_s \vec{u}\,\vec{u}\right) = -\varepsilon_s \nabla p + \varepsilon_s \rho_s g - (\nabla \cdot \varepsilon_s \boldsymbol{\tau}_s) + \beta\left(\vec{u}_g - \vec{u}_s\right) \tag{4.18}$$

To describe the solids phase rheology, the widely used KTGF is adopted: in this framework in addition to the mass and momentum conservation equations the granular temperature θ, accounting for frictional stresses due to particle–particle and particle–wall collisions, needs to be solved by:

$$\frac{2}{3}\left[\frac{\partial(\varepsilon_s \rho_s \theta)}{\partial t} + \nabla \cdot \left(\varepsilon_s \rho_s \theta \vec{u}\right)\right] = -(p_s \boldsymbol{I} + \varepsilon_s \boldsymbol{\tau}_s) : \nabla \vec{u}_s - \nabla \cdot \left(\varepsilon_s \vec{q}_s\right) - 3\beta\theta - \gamma \tag{4.19}$$

In this equation, $\boldsymbol{I}$ is the unit tensor, $\vec{q}_s$ is the pseudo-Fourier fluctuating kinetic energy flux, and γ is the dissipation rate of granular energy due to inelastic particle–particle collisions. In the KTGF, collisions are assumed binary and quasi-instantaneous and do not take long-term and multiple particle contact into account (which is the case in the dense part of the fluidized bed). To correct for this shortcoming, the solids phase viscosity μ_s and the solids phase pressure p_s are split up into a kinetic part and a frictional part, which are treated separately (after Johnson and Jackson, 1987; Srivastava and Sundaresan, 2003):

$$p_s = p_{s,\mathrm{kin}} + p_{s,f} \tag{4.20}$$

$$\mu_s = \mu_{s,\mathrm{kin}} + \mu_{s,f} \tag{4.21}$$

$$p_{s,f} = p_c(\varepsilon_s) = \begin{cases} F\dfrac{(\varepsilon_s - \varepsilon_{s,\min})^r}{(\varepsilon_{s,\min} - \varepsilon_s)^s} & \text{for } \varepsilon_s > \varepsilon_{s,\min} \\ 0 & \text{for } \varepsilon_s > \varepsilon_{s,\min} \end{cases} \tag{4.22}$$

$$\mu_{s,f} = \frac{p_c(\varepsilon_s)\sqrt{2}\sin(\phi_I)}{2\varepsilon_s\sqrt{\boldsymbol{D}_{i,j}:\boldsymbol{D}_{i,j} + \Psi\theta_s/d_p^2}} \tag{4.23}$$

In these equations, F, r, s, and Ψ are constants, ϕ_I represents the internal angle of friction, and $\boldsymbol{D}_{i,j}$ represents the rate of strain tensor. An overview of all closures is given in Table 4.4 (De Jong et al., 2012d). For a detailed description of the model, the reader is referred to Laverman (2010).

2.2.3 Modeling of Permeable Membranes

The incorporation of gas permeable membranes in the models described above, either in the reactor walls or submerged in the bed itself, is required to investigate the hydrodynamics in FBMRs. While permeable walls can be implemented in a relatively straightforward fashion (if one only considers the hydrodynamic effect of adding or extracting gas and not the selective extraction of a single chemical species from a gas-phase mixture), the incorporation of submerged, cylindrical membranes is more involved. For this reason, De Jong et al. (2012b) presents a novel hybrid computational method, combining the DPM and the IBM, that allows the inclusion of arbitrary shapes onto a fixed Cartesian grid such as used by DPM. The advantage of continuing to use a Cartesian grid is that calculation times are small compared to models using unstructured or conformal grids. In contrast to continuum models (i.e., two-fluid or multifluid models), the DPM gives sufficient detail to understand the underlying phenomena, yet allows for much more particles than with fully resolved simulations.

The gas-phase motion is described with the volume-averaged Navier–Stokes equations, similar to those described in the DPM model description, but with one additional source term $\vec{F}^{\mathrm{IBM}}$that accounts for the presence of the membranes, hence the addition or extraction of gas at an arbitrary boundary:

Table 4.4 Equations for the TFM model

Granular temperature	$\theta=\frac{1}{3}\left\langle \vec{C}\cdot\vec{C}\right\rangle$
Solids phase pressure	$p_s=p_{s,kc}+p_{s,f}$
Solids phase viscosity	$\mu_s=\mu_{s,kc}+\mu_{s,f}$
Solids phase frictional pressure	$p_{s,f}=\begin{cases} F\frac{(\varepsilon_s-\varepsilon_{s,\min})^r}{(\varepsilon_{s,\min})^s} & \varepsilon_s>\varepsilon_{s,\min} \\ 0 & \varepsilon_s<\varepsilon_{s,\min}\end{cases}$
Frictional stress tensor	$\mu_{s,f}=\frac{p_c(\varepsilon_s)\sqrt{2}\sin(\phi_I)}{2\varepsilon_s\sqrt{\boldsymbol{D}_{i,j}:\boldsymbol{D}_{i,j}+\Psi\theta_s/d_p^2}}$
Rate of strain tensor	$\boldsymbol{D}_{i,j}=\frac{1}{2}\left(\left(\nabla\vec{u}_s\right)+\left(\nabla\vec{u}_s\right)^T\right)-\frac{1}{3}\nabla\cdot\vec{u}_s\boldsymbol{I}$
Partial slip boundary condition	$\begin{cases}\left(I-\vec{n}\,\vec{n}\right)\cdot\varepsilon_s\boldsymbol{\tau}_s\cdot\vec{n}=\frac{\alpha_{wall}\pi\varepsilon_s\rho_s g_0\sqrt{\theta}}{2\sqrt{3}\varepsilon_0}\vec{u}_s \\ \varepsilon_s\vec{q}_s\cdot\vec{n}=-\vec{u}_s\cdot\varepsilon_s\tau_s\cdot\vec{n}+ \\ \frac{\sqrt{3}\pi\left(1-e_{n,wall}^2\right)\varepsilon_s\rho_s g_0\sqrt{\theta}}{4\varepsilon_0}\theta.\end{cases}$
Particle pressure	$p_s=[1+2(1+e_E)\varepsilon_s g_0]\varepsilon_s\rho_s\theta$
Newtonian stress tensor	$\boldsymbol{\tau}_s=-\mu_s\left[\nabla\vec{u}+\left(\nabla\vec{u}\right)^T-\frac{2}{3}\vec{I}\left(\nabla\vec{u}\right)\right]$

Continued

Table 4.4 Equations for the TFM model—cont'd

Bulk viscosity	$\lambda_s = \frac{4}{3}\varepsilon_s \rho_s d_p g_0 (1 + e_n)\sqrt{\frac{\theta}{\pi}}$
Shear viscosity	$\mu_s = \left\{ \begin{array}{c} 1.01600 \frac{5}{96}\pi \rho_s d_p \sqrt{\frac{\theta}{\pi}} \\ \left(1 + \frac{\frac{\frac{8}{5}(1+e_n)}{2}\varepsilon_s g_0 \left(1 + \frac{8}{5}\varepsilon_s g_0\right)}{\varepsilon_s g_0} + \frac{4}{5}\varepsilon_s \rho_s d_p g_0 (1 + e_n)\sqrt{\frac{\theta}{\pi}} \right) \end{array} \right.$
Pseudo-Fourier fluctuating kinetic energy flux	$\vec{q}_s = -\kappa_s \nabla\theta$
Pseudothermal conductivity	$\kappa_s = \left\{ \begin{array}{c} 1.02513 \frac{75}{384}\pi \rho_s d_p \sqrt{\frac{\theta}{\pi}} \\ \left(1 + \frac{\frac{\frac{12}{5}(1+e_n)}{2}\varepsilon_s g_0 \left(1 + \frac{12}{5}\varepsilon_s g_0\right)}{\varepsilon_s g_0} + 2\varepsilon_s \rho_s d_p g_0 (1 + e_n)\sqrt{\frac{\theta}{\pi}} \right) \end{array} \right.$
Dissipation of granular energy due to inelastic particle–particle collisions	$\gamma = 3\left(1 - e_n^2\right)\varepsilon_s^2 \rho_s g_0 \theta \left[\frac{4}{d_p}\sqrt{\frac{\theta}{\pi}} - (\nabla \cdot \vec{u}_s)\right]$

$$\frac{\partial(\varepsilon_f \rho_f)}{\partial t} + \left(\nabla \cdot \varepsilon_f \rho_f \vec{u}\right) = 0 \tag{4.24}$$

$$\frac{\partial\left(\varepsilon_f \rho_f \vec{u}\right)}{\partial t} + \left(\nabla \cdot \varepsilon_f \rho_f \vec{u}\,\vec{u}\right) = -\varepsilon_f \nabla p + \varepsilon_f \rho_f g - \left(\nabla \cdot \varepsilon_f \boldsymbol{\tau}_f\right) - \vec{S}_{f \to s} + \vec{F}^{\mathrm{IBM}} \tag{4.25}$$

The IBM due to De Jong et al. (2012b) builds on the work by Uhlmann (2005) and takes into account the presence of—and gas permeation through—cylindrical membranes inside the computational domain; the transport of gas through the membrane structure itself is not resolved here. Lagrangian force points, equally distributed over the surface of the membrane, are employed to enforce the Dirichlet condition at the surface of the submerged membrane. A schematic representation of these force points is shown in Fig. 4.16.

Each force point exerts a force on the gas phase such that the interpolated preliminary velocity u^{p} of the gas phase is equal to the specified desired gas velocity u^{d} at the position of that force point. This force ($\vec{F}^{\mathrm{IBM}}$) is included as a source term in the gas-phase momentum equation. Euler discretization in time of the momentum equation leads to

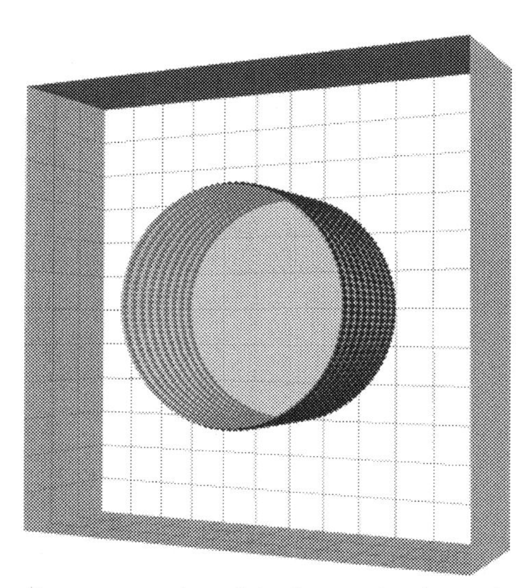

Figure 4.16 Schematic representation of the force points located on the surface of the cylindrical membrane. *Reprinted from De Jong et al. (2012b) with permission from Elsevier.*

$$\left[\varepsilon_f \rho_f \vec{u}\right]^{n+1} = \text{RHS}^n - \Delta t\, b^n \vec{u}^{n+1} + \Delta t \left[\vec{F}^{\text{IBM}}\right]^n \tag{4.26}$$

$$\begin{aligned} \text{RHS}^n = \left[\varepsilon_f \rho_f \vec{u}\right]^n + \Delta t \cdot (-[\varepsilon_f \nabla p]^n + [\varepsilon_f \rho_f g]^n = -\nabla[\varepsilon_f \boldsymbol{\tau}_f]^n \\ -\left[\nabla \cdot \varepsilon_f \rho_f \vec{u}\vec{u}\right]^n + \vec{a}^n) \end{aligned} \tag{4.27}$$

using a linear-implicit treatment of the gas–particle drag force. Note that the pressure term and density are evaluated at time level n instead of $n+1$. In case no force would be applied (read $\vec{F}^{\text{IBM}} = 0$), the solution of $\vec{u}^{n+1}$ would equal the preliminary velocity field $\vec{u}^p$. Applying this condition yields

$$\vec{u}^{\text{p}} = \frac{\text{RHS}^n}{\left[\varepsilon_f \rho_f\right]^{n+1} + \Delta t b^n} \tag{4.28}$$

However, we do apply a force and, therefore, the solution of $\vec{u}^{n+1}$in Eq. (4.27) will be the desired velocity $\vec{u}^d$. Combining Eqs. (4.27) and (4.28) results in the relation between the force density $\left[\vec{F}^{\text{IBM}}\right]^n$ and the velocity at the old time level n ($\vec{u}^{\text{p}}$)

$$\vec{u}^{\text{d}} = \vec{u}^{\text{p}} + \frac{\Delta t}{\left[\varepsilon_f \rho_f\right]^{n+1} + \Delta t b^n} \left[\vec{F}^{\text{IBM}}\right]^n \tag{4.29}$$

The Eulerian momentum density is subsequently mapped to the Lagrangian force points using a distribution function

$$\left(\varepsilon_f \rho_f \vec{u}\right)_{\text{fp}}^{\text{p}} = \sum_{\text{fp}}^{N} \delta\left(\vec{x} - \vec{X}_{\text{fp}}\right) \left(\varepsilon_f \rho_f \vec{u}\right)^{\text{p}} \tag{4.30}$$

where $\vec{x}$ and $\vec{X}_{\text{fp}}$ represent, respectively, the Eulerian velocity node and the Lagrangian force point positions. Likewise, the macroscopic density at the force points $E_{\text{fp}}^{\text{p}} R_{\text{fp}}^{\text{p}}$ and the interphase momentum transfer coefficient at the force point B_{fp}^{p} are obtained by interpolation of $\varepsilon_f \rho_f$ and β at the Eulerian nodes, using a fourth-order polynomial interpolation given by Deen et al. (2004). Substituting these Lagrangian values into Eq. (4.29) results in the equation for the force density at the Lagrangian force point

$$F_{\text{fp}}^{\text{p}} = \varphi\left(E_{\text{fp}}^{\text{p}} R_{\text{fp}}^{\text{p}} + \Delta t B_{\text{fp}}^{\text{p}}\right)\left(\vec{u}_{\text{fp}}^{\text{d}} - \vec{u}_{\text{fp}}^{\text{p}}\right) \tag{4.31}$$

The normalization factor φ signifies the relative range of influence of each force point. For a cylinder with diameter D and height H, φ is defined as the volume per force point divided by the total volume of a grid cell

$$\varphi = \frac{\pi DH}{N_{\mathrm{fp}}\sqrt[3]{\Delta x \Delta y \Delta z}} \tag{4.32}$$

After calculation of the Lagrangian force density $\vec{F}_{\mathrm{fp}}^{\mathrm{p}}$, mapping needs to be carried out again for each component of the force density separately and needs to be summed over all Lagrangian force points N

$$\left[\vec{F}^{\mathrm{IBM}}\right]^n = \sum\nolimits_{\mathrm{fp}}^{N} \delta\left(\vec{x} - \vec{X}_{\mathrm{fp}}\right)\vec{F}_{\mathrm{fp}}^{\mathrm{p}} \tag{4.33}$$

Mass is added to (or extracted from) the system by means of inflow (outflow) cells in the front and rear wall within the cylinder perimeter. The IBM force is added for each force point separately. Because of its explicit nature, after a converged pressure correction, the solution shows a small deviation from the prescribed conditions. This numerical error can be minimized by iterating until the solution has converged. To decrease the number of iterations, only a correction to the solution obtained in the previous time step is calculated. This process is schematically summarized in Fig. 4.17.

The incorporation of the IBM has been verified for creeping flows and for higher *Re* flows by comparison of the drag exerted on an array of cylinders in the domain. Only for very high cylinder packings, larger than 0.5, deviations have been found for time steps of 10^{-6} s; however, for FBMRs, such a packing fraction will not be achieved.

In order to verify the permeation behavior of the immersed boundary cylinders, the case of a single gas-permeating cylinder in the center of the computational domain is considered, where gas is added (or extracted) equally through the frontal surface of all grid cells that are situated well within the cylinder perimeter. Because there is no analytical solution for this case, it is difficult to validate the results in a quantitative way. Three test cases (no permeation, gas addition, and gas extraction) were simulated with one cylinder (0.01 m in diameter at x/z position of 0.015/0.015 m) in a $0.09 \times 0.003 \times 0.03\text{-m}^3$ domain with $60 \times 1 \times 180$ computational cells at a gas velocity of 0.65 m/s, with a loose packing of static particles (packing fraction 0.35, see Fig. 4.18D) surrounding the membrane tube. Relatively high permeation ratios of 200% relative to the background fluidization velocity were chosen for these test cases. Qualitative results of these simulations are presented in Fig. 4.18. These simulations were also carried out without the presence of an array of static particles. The results were very similar to the results shown in Fig. 4.18.

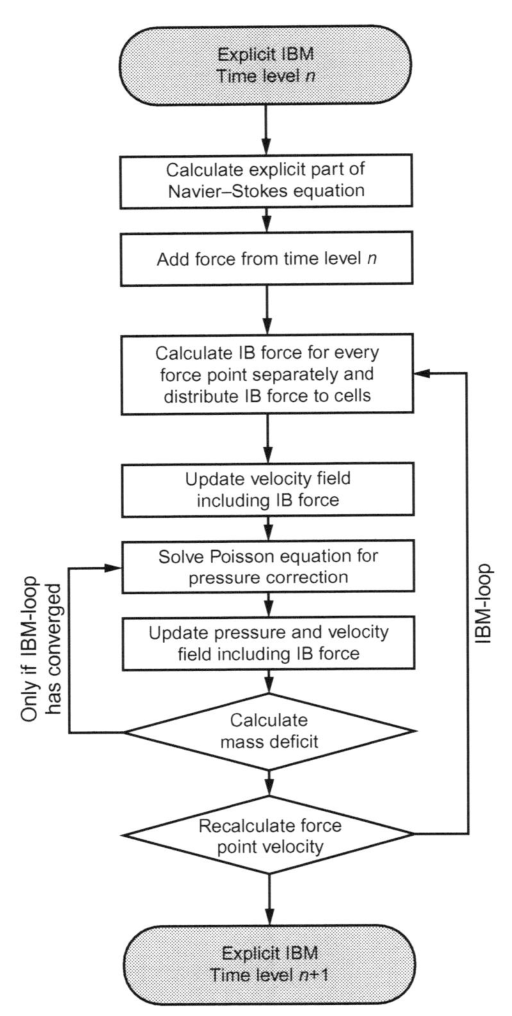

Figure 4.17 Schematic representation of the implementation of the IBM into the flow solver of the DPM code. *Reprinted from De Jong et al. (2012b) with permission from Elsevier.*

Fig. 4.18A shows the flow pattern without permeation through the membrane tube but with the no-slip condition imposed. As expected, the gas flows around the cylinder. Note that inside the cylinder a flow is induced as well; this phenomenon is well known (Uhlmann, 2005). We add force to the cells surrounding the force points on the cylinder surface, but do not enforce any restriction on these cells, thus allowing for internal flow in the submerged object (membrane). Usually, these gas velocities inside the

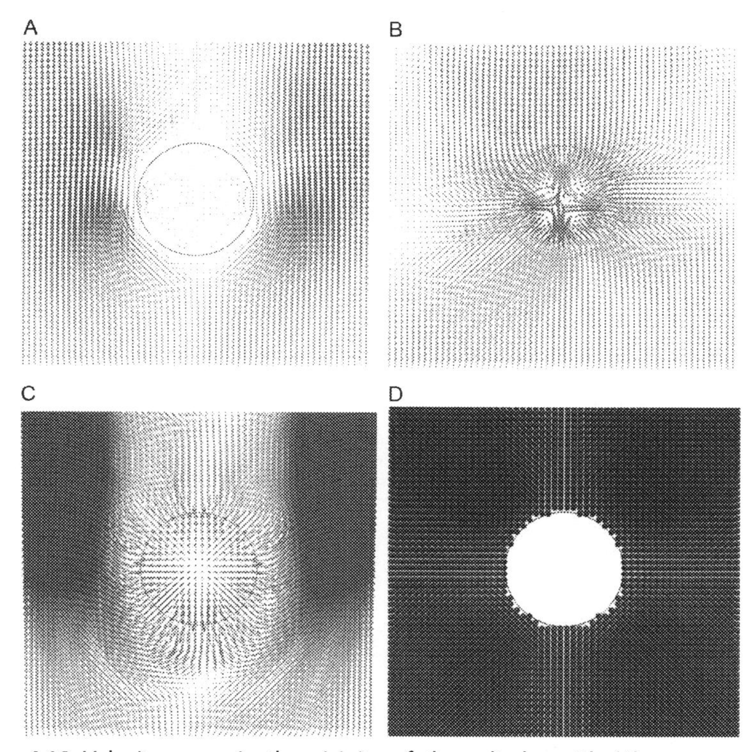

Figure 4.18 Velocity maps in the vicinity of the cylinder with (A) no permeation, (B) 200% gas extraction (relative to the background fluidization velocity), (C) 200% gas addition through the membrane tube, and (D) the particle distribution of the static packing. The scale of the velocity vectors ranges from 0 m/s (blue (black in print version)) to 3 m/s (red (gray in print version)). *Reprinted from De Jong et al. (2012b) with permission from Elsevier.*

cylinder are much smaller than the exterior velocities. Moreover, as can be seen from this figure, this phenomenon does not affect the boundary condition imposed on the membrane surface.

Table 4.5 displays the mass balance over the membrane and over the entire system. It can be concluded from here that the error in the mass balance over the system Δ_s is very small. The error in the mass balance over the membrane Δ_m is approximately a factor 10–100 larger than Δ_s; however, this error remains relatively small.

Fig. 4.18B and C show the cases of gas extraction and gas addition. Note that because the extraction rate is very high, an equal amount of gas is taken from the outlet (top) of the system, as is fed via the inlet (bottom). Since the simulation domain is higher than actually shown in Fig. 4.18, the flow from the top already has a parabolic flow profile, while the gas entering from the

Table 4.5 Error in the Mass Balance Over the Membranes Δ_m (Relative to the Permeation Flow or, in Case of Zero Permeation, Relative to the Background Gas Velocity) and Error in the Mass Balance Over the Entire System Δ_s (Relative to the Total Inflow)

$N_{p,static}$ (–)	Permeation (%)	Δ_m (%)	Δ_s (%)
0	−200	7×10^{-4}	7.3×10^{-5}
0	0	$\ll1\times10^{-4}$	$<0.5\times10^{-5}$
0	200	7×10^{-4}	4.4×10^{-5}
13,180	−200	22×10^{-4}	8.7×10^{-5}
13,180	0	$\ll1\times10^{-4}$	1.5×10^{-5}
13,180	200	22×10^{-4}	0.5×10^{-5}

The number of static particles N_p and permeation (relative to the background fluidization velocity) are varied, the prescribed accuracy is kept constant at 5%.

bottom has a flat flow profile. It can be clearly seen from the figures that also with permeation, there is a flow induced inside the cylinder.

For transient simulations, the IBM loop needs a convergence criterion. For this criterion, the maximum deviation of all force points N_{fp} (deviation of the actual gas velocity at the force points from the desired gas velocity) is used. It is defined as follows:

$$Deviation_{fp} = 100\cdot\frac{\left|\left|\vec{u}^{d} - \vec{u}\right|\right|}{\left|\left|\vec{u}^{d}\right|\right|} \tag{4.34}$$

or in cases without permeation relative to the background fluidization velocity

$$Deviation_{fp} = 100\cdot\frac{\left|\left|\vec{u}^{d} - \vec{u}\right|\right|}{\left|\left|\vec{u}_{bg}\right|\right|} \tag{4.35}$$

Because the accuracy (i.e., the maximum deviation allowed) of the IBM is simply enforced as the convergence criterion for the flow solver, it is more relevant to compare the number of iterations needed to achieve the prescribed accuracy. Therefore, several test cases have been conducted using two cylinders of 0.01 m in diameter at x/z positions of 0.01/0.01 m and 0.02/0.03 m, respectively, in the same computational domain as described previously. The results are summarized in Table 4.6. The last entry of this table is a simulation using different simulation parameters: in this case, with nine membrane tubes, it coincides with the settings described in Section 2.4

Table 4.6 Average Number of Iterations N_{iter} Needed for Several Test Simulations and Average Deviation of the Gas Velocity of the Force Points Δu,fp with Varying Accuracy, Number of Particles N_p and Permeation (Relative to the Background Fluidization Velocity)

Prescribed Deviation (%)	N_p (−)	Permeation (%)	N_{iter} (−)	Δu,fp (%)
0.1	0	0	655	0.016
1	0	0	2	0.097
2	0	−50	2	0.003
2	0	50	2	0.002
0.1	25,000	0	10,500	0.022
1	25,000	0	1000	0.214
2	25,000	−50	2	0.024
2	25,000	50	3	0.001
5	200,000	0	2	2.64

and is added here to illustrate the influence of a larger system geometry on the number of iterations and the average deviation.

For the test cases with a prescribed accuracy of 0.01% (for both cases with and without particles), the results are not available, because these simulations never converged. This fact implies that there is a limit to the accuracy of the IBM. From N_{iter} in Table 4.6, it can clearly be concluded that the number of iterations, needed to converge to the prescribed maximum deviation of the gas velocity at the force points, increases dramatically as the prescribed deviation is decreased. When particles are added to the system, this effect becomes even more pronounced, because of the varying porosity, pressure, and gas velocity in the vicinity of the membranes. Moreover, it can be concluded that the average deviation of gas velocity Δu,fp depends on the prescribed accuracy, although this dependency is not linear; it varies between approximately 10% and 50% relative to the prescribed accuracy. It is believed that the static pressure induced by the presence of the particles has an influence on this parameter.

2.3. Hydrodynamics in FBMRs with Permeable Membrane Walls

Investigations have been performed into the effect of permeable membrane walls on the hydrodynamics in FBMRs. First of all, an experimental investigation on the fluidization behavior of an FBMR with wall-mounted membranes is outlined. The results are then reinterpreted for micro-FBRs. The sections that follow describe the numerical simulations of these systems.

2.3.1 Experimental Investigation on the Fluidization Behavior of FMBRs with Permeable Membrane Walls

An investigation has been performed using the setup described in Section 2.1.1, with a bed width, depthm and height of 0.3, 0.015, and 1 m respectively, using Geldart B particles (400–600 μm glass beads) and air as a fluidization agent.

The minimum fluidization velocity u_{mf} was determined by slowly decreasing the fluidization velocity. Based on the pressure drop, u_{mf} was determined to be 0.25 m/s. All experiments reported here have been performed with a total gas feed corresponding to $u/u_{\mathrm{mf}}=2.6$.

Two sets of experiments have been carried out. In the first set of experiments, the background fluidization velocity was kept constant, while the amount of secondary gas added or extracted via the membranes was varied. In the second set of experiments, the amount of secondary gas added/extracted via the membranes was varied, while simultaneously the background fluidization velocity was adjusted so that the outlet flow rate remained constant (see Table 4.7). Measurement 1 is the reference case, without in- or outflow via the membranes. Although the fluxes of gas through permselective membranes are much smaller than the cases used in this study, these values were chosen to emphasize the effect of permeation and to anticipate future improvements in membrane performance. On the other hand, the selected permeation cases are much more realistic for the nonselective porous membranes.

For the discussion of the effect of gas permeation on bed hydrodynamics, we first focus on the solids circulation patterns, followed by the bubble properties.

2.3.1.1 Solids Phase Circulation Patterns

The time-averaged solids circulation pattern and the time-averaged lateral profile of the axial solids phase velocity at different heights in the bed are shown in Fig. 4.19 for all series where the background fluidization velocity has been kept constant; i.e., 100% − 40%, 100% − 20%, the reference, 100% + 20% and 100% + 40%. In the reference series with no secondary gas extraction or addition, the characteristic pattern for fluidized beds with an upward-directed solids flow through the core and a downward solids flow along the

Table 4.7 Measurement Series

Measurement Number/Name		Background Gas Flow/Velocity		Total Membrane Flow/Velocity		Number of Pictures	
		(%)	(m/s)	(%)	(m/s)	For DIA (–)	For PIV (–)
1	Reference	100	0.65	0	0	2700	2160
2	80% + 20%	80	0.52	20	0.065	2700	2160
3	60% + 40%	60	0.39	40	0.13	2700	2160
4	100% + 20%	100	0.65	20	0.065	2700	2160
5	100% + 40%	100	0.65	40	0.13	2700	2160
6	120% – 20%	120	0.78	–20	–0.065	1350	2160
7	140% – 40%	140	0.91	–40	–0.130	1350	2160
8	100% – 20%	100	0.65	–20	–0.065	1350	2160
9	100% – 40%	100	0.65	–40	–0.130	1350	2160

walls of the fluidized bed can be clearly discerned. This well-known solids circulation pattern is also clearly visible in the lateral profiles of the axial solids phase velocity at different heights: a broad region in which particles move upward in the center of the fluidized bed, and near the walls a small region where the particles move downward. It is interesting to notice the local minimum in the axial solids phase velocity in the center at 10 cm above the bottom distributor plate, corresponding to the well-known average bubble trajectories from the walls toward the center in the lower sections of the fluidized bed (see also Laverman et al., 2008).

When comparing the cases with gas extraction to the reference case, a striking difference is the stagnant regions near the membranes in case of gas extraction. It is already appearing in the 100% – 20% case, but becomes even more pronounced when 40% of the background fluidization gas is extracted. These stagnant zones have two consequences: the first consequence is that the bed height is reduced, implying a smaller number of bubbles or smaller bubbles present inside the fluidized bed. Second, the velocity plot shows that the peak of upward moving solids has become steeper, while the downward directed "peak" for the downward moving solids has become less pronounced and has shifted somewhat toward the center of the bed. The reason for these phenomena is that the stagnant zones near the membranes

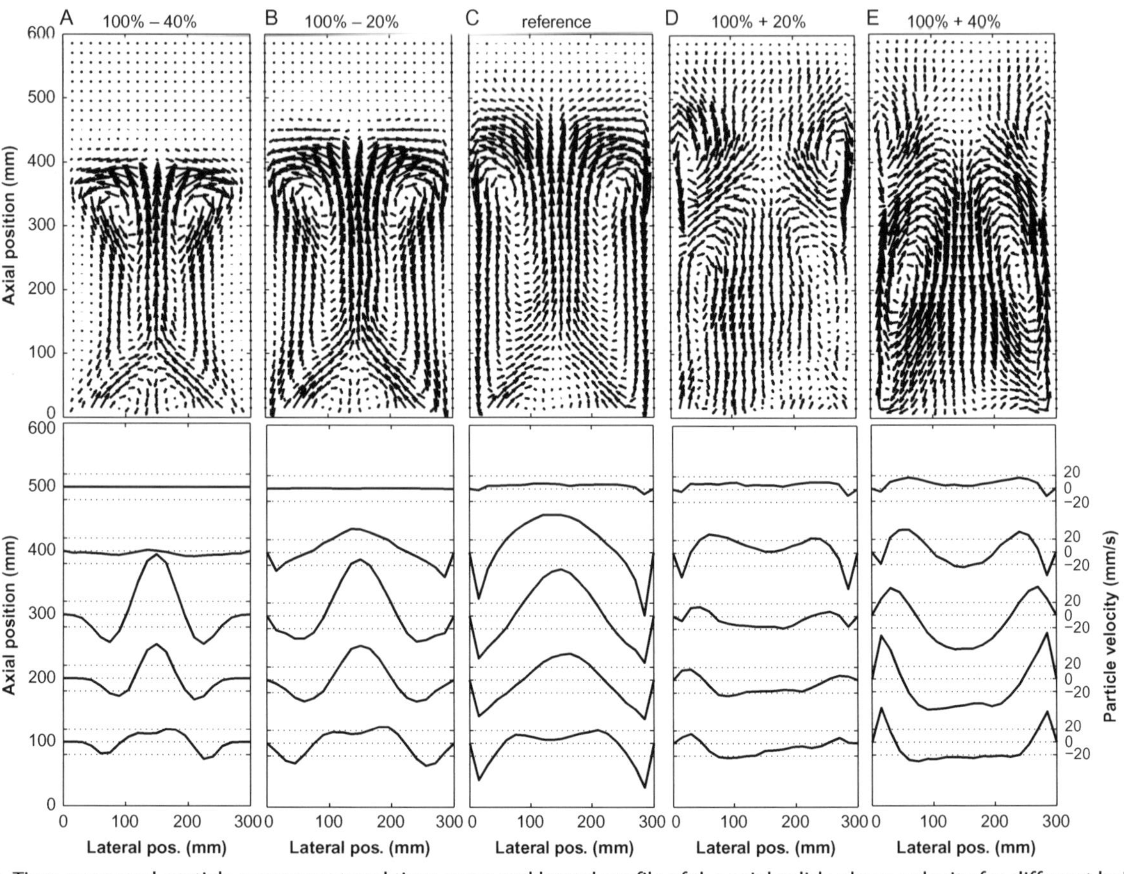

Figure 4.19 Time-averaged particle movement and time-averaged lateral profile of the axial solids phase velocity for different heights in the fluidized bed with (A) 40% gas extraction, (B) 20% gas extraction, (C) the reference series with no gas addition/extraction, (D) 20% gas addition, and (E) 40% gas addition. In all experiments, the background fluidization velocity was kept constant ($u_0 = 100\%$). *Reprinted from De Jong et al. (2011) with permission from Elsevier.*

leave less space for bubbles to rise and for particles to recirculate to the bottom of the bed, resulting in narrower vortices in the solids circulation.

In contrast to gas extraction, gas addition via the membranes has an even more distinctive effect on the particle circulation pattern: gas addition inverts the circulation pattern. The inversion of the particle circulation is interesting; this phenomenon shows that there is a competition between the background gas velocity and the additional gas entering via the membranes to drag the particles along. Already in the 100% + 20% series, this phenomenon starts to become apparent, but is even more pronounced for the 100% + 40% series. Usually particles would move downward near the walls. However, due to the gas addition, particles near the wall (in the first 30 cm) are dragged upward instead, causing the particles to move downward in the center of the bed. This phenomenon is also illustrated by the lowest three lateral profiles of the axial solids phase velocity profiles; the upward-directed peak is now near the wall, while the velocity in the center of the bed is slightly negative. Above the membrane (above 30 cm), the particles are pushed toward the center, and continue their way as usual: upward via the center and downward via the sides. This division in a part with membrane and a part without membrane results in four vortices inside the fluidized bed, each one rotating differently than its neighbor.

In the previous cases, the background fluidization velocity was kept constant, while the amount of gas added or extracted via the membranes was varied. An additional set of experiments was carried out where the outlet velocity was kept constant by adjusting the background velocity to the amount permeated through the membranes: 140% − 40%, 120% − 20%, 80% + 20%, and 60% + 40%. In Fig. 4.20, the particle velocity profiles of these series are compared to the reference case.

First, it can be observed that in these cases the bed height remains approximately the same. Additionally, it can be seen that also in these cases there is a constant "competition" between the gas forcing particles upward and the effect of gas permeation through the membranes. For the 140% − 40% and 120% − 20% series, the effect of gas extraction via the membranes (i.e., stagnant zones) is the same, but, the stagnant zones are smaller than for the 100% − 40% and 100% − 20% cases shown in Fig. 4.20, because of the increased background fluidization flow rate.

In contrast to the cases of gas extraction with increased background fluidization gas velocity, where the effect of stagnant zones is decreased by the additional background fluidization flow rate, for the cases of gas addition via the membranes, the decreased background fluidization velocity increases the

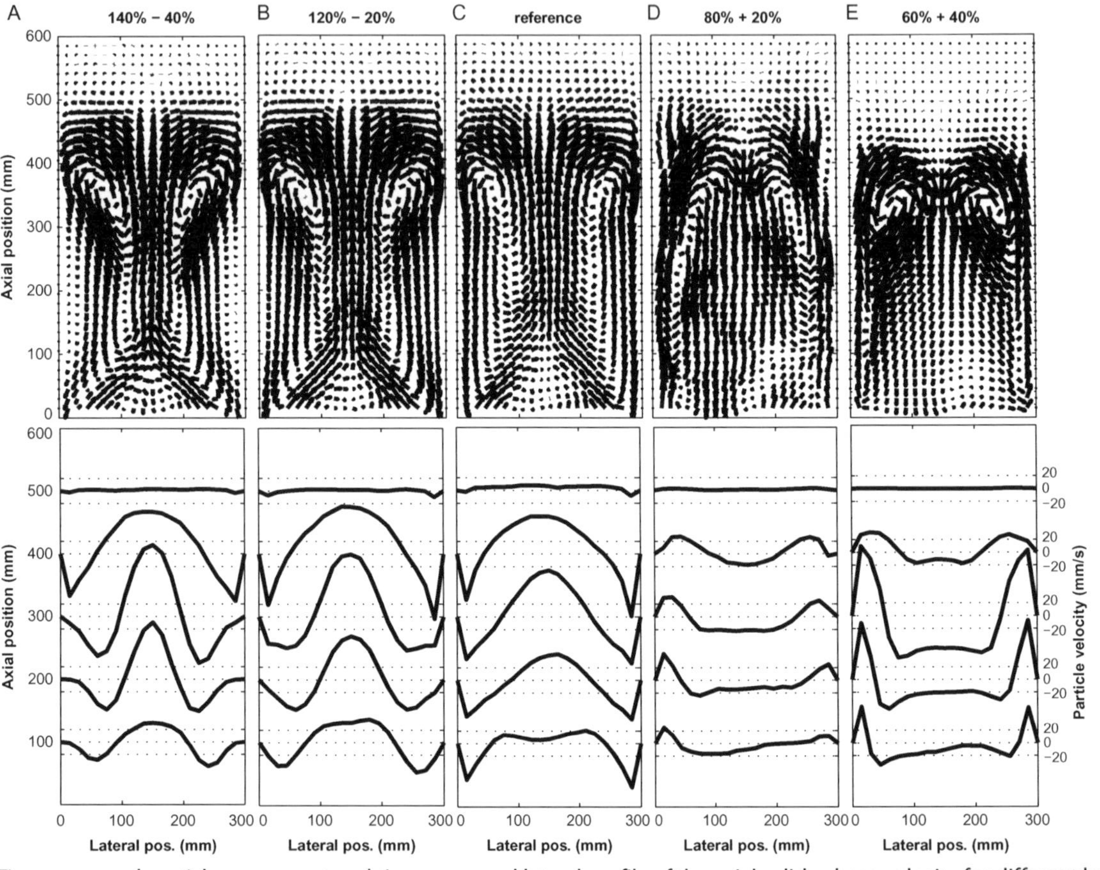

Figure 4.20 Time-averaged particle movement and time-averaged lateral profile of the axial solids phase velocity for different heights in the fluidized bed with (A) 40% gas extraction, (B) 20% gas extraction, (C) the reference series with no gas addition/extraction, (D) 20% gas addition, and (E) 40% gas addition. In all experiments, the outflow velocity was kept constant by adjusting the background fluidization to the amount of gas inflow or outflow via the membranes. *Reprinted from De Jong et al. (2011) with permission from Elsevier.*

effect of the gas addition. For the 100% + 20% and 100% + 40% series, there was a competition between the background fluidization gas and the additionally added gas, trying to drag particles along with the gas stream. The decrease of the background fluidization flow in the 80% + 20% and 60% + 40% series changes the competition in favor of the gas addition via the membranes. The 80% + 20% series already shows clearly the inversion of the particle circulation, and in the 60% + 40% series, this phenomenon is even more pronounced, which can be seen particularly well in the two strong vortices at the top of the fluidized bed.

The findings described above can be schematically summarized as depicted in Fig. 4.21. In all cases, the magnitude of the effects depends on the background fluidization velocity and amount of gas extraction or gas addition. It can be expected that the change in particle behavior has a pronounced effect on the bubble properties and bubble size distribution, which is discussed next.

2.3.1.2 Bubble Properties

An empirical relation used frequently in literature to describe the bubble diameter d_b as a function of the bed height h was given by Darton et al. (1977) and modified to fit pseudo-2D beds by Shen et al. (2004).

$$d_b = 0.89\left[(u_0 - u_{mf})\left(h + \frac{3A_0}{t}\right)\right]^{2/3} g^{-1/3} \tag{4.36}$$

In this equation, A_0 represents the catchment area and t the depth of the bed. Fig. 4.22 shows a graph of the equivalent bubble diameter of the reference series as a function of the bed height; the reference without gas

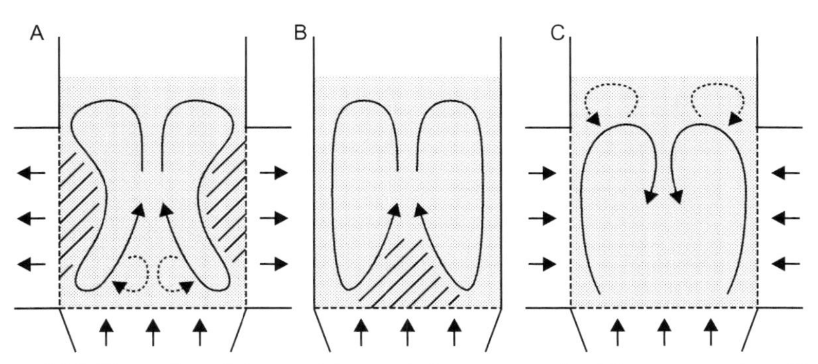

Figure 4.21 Illustration of the particle circulation patterns for (A) gas extraction, (B) the reference, and (C) gas addition. *Reprinted from De Jong et al. (2011) with permission from Elsevier.*

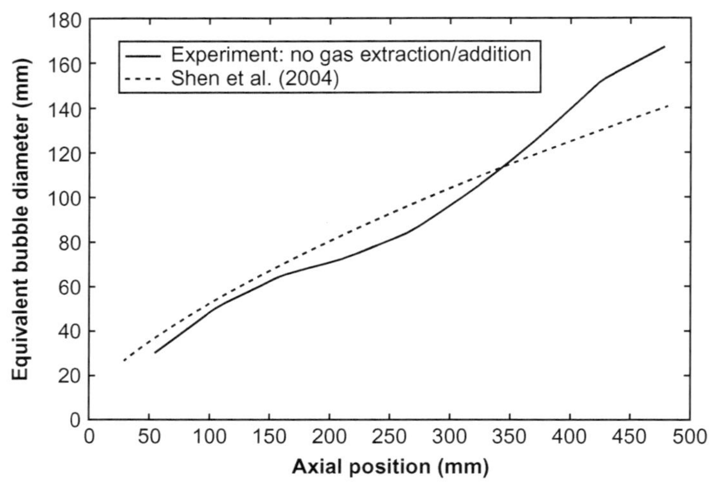

Figure 4.22 Comparison of the equivalent bubble diameter as a function of the bed height for the reference case with literature correlation. *Reprinted from De Jong et al. (2011) with permission from Elsevier.*

extraction or addition compares well to the relation proposed by Shen et al. (2004). The overprediction of the bubble diameter in the lower part of the fluidized bed by the proposed correlation can be attributed to the fact that Shen et al. (2004) used a larger setup (0.68×0.07 m^2 in width and depth) than in this experiment, so that the smallest bubbles were not accounted for in the correlation. The deviation from the predicted trend in the upper part of the fluidized bed is believed to have two causes: first, because the bubble diameter has grown to almost two-thirds of the bed width, the deviation can be caused by wall effects. Second, it can be the result of the relatively small number of bubbles detected in that region, resulting in a relatively large error in that area.

Subsequently, we will first compare the cases where the background fluidization velocity was kept constant and the amount of gas added or extracted was varied. The equivalent bubble diameter as a function of the height for these series is shown in Fig. 4.23A. In the lower part of the fluidized bed, the bubbles remain approximately the same size, irrespective of the amount of gas extraction or addition. Only from a height of approximately 20 cm, a difference becomes apparent. However, unlike what would be expected intuitively, extracting gas leads to larger bubbles, while adding gas results in smaller bubbles.

In particular, the experimental series in which gas is added via the membranes deviates substantially from the reference case above a height of 30 cm. Note that the largest bubbles for the cases of gas extraction appear at 40 cm

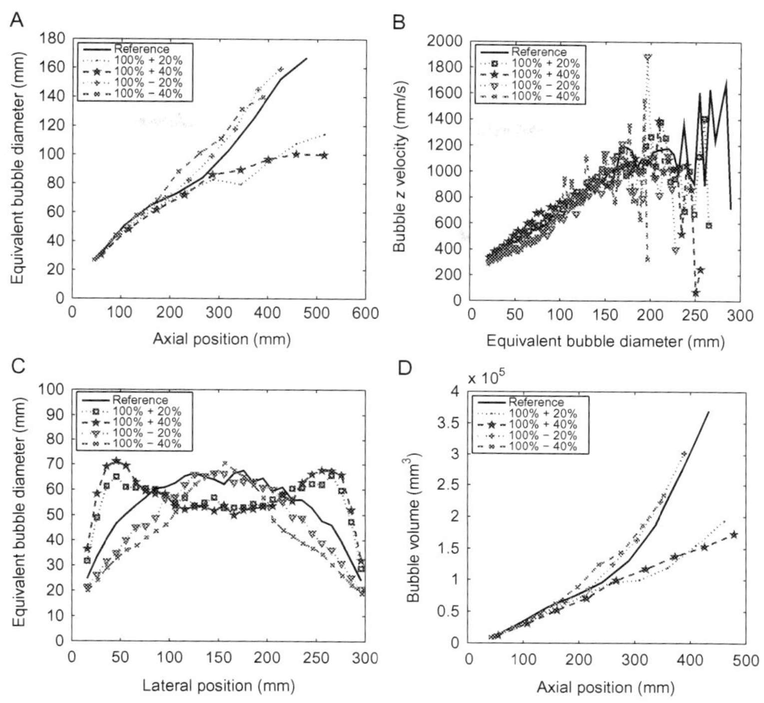

Figure 4.23 Effect of gas extraction and addition on (A) equivalent bubble diameter as a function of the bed height, (B) vertical bubble velocity as a function of the bubble diameter, (C) average bubble diameter as a function of the lateral position, and (D) bubble holdup as a function of the bed height. Data are shown for the series with constant background fluidization velocity. *Reprinted from De Jong et al. (2011) with permission from Elsevier.*

height, the ones for the reference case at about 46 cm, and the bubbles for the cases with gas addition appear even at 52 cm height, reflecting the difference in fluidized bed height.

The bubble rise velocity (Fig. 4.23B) as a function of the equivalent bubble diameter appears to be quite similar for all cases. The graphs of the lateral profile of the equivalent bubble diameter and the axial profile of the bubble holdup (Fig. 4.23C and D) provide more insight into the bubble behavior.

There is a significant difference in the average lateral position of the bubbles. The reference case shows an almost parabolic distribution, as expected, because bubbles are formed over the entire width of the fluidized bed and move toward the center due to bubble coalescence. The 100% − 20% and 100% − 40% series show a similar distribution, although bubbles reside more in the center (which is in line with the conclusions drawn from Fig. 4.19). The 100% + 20% and 100% + 40% series reveal a very different bubble distribution: in these cases, the large bubbles are situated much closer to the walls. In the center, a significant decrease in bubble diameter is visible, indicating that the movement of the bubbles is reversed, i.e., while bubbles are rising and growing, they are moving away from the center and toward the membranes. This is in line with the particle movement seen in Fig. 4.19.

Not only the location but also the bubble volume is different for these cases. The series with gas extraction show a slightly larger bubble volume, although this difference is very small. However, the series with gas addition reveal that—in particular in the top section of the bed—the bubble volume is much smaller compared to the reference case. Now the question remains why for the 100% + 20% and 100% + 40% series, both the average bubble diameter as well as the average bubble volume are lower than the reference case. This phenomenon is caused by a combination of particle movement and bubble detection (see Fig. 4.24): large gas voids near the walls are likely to be part of the freeboard of the fluidized bed and are therefore no longer defined as bubbles. This is caused by particles near the freeboard that are—in contrast to the reference case—moving away from the wall toward the center of the fluidized bed, and as a consequence, there are much fewer large bubbles surrounded by solids phase.

Subsequently, the effect of gas extraction or addition on the global gas holdup in the fluidized bed has been investigated. The results for all series

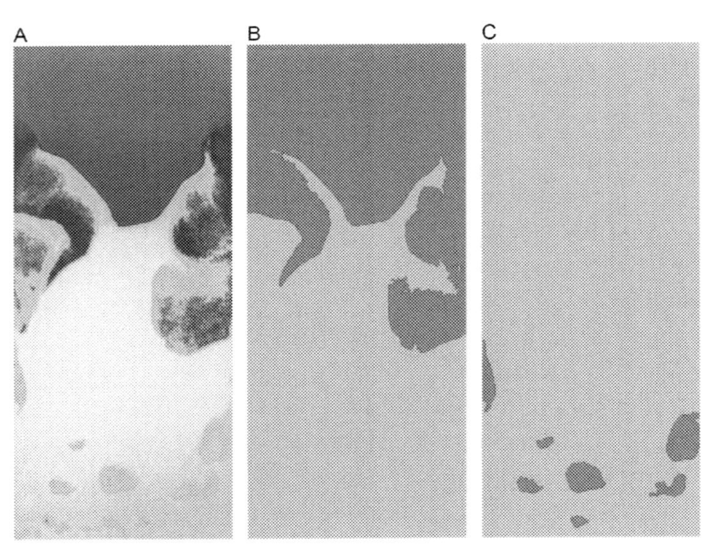

Figure 4.24 Snapshots of (A) an original PIV image, (B) the freeboard detection, and (C) the subsequent bubble detection by DIA. *Reprinted from De Jong et al. (2011) with permission from Elsevier.*

Table 4.8 Effect of Gas Extraction and Addition on the Total Gas Holdup

U_0 constant	100% − 40%	100% − 20%	Reference	100% + 20%	100% + 40%
Gas holdup (%)	44.9 ± 0.9	45.9 ± 1.2	46.1 ± 1.8	46.7 ± 1.8	47.4 ± 1.8
U_0 adjusted	140% − 40%	120% − 20%	Reference	80% + 20%	60% + 40%
Gas holdup (%)	46.5 ± 1.9	46.8 ± 1.9	46.1 ± 1.8	45.8 ± 1.2	45.1 ± 0.6

Also the standard deviation for the temporal fluctuation is provided.

have been summarized in Table 4.8. The series with constant background fluidization velocity show a slight increase in the global gas holdup going from gas extraction to gas addition. For the series with adjusted background fluidization velocity, this is exactly opposite, indicating that the total amount of fluidization gas is most important. However, the difference in gas holdup is relatively small, in particular when compared to the standard deviation.

Just as before for the particle movement, adjusting the background fluidization velocity to keep the outlet velocity constant makes the observed phenomena related to bubble behavior more pronounced. Fig. 4.25A shows an increase in equivalent bubble diameter for the series with gas extraction, and Fig. 4.25D shows an increase in bubble volume as a function of the bed height. The explanation for these findings can be discerned from Fig. 4.25C, which shows a substantial increase in average equivalent bubble diameter in the center of the bed for the 120% − 20% and 140% − 40% series. As shown schematically in Fig. 4.20, the bubbles (and particles) are forced through the center, resulting in vertically stretched—and therefore larger—bubbles. In Fig. 4.26, three snapshots are given that illustrate this. The presence of vertically elongated bubbles at this fluidization velocity indicates that a significant amount of the gas flow is bypassing the emulsion phase by means of throughflow (Grace and Clift, 1974; Grace and Harrison, 1969). In reactive systems, this increased throughflow decreases the gas–emulsion contact with a negative impact on the reactor performance.

The effect of gas extraction and gas addition on the bubble behavior is schematically depicted in Fig. 4.27.

2.3.2 One-to-One Comparison of Experiments and Simulations

Section 2.3.1 discusses the experimental results from a 30-cm wide fluidized bed with wall-mounted membranes. This section continues with a direct comparison of a number of experiments using a bed with DPM and TFM simulations.

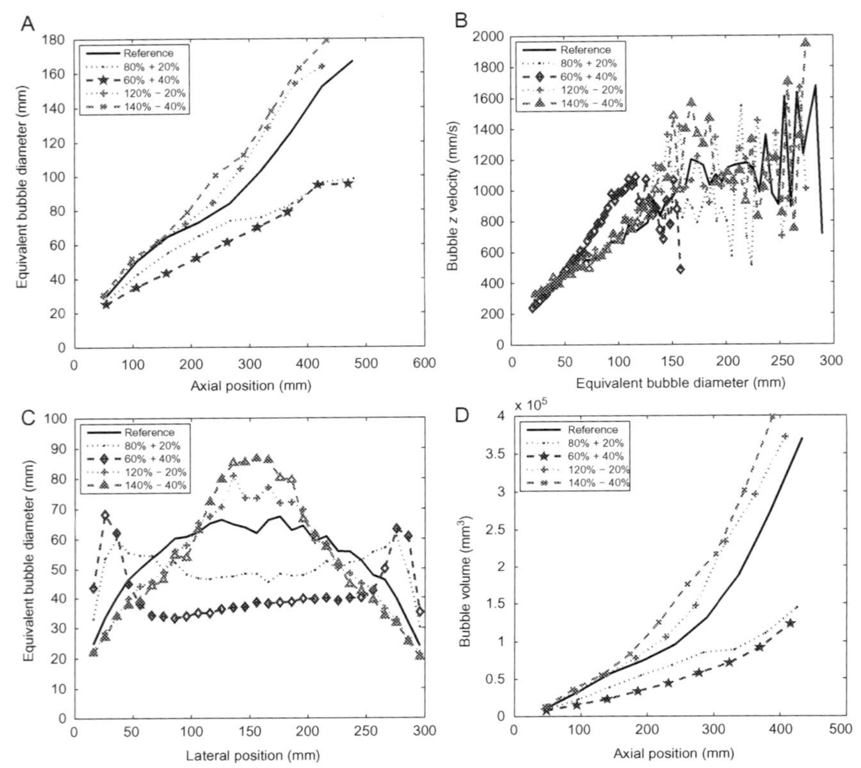

Figure 4.25 (A) Equivalent bubble diameter as a function of the height in the bed, (B) vertical bubble velocity as a function of the bubble diameter, (C) equivalent bubble diameter as a function of the lateral position, and (D) bubble holdup as a function of the bed height. Data are shown for the series with constant outlet velocity by adjusting the background velocity to the amount of gas extracted or added via the membranes. *Reprinted from De Jong et al. (2011) with permission from Elsevier.*

2.3.2.1 Experiment and Simulation Outline

A pseudo-2D setup was used to investigate the effect of gas permeation on the fluidized bed hydrodynamics using a small-scale fluidized bed. The bed width, depth, and height were 4, 1, and 50 cm, respectively. These dimensions have been chosen to have the setup as large as possible, yet being able to perform DPM simulations with the resulting number of particles in this domain. Apart from the dimensions, the experimental setup and measurement procedures have been kept identical to that described in Section 2.3.1, i.e., glass beads with diameter 400–600 μm have been used, which results in a measured minimum fluidization velocity of 0.25 m/s.

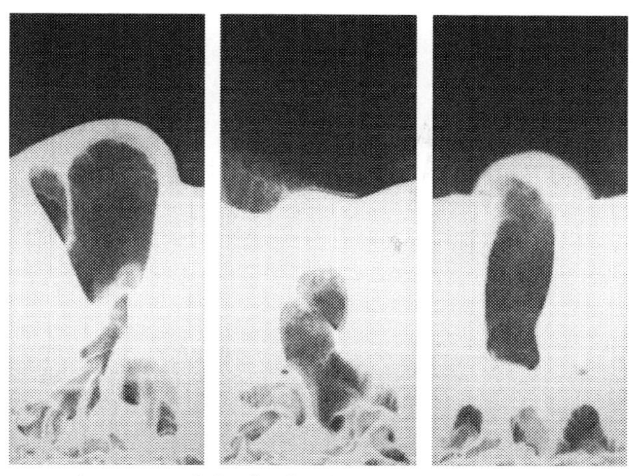

Figure 4.26 Typical pictures of the fluidized bed of the 140% – 40% series. *Reprinted from De Jong et al. (2011) with permission from Elsevier.*

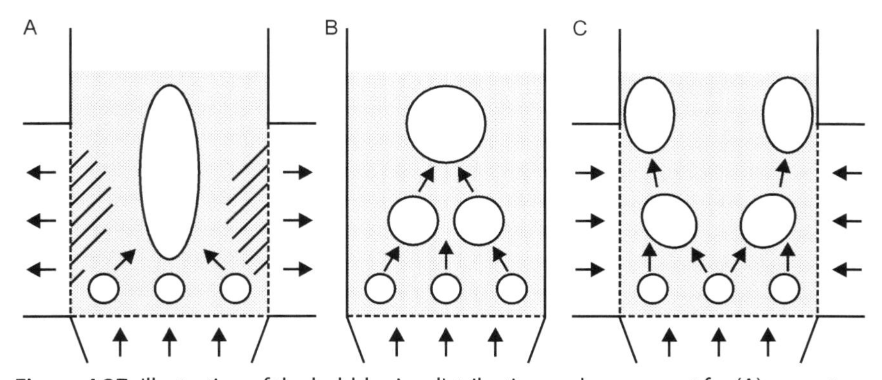

Figure 4.27 Illustration of the bubble size distribution and movement for (A) gas extraction, (B) the reference, and (C) gas addition. *Reprinted from De Jong et al. (2011) with permission from Elsevier.*

The reference experiment has been performed with a total gas feed corresponding to $u/u_{mf}=2.6$. Similar to the experiments discussed in Section 2.3.1, two sets of experiments and simulations have been carried out, one keeping the background gas velocity constant and the other one keeping the total gas outlet velocity constant (see Table 4.9). Experimental and numerical details can be found in the work due to De Jong et al. (2012d).

2.3.2.2 Solids Circulation Patterns

Fig. 4.28 displays the solids flux obtained from experiments, DPM and TFM simulations during gas extraction, without permeation and with gas

Table 4.9 Overview of the Experimental Series Researched in This Chapter

Measurement Series	Background Gas Velocity (m/s)	Membrane Gas Velocity (m/s)	Number of Images for DIA	Number of Images for PIV	Simulation Time (s) DPM	Simulation Time (s) TFM
Reference	0.650	0.0000	3322	6644	35	50
100% − 20%	0.650	−0.0650	3000	6000	31	50
100% − 40%	0.650	−0.1300	3000	6000	26	50
100% + 20%	0.650	+0.0650	3000	6000	34	50
100% + 40%	0.650	+0.1300	2167	4334	20	50
120% − 20%	0.780	−0.0650	3172	6344	30	50
140% − 40%	0.910	−0.1300	3172	6344	28	50
80% + 20%	0.520	+0.0650	3259	6518	35	50
60% + 40%	0.390	+0.1300	2791	5582	31	50

addition, respectively. In this case, the results for experiments and simulations constant flow rate at the reactor outlet (and thus varying inflow depending on the permeation rate) are displayed.

Fig. 4.28 portrays the solids circulation patterns of experimental PIV results, DPM and TFM for gas extraction (top), no permeation (middle) and gas addition (bottom). During gas extraction, the solids motion near the membranes is slightly reduced. This is, of course, only the case in the bottom 4 cm of the bed, where the membranes are located. During gas addition (bottom row), the solids circulation pattern is inverted, with solids moving downward via the center of the bed and upward via both sides. This trend is not only obtained experimentally but also captured in both DPM and TFM simulations. The fact that the DPM simulation results lack complete symmetry can be attributed to the relatively short simulation time. Qualitatively, the solids circulation patterns are in agreement with the same trend as seen in the previous Section 2.3.1.1 for a 30-cm wide bed; for gas addition, the solids circulation completely inverts, resulting in a solids up flow at the walls and a solids downflow through the center of the bed.

Another noticeable difference among experiment, DPM simulation, and TFM simulation is the bed expansion (see Fig. 4.28). Although the experiment contains approximately 5% more particles (4.3 cm fixed bed height vs. 4.1 cm for the simulations), this cannot explain the difference we observe in these graphs. In all three cases shown, the experimental bed expansion is

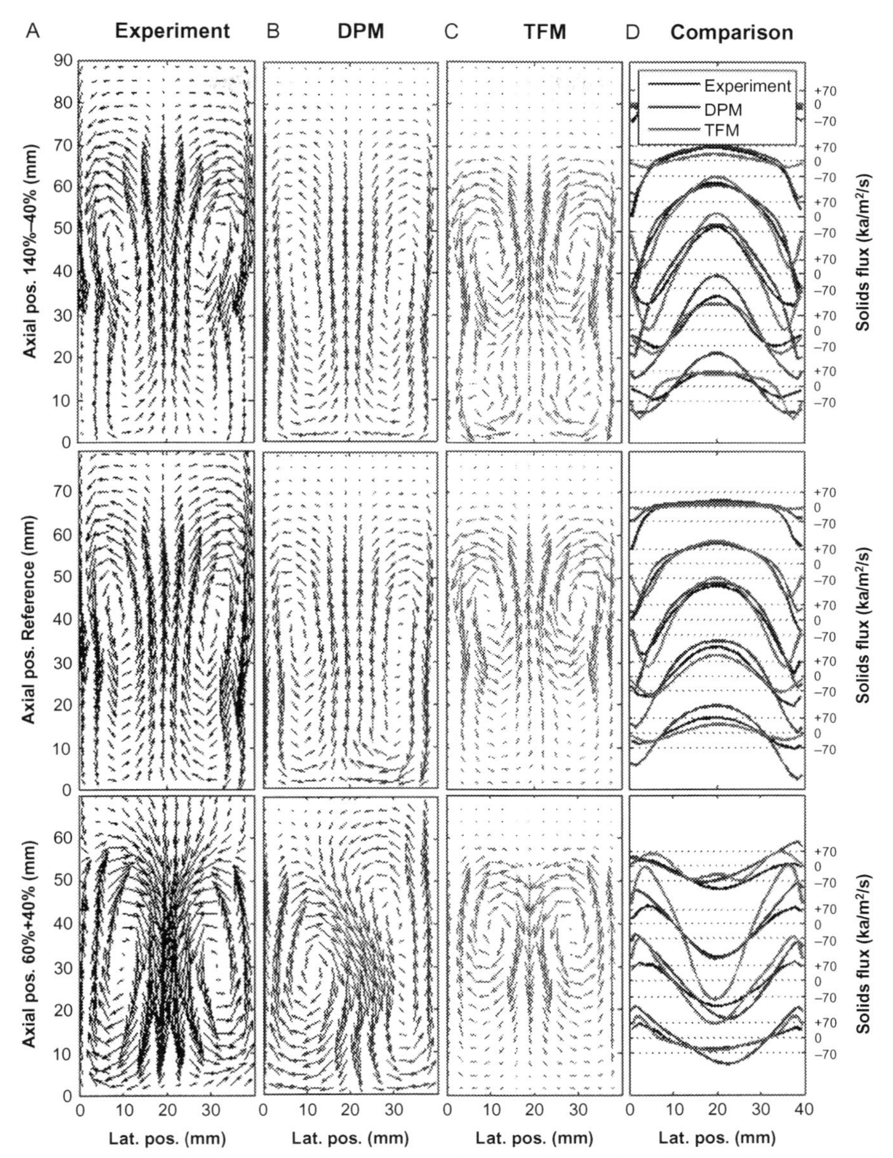

Figure 4.28 Comparison of (A) experimental, (B) DPM, and (C) TFM solids fluxes (and their lateral profiles) for constant outflow in the 4-cm wide bed with 40% gas extraction (top), no permeation (middle), and 40% addition (bottom). *Reprinted from De Jong et al. (2012d) with permission from Elsevier.*

clearly larger than the bed expansion for the numerical cases. The TFM has the smallest bed expansion.

However, even in the graphs displaying the lateral profiles of the axial solids fluxes, the direct comparison between the DPM and the TFM models is difficult. For that reason, a quantitative method, in which all solids fluxes are compared at once, is used. For both models, an average deviation from the experimental data is calculated based on the absolute differences between each solids flux vector, that is, the following equation is employed:

$$\Delta\phi = 100 \cdot \sum_{N_i} \frac{\sqrt{\left(v_{x,\text{sim}} - v_{x,\text{exp}}\right)^2 + \left(v_{y,\text{sim}} - v_{y,\text{exp}}\right)^2}}{\sqrt{\left(v_{x,\text{exp,max}}^2 + v_{y,\text{exp,max}}^2\right)}} \tag{4.37}$$

The resulting deviations (a combination of over prediction and under prediction) for all series are summarized in Table 4.10. Although the deviations per series vary to some extent, the average deviations for the DPM model and the TFM model do not differ very much: their average compared to the experiments is 18.0% and 19.3%, respectively. This comes not as a surprise in view of the results from Fig. 4.28D, in which we can see that the TFM differs from experiments in particular in the center of the bed, and the DPM differs more near the walls. Table 4.10 seems to show a general trend that the error increases when the gas flow decreases; the average error for the 100% − 40%, 100% − 20%, 80% + 20%, and 60% + 40% cases is 27.1%, while for the 100% + 20%, 100% + 40%, 140% − 40%, and 120% − 20% cases, it is only 12.3%. It is known that the description of the friction in the TFM is nontrivial. This is believed to be the cause for the larger error at decreased gas flow, because friction plays a more dominant role in dense systems. The deviation between the DPM and the experiments

Table 4.10 Deviation in Solids Flux $\Delta\Phi_s$, Between the Simulation (Either DPM or TFM) and the Experimental Data

U_0 constant	100% − 40%	100% − 20%	Reference	100% + 20%	100% + 40%
DPM	32.6	20	10.7	6.9	16.3
TFM	29.8	15.8	10.2	15.3	14.2
U_0 varying	140% − 40%	120% − 20%	Reference	80% + 20%	60% + 40%
DPM	11.5	12.5	10.7	29.7	22.1
TFM	11.5	10.4	10.2	40	26.9

in the vicinity of the walls is most likely due to the extremely rough surface (sintered steel) of the membranes in the experiment.

2.3.2.3 Bubble Properties

Fig. 4.29 shows the bubble properties of experiments, DPM, and TFM simulations for different permeation ratios. The experimental results as well as the DPM and TFM simulation results again follow the same qualitative trend as seen for the 30-cm wide bed; because of the inverted solids circulation pattern during gas addition, bubbles split between the left and the right part of the bed, which results in a smaller average bubble diameter. On the other hand, gas extraction leads to slightly larger bubbles in the center of the bed due to semistagnant zones of particles in the vicinity of the membranes. Again, this can also be seen in Fig. 4.29 not only for the experiments but also for the DPM and TFM simulations.

However, from the graphs indicating the bubble size, it can be concluded that quantitatively neither the DPM nor the TFM gives a perfect match with the experiments; only the reference case shows a reasonable agreement between all three cases (except for the region near the freeboard). For gas addition, the DPM systematically overpredicts the bubble size, while the TFM has the tendency to underpredict the bubble size. From the graphs showing the number of bubbles, it can be seen that the DPM predicts more bubbles in the bottom zone of the bed. However, for the central region, the number of bubbles compares well to the experiments.

For gas extraction, on the other hand, both models predict smaller bubbles near the walls, and slightly larger bubbles in the center of the bed compared to the experiments. The bubble size can be correlated to the number of bubbles; the DPM, but in particular the TFM, predicts many more bubbles in the bottom central region. Since a larger density of bubbles increases the likelihood of coalescence, the peak in bubble size can be explained. However, with increasing axial position, this peak vanishes, and so does the peak of the number of bubbles for both models. In general, the experiments show a wider distribution of both number of bubbles and bubble size compared to the DPM and TFM simulations results, because the models overpredict the influence of the stagnant zones of solids near the membranes.

2.3.3 Extension to Membrane-Assisted MicroFluidized Beds

In the previous Sections 2.3.1–2.3.2, it has been demonstrated that in FBMRs, using membrane walls with a high permeation flux, densified zones

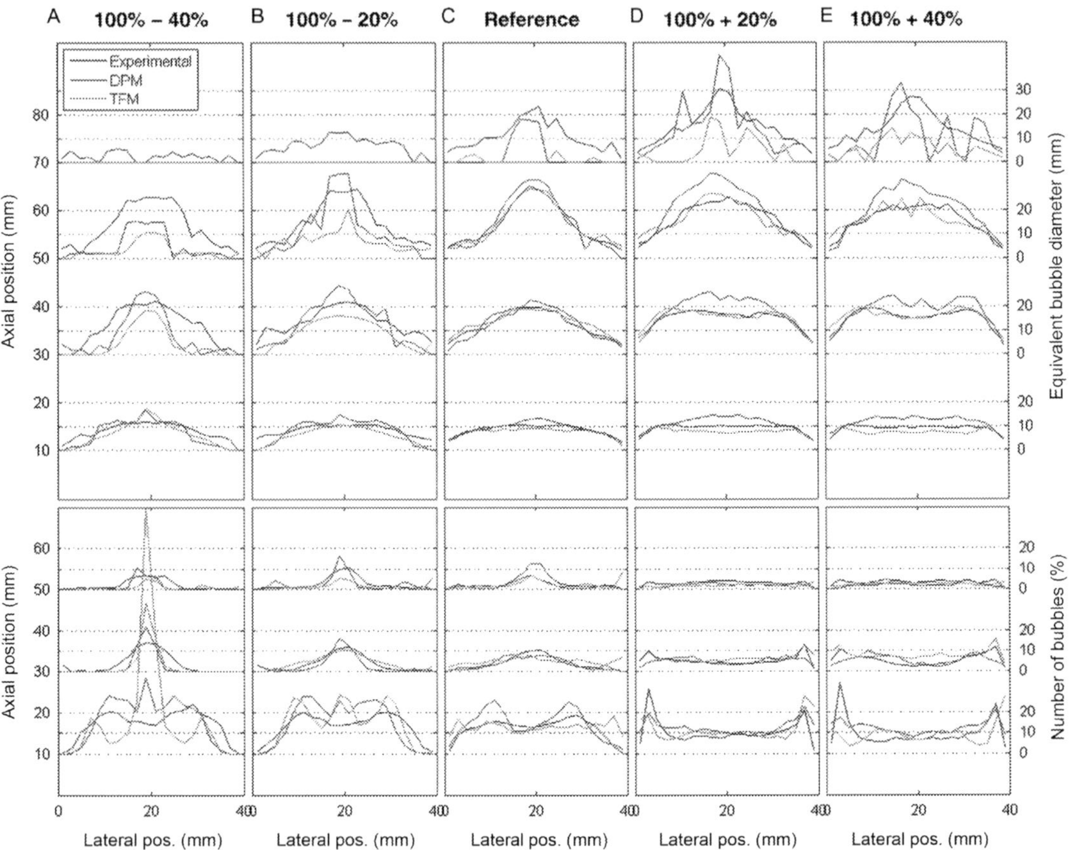

Figure 4.29 Lateral profiles of the average bubble size (top) and the average number of bubbles per frame (bottom) at several axial positions in the fluidized bed for different permeation ratios in the experimental and numerical (DPM and TFM) series. (A) 40% gas extraction; (B) 20% gas extraction; (C) reference case; (D) 20% gas addition; (E) 40% gas addition. Background velocity is constant 100%. *Reprinted from De Jong et al. (2012d) with permission from Elsevier.*

close to the membrane walls are formed, which would most probably induce strong bed-to-membrane mass transfer limitations.

For this section, we shift our focus to microstructured FBMRs. It will be investigated whether it is indeed possible to overcome the external mass transfer limitations by using smaller compartments even in case a large amount of gas is extracted through the membranes. Is it possible to operate smaller sized reactors in the turbulent regime in order to circumvent the detrimental formation of densified zones. And finally, is it possible to operate these reactors with state-of-the-art membranes without inducing mass transfer limitations (due to densified zones)?

To answer these questions and give guidelines for the design and operation of FBMRs, the results are reported of a detailed experimental investigation on the hydrodynamics of small membrane-assisted fluidized beds operated in both the bubbling and the turbulent fluidization regimes where the gas is added or extracted through two opposed vertical membranes confining the fluidized suspension.

2.3.3.1 Experimental Setup and Method

An experimental setup consisting of a pseudo-2D membrane-assisted microstructured fluidized bed ($8 \times 1 \times 60$ cm) has been constructed, which is in further aspects comparable to the setup described in the previous Section 2.1.1. For the used glass beads, the transition between bubbling regime and turbulent fluidization regime occurs at $u_c/u_{mf}=7.6$. The particle size may be considered as large for the small fluidized bed, this particle size was selected on the basis of CFD simulation results that have indicated that because of the effect of the particle size, bigger particles would be preferred in membrane-assisted fluidized beds (providing that the particle size is small enough to avoid internal mass transfer limitations).

All experimental conditions used have been summarized in Table 4.11 in which both the membrane length at both sides and the static bed height were kept at 80 mm.

2.3.3.2 Avoiding Densified Zones in Bubbling Microstructured FBMRs

In these first experiments, we verified whether also small membrane-assisted fluidized beds suffer from the formation of densified zones. In Fig. 4.30, the results of experiments with gas addition and gas extraction on the solids holdup and solids circulation patterns are reported for a case of constant inflow. In the top row images (Fig. 4.30A), it can be noted that for the reference case, the

Table 4.11 Operating Conditions Used in the Experiments: Both Membrane Height and Static Bed Height Were Kept as 80 mm at Both Sides

	Background		Gas Velocity at the Membrane		Total Outlet Superficial Gas Velocity	
Measurement[a] Label	**%**	**(m/s)**	**%**	**(m/s)**	**%**	**(m/s)**
Bubbling						
Reference[b]	100	0.75	0	0	100	0.75
80% + 20%	80	0.6	10	0.075	100	0.75
60% + 40%	60	0.45	20	0.15	100	0.75
120% − 20%	120	0.9	10	−0.075	100	0.75
140% − 40%	140	1.05	20	−0.150	100	0.75
100% + 20%[b]	100	0.75	10	0.075	120	0.9
100% + 40%[b]	100	0.75	20	0.15	140	1.05
100% − 20%[b]	100	0.75	10	−0.075	80	0.6
100% − 40%[b]	100	0.75	20	−0.150	60	0.45
Reference[b]	100	2	0	0	100	2
80% + 20%	80	1.6	10	0.2	100	2
Turbulent						
60% + 40%	60	1.2	20	0.4	100	2
120% − 20%	120	2.4	10	−0.20	100	2
140% − 40%[c]	140	2.8	20	−0.40	100	2
100% + 20%[b]	100	2	10	0.2	120	2.4
100% + 40%[b]	100	2	20	0.4	140	2.8
100% − 20%[b]	100	2	10	−0.20	80	1.6
100% − 40%[b]	100	2	20	−0.40	60	1.2

[a]80% + 20% indicates that 80% of the total gas fed is fed via the bottom distributor while 20% is added via the left and right walls.
[b]Results of these experiments are reported in the paper.
[c]This case is also repeated for 160-mm membrane length.
The reference experiment indicates no gas addition or extraction from the left and right walls.

solids holdup is higher near the walls, as expected by the fact that bubbles are growing and moving toward the center of the bed, while the solids dragged by bubbles in the center of the bed are descending near the walls of the bed. For the gas extraction cases, a higher solid holdup close to the membrane walls

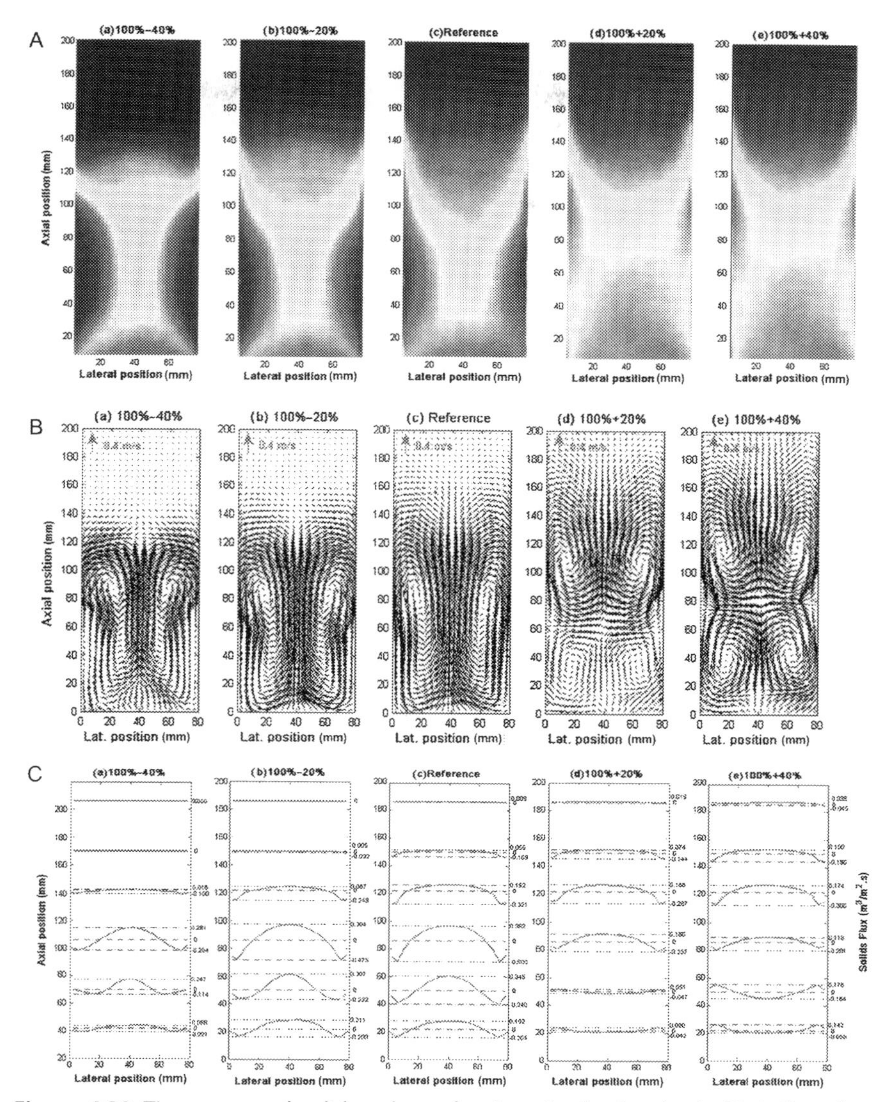

Figure 4.30 Time-averaged solids volume fraction distribution (scale [0–0.6]) with a constant inflow (A). Time-averaged solids flux circulation pattern with a constant inflow (B) for different gas addition and extraction ratios, and lateral profiles of the axial solids flux for different heights in the fluidized bed (C). Bubbling fluidization regime, superficial gas velocity via the bottom distributor is 0.75 m/s. *Reprinted from Dang et al. (2014) with permission from Elsevier.*

compared to the reference case is observed. These densified zones are more enlarged if more gas is extracted (see 100% − 40% case). It is interesting to see that the extraction of gas has two more effects on the hydrodynamics of the bed: first, the averaged expansion of the bed height is decreased, and second, the number of bubbles is decreased as well, due to the formation of elongated bubbles (as explained in the following sections, see also Fig. 4.31).

In contrast to gas extraction, the addition of gas from the left and right walls (100% + 20% and 100% + 40% cases) results in increased gas holdup close to the walls and a higher solids holdup toward the center of the bed. This ultimately results in the inversion of the solids circulation pattern inside the fluidized bed. It is interesting to note that this effect is pronounced

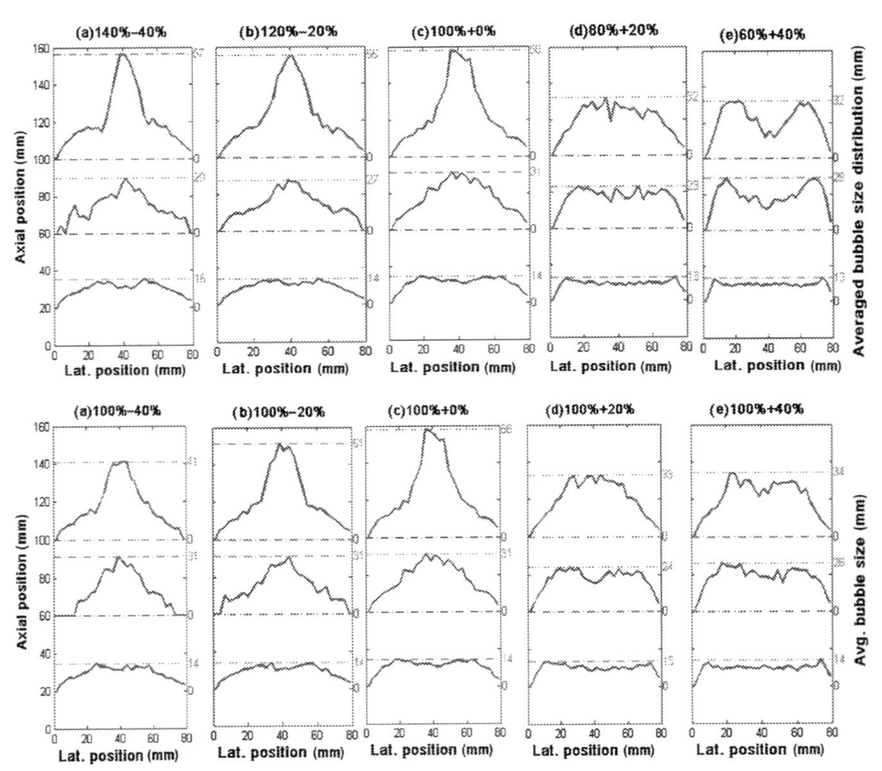

Figure 4.31 Number average of bubble size distribution in lateral direction at different bed heights of constant inflow experiments (top) and constant outflow (below). The bubble size has been detected at three different axial positions (20, 60, and 100 mm) at which the bubbles cross-in and noted by the black dash-lines. The red (gray in print version) dash-lines present the maximum averaged equivalent bubble size. *Reprinted from Dang et al. (2014) with permission from Elsevier.*

in the zones of the bed where the membranes are used, while at higher axial positions in the bed (where membranes are not used) the distribution (and circulation pattern) of solids becomes similar to the reference case with high solids concentration near the walls.

A major difference that has been observed is that the densified zones still move in contrast to what has been found for larger beds. Therefore, the microstructured FBMRs surely enhance the heat and mass transfer characteristics.

For the gas addition experiments (100% + 20% and 100% + 40%), two distinct solids circulation patterns can be distinguished. Similar to the cases of gas extraction, within the membrane region, solids are pushed toward the center by the gas added through the membranes, while solids also change their flow pattern with upward flows near the membrane walls and downward in the center region. Above this position (80 mm in this experiment), there are no membranes operated and a solids flow profile similar to the reference case can be observed. The experimental results also show how the effect of gas addition through membranes is more pronounced when more gas is added to the system through the membranes (higher gas addition velocities).

The answer to the first question (whether densified zones can be avoided in microstructured fluidized bed with gas extraction) is negative. Although the smaller reactors operated in bubbling fluidization regime do not present dead-zones as observed in bigger reactors, still some densified zones can develop (strongly dependent on the gas extraction velocity) that can induce bed-to-wall mass transfer limitations with an adverse effect on the performance of the reactor.

2.3.3.3 Avoiding Densified Zones in Turbulent Microstructured FBMRs

Turbulent fluidization is often preferred due to increased bubble breakup with improved mass transfer characteristics. The interesting question is whether turbulent fluidization can avoid the formation of densified zones that affect the membrane reactors when operated in the bubbling regime. Fig. 4.32 shows the results of the hydrodynamic study in the turbulent fluidization regime. To operate in the turbulent fluidization regime, both the fluidization velocity and the permeation gas velocity are increased while the bed mass and the membrane area were kept the same as before in the bubbling regime (80 mm height, this is the limitation of the current experimental setup). The time-averaged solids distribution in Fig. 4.32A shows that the averaged-bed height in the turbulent regime is much more expanded than for the bubbling regime, while obviously the bed itself is more dilute. The extraction of gas (100% − 20% and 100% − 40%) results also in this case in a

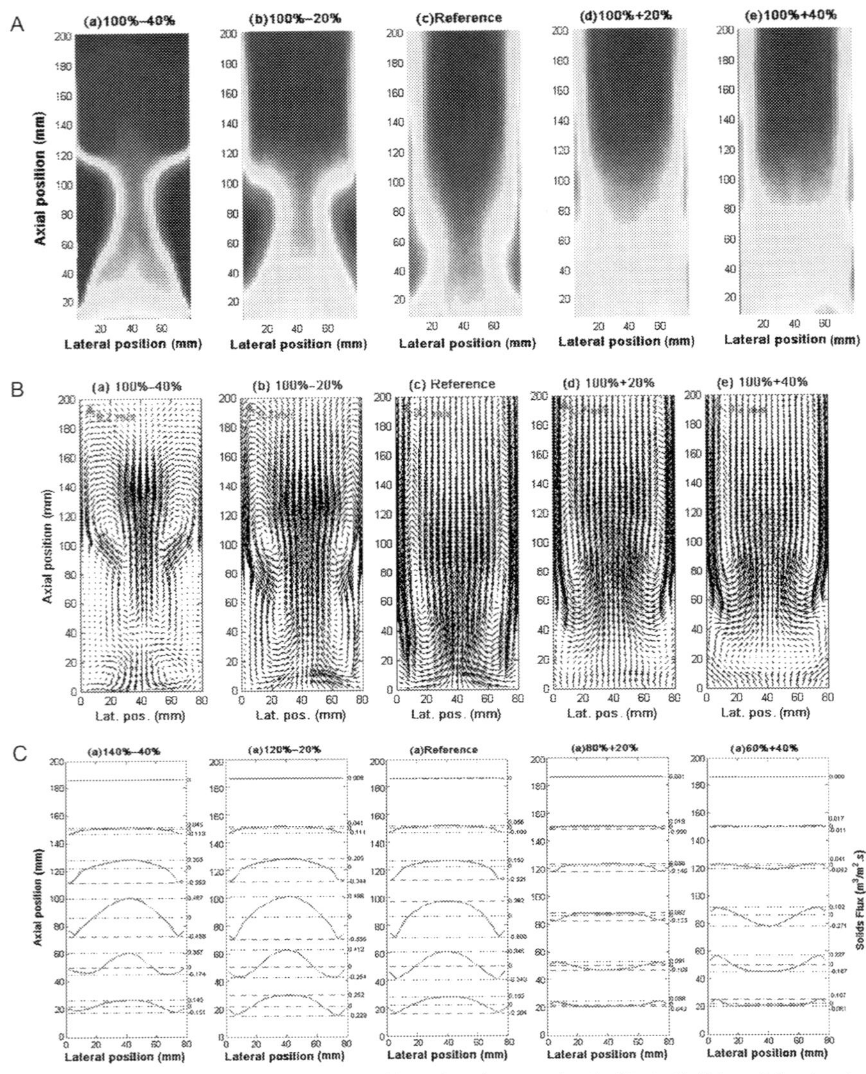

Figure 4.32 Time-averaged solids holdup distribution (scale [0–0.6]) (A), solids circulation pattern (B), and lateral profiles of the axial solids flux for different heights (C) in the fluidized bed with a constant inflow, for different gas extraction and addition ratios. Turbulent fluidization regime, superficial gas velocity used from the bottom distributor is 2.0 m/s. *Reprinted from Dang et al. (2014) with permission from Elsevier.*

very dense phase located at the membrane walls (due to the high extraction velocities) and a narrow dilute channel in the center.

The time-averaged solids flux in the turbulent regime is illustrated in Fig. 4.32B. The reference case presents the broad upflow in the center of the bed and narrow downflow at the left and right walls (reference case), which could improve solids mixing and reduce gas back-mixing. The lateral profiles of the axial solids fluxes at different heights show a flattened velocity profile compared to the parabolic-shaped velocity profile observed in the bubbling fluidization regime.

Extraction of gas creates much extended densified zones on the left and right walls toward the center bed. These densified zones are already present when the gas extraction ratio is low (100% – 20%), and their extent increases as the extraction gas velocity increases (100% – 40%). This phenomenon creates a "channeling fluidization" in the center where reactants in the gas phase move quite fast with a short residence time and low contact time with solids which would ultimately result in more bypassing of reagents and decreased performance of the reactor.

In contrast to the findings in the bubbling regime, the addition of gas in the turbulent regime shows that the solids circulation pattern is almost the same (100% + 20%) as found for the reference case with only slightly reduced solids mixing just above the bottom distributor when higher gas addition velocities are used (100% + 40%). The background gas velocity has a significant influence on the solids circulation pattern; increasing the background gas velocity to reach the turbulent regime can avoid the inversion of the solids circulation pattern and the bypassing of gas via the membrane walls as was observed in the bubbling fluidization regime.

For the case of gas addition, the turbulent fluidization is surely to be preferred because of decreased gas bypassing near the membrane walls, while for the case of gas extraction still densified zones are formed. However, it seems that the formation of these densified zones is strongly related to the very high gas extraction velocity at the membranes. In the turbulent fluidization experiments, the gas extraction velocity is much higher than in the bubbling fluidization experiments because the relative amount of gas added/withdrawn was kept the same but at a much higher superficial gas velocity.

2.3.3.4 Avoiding Densified Zones with High-Flux Membranes

In this case, we will focus our attention to the turbulent fluidization regime with gas extraction. For the bubbling regime and for the gas addition cases, the main conclusion will also hold (for compactness these figures are not shown). Fig. 4.33 shows the results of the hydrodynamic study with 40% gas extraction with the original membrane length (80 mm) and with double membrane length (160 mm). The static bed height is still kept at 80 mm. while the expanded bed height is more than 160 mm for these experiments.

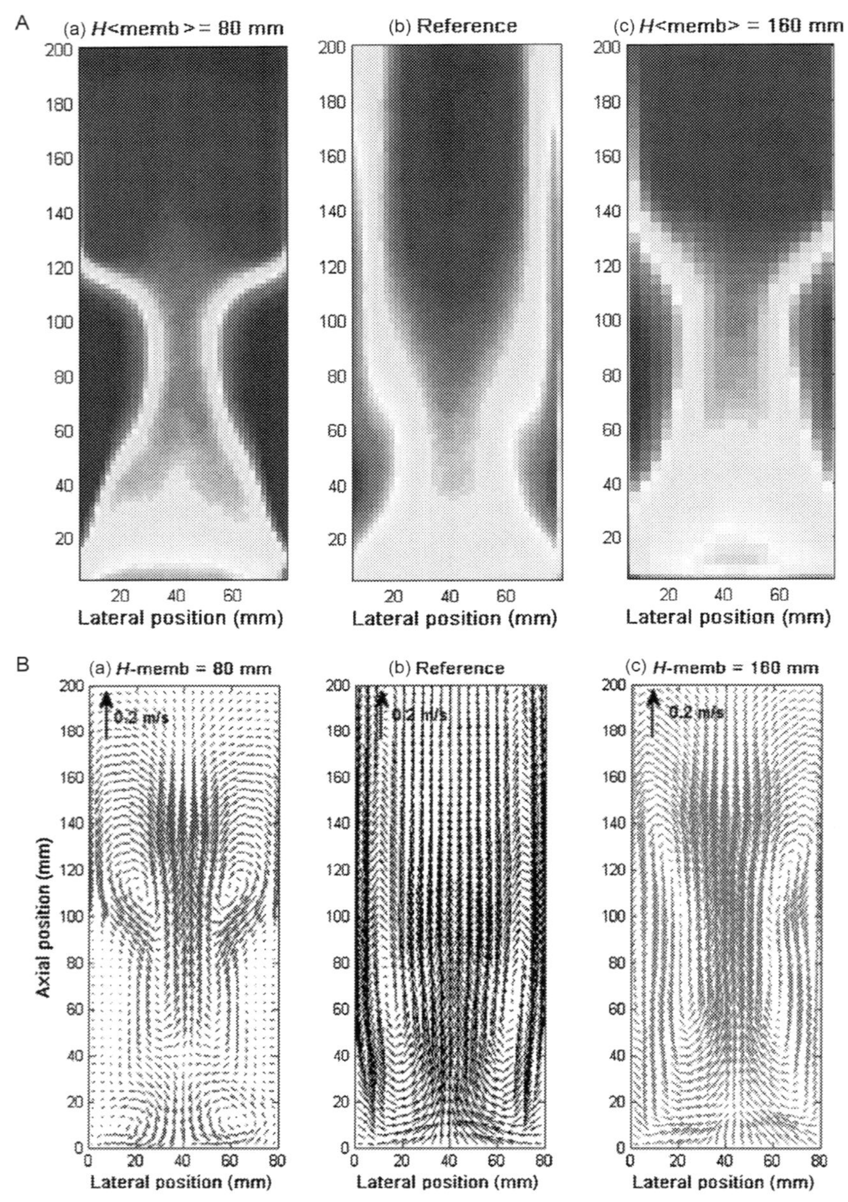

Figure 4.33 Influence of the membrane areas on the solid holdup (A) and time-averaged solids circulation pattern (B) in turbulent fluidization regime with 40% of gas extraction, (a) 80-mm membrane length, (b) no gas extraction, and (c) 160-mm membrane length. *Reprinted from Dang et al. (2014) with permission from Elsevier.*

Doubling the membrane length results in halving the gas extraction velocity through the membrane, which significantly reduces the development of the semistagnant zones and improves the solids mixing.

Reducing the ratio between the membrane flux velocity and the background velocity decreases the extent of semistagnant zones. This means that by using state-of-the-art membranes, it is possible to operate the small reactor in the turbulent regime with decreased densified zones, and thus reduced mass transfer limitations. This underlines that it is more important to produce membrane with the same flux as current state-of-the-art membranes but with increased resistance (and thus lifetime) rather than further increasing the membrane flux (compromising the membrane selectivity or stability) that may induce mass transfer limitations even in fluidized beds in the turbulent fluidization regime.

It is interesting to have a more precise quantification of the extent of densified zones at different operating conditions, where the 3D solids holdup has been used as "densification" criterion. The part of the bed with a solid holdup above a threshold value of 0.51 has been classified as densified region, corresponding to 85% of the maximum solids holdup in packed beds.

The extent of the densified zones (time-averaged) is subsequently quantified as in following equation:

$$E = \frac{\sum A_{\text{densified zones}}}{A_{\text{total}}} \tag{4.38}$$

where $A_{\text{densified zones}}$ is the area with a (time-averaged) solid holdup above the threshold value. It should be noted that, in case of turbulent fluidization, the densified zone is found above the fixed membrane height (above 80 mm membrane height) which will also be taken into account.

Fig. 4.34 shows the extent of the stagnant zones as a function of the permeation velocity at different fluidization gas velocities in both bubbling fluidization (A) and turbulent fluidization regimes (B). It is pointed out that with the selected threshold value even in the reference case without gas extraction, the presence of small semistagnant zones is found, e.g., for the reference case in the bubbling regime, the extent of the densified zones is approximately 8% of the bed volume. Increasing the permeation gas velocity results in much extended densified zones (that even reach values higher than 30% as shown in Fig. 4.34). However at the same permeation gas velocity, increasing the fluidization gas velocity will decrease the extent of densified zones; for example, the densified zones decrease from 32% to 23% for 40% of

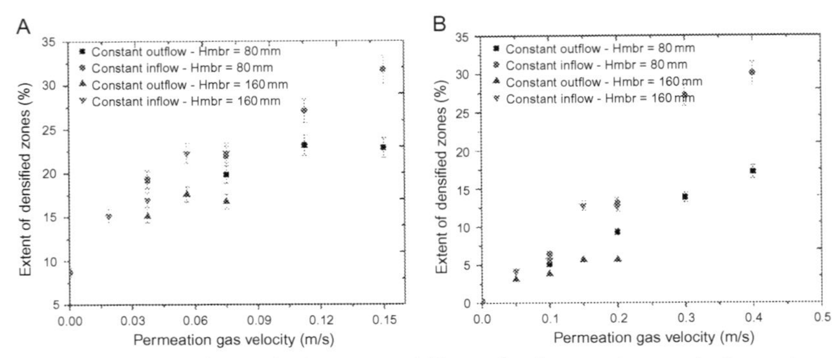

Figure 4.34 Extent of densified zones at different background gas velocity and membrane areas against permeation gas velocity for bubbling fluidization (A) and turbulent fluidization (B). *Reprinted from Dang et al. (2014) with permission from Elsevier.*

gas permeation when increasing the fluidization velocity to keep the outflow constant.

A similar effect (increasing the fluidization gas velocity will decrease the extent of densified zones) has been found for the turbulent fluidization regime (Fig. 4.34B). However, the slopes of the curves representing the extent of the densified zones as a function of the permeation gas velocity are lower than for the bubbling regime. This means that it is preferred to operate in the turbulent fluidization regime, because with the same permeation gas velocity (i.e., membrane permeance), the extent of densified zones is much decreased.

2.3.3.5 Simulations of Bubbling Membrane-Assisted Microfluidized Beds

A study on the performance of membrane-assisted microfluidized beds with different membrane areas has been performed by Tan et al. (2014). Their work simulates one compartment of 3 mm in width and 1.4 mm in depth with two opposing vertical walls acting as membranes for gas addition or extraction. The height of the membranes and the simulated domain are 5 and 18 mm, respectively. Considering the small size of the simulation domain, monodisperse Geldart A type particles with a diameter of 75 μm were selected and air was used as a fluidizing agent. The outlet superficial gas velocity was kept constant at 0.03 or 0.05 m/s by adjusting the background gas velocity according to the selected gas addition or extraction ratio (i.e., 40% and 20%) applied via the membranes. According to the study on the critical velocities of the onset of turbulent fluidization regime in microfluidized beds (Wang et al., 2011), these fluidized beds were all simulated under the bubbling fluidization regime.

All the simulation series with different membrane areas and average gas permeation velocities are listed in Table 4.12. All the simulations were performed for 7 s and a time-averaged analysis has been carried out from 2 to 7 s.

Table 4.12 Summary of the Simulation Series with "*a*" Denoting the Left/Right Membrane Wall Configuration of 14 mm^2 and "*b*" denoting the Front/Back Membrane Wall Configuration with Area 30 mm^2

		Background Gas Flow/Velocity		Total Membrane Flow/Velocity		
Simulations		**(%)**	**(m/s)**	**(%)**	**(m/s)**	**Membrane Area (Position) (mm^2)**
1-0 Reference		100	0.03	0	0	–
1-1 80% + 20%	*a*	80	0.024	20	0.0018	14 (left/right wall)
1-2 60% + 40%	*a*	60	0.018	40	0.0036	
1-3 120% − 20%	*a*	120	0.036	−20	−0.0018	
1-4 140%-40%	*a*	140	0.042	−40	−0.0036	
1-5 80% + 20%	*b*	80	0.024	20	0.00084	30 (front/back wall)
1-6 60% + 40%	*b*	60	0.018	40	0.00168	
1-7 120% − 20%	*b*	120	0.036	−20	−0.00084	
1-8 140% − 40%	*b*	140	0.042	−40	−0.00168	
2-0 Reference		100	0.05	0	0	–
2-1 80% + 20%	*a*	80	0.04	20	0.003	14 (left/right wall)
2-2 60% + 40%	*a*	60	0.03	40	0.006	
2-3 120% − 20%	*a*	120	0.06	−20	−0.003	
2-4 140% − 40%	a	140	0.07	−40	−0.006	
2-5 80% + 20%	*b*	80	0.04	20	0.0014	30 (front/back wall)
2-6 60% + 40%	*b*	60	0.03	40	0.0028	
2-7 120% − 20%	*b*	120	0.06	−20	−0.0014	
2-8 140% − 40%	*b*	140	0.07	−40	−0.0028	

2.3.3.6 Simulation Results

In this section, all simulation results have been analyzed to understand the potential impact of the flow phenomena/structures on the future reactor design of microstructured membrane-assisted FBRs.

In Fig. 4.35, the solids circulation patterns can be discerned for membrane-assisted microfluidized beds, which were operated in the bubbling fluidization regime with the outlet superficial gas velocity of 0.05 m/s and gas extraction or gas addition via membrane assumed built in the left and right walls (case (a) with a membrane area of 14 mm^2): 140% − 40%, 120% − 20%, the reference, 80% + 20% and 60% + 40% cases. This figure shows a good qualitatively correspondence to the results mentioned earlier in this chapter (see also De Jong et al., 2011, 2012d) for experimental and simulation studies using larger systems.

The results from the cases where gas extraction or gas addition was imposed via membranes built in the front and back walls (case (b) with a membrane area of 30 mm^2) with the same relative flow rates (see Table 4.12) show very similar, but much less pronounced effects of gas permeation as what has been observed in Fig. 4.35. In Fig. 4.36, a solids circulation pattern very similar to the reference case can be discerned for the cases with gas extraction while a more complicated solids circulation pattern for the case with gas addition is indicated. Even though the well-known solids circulation pattern seen in the reference case can be

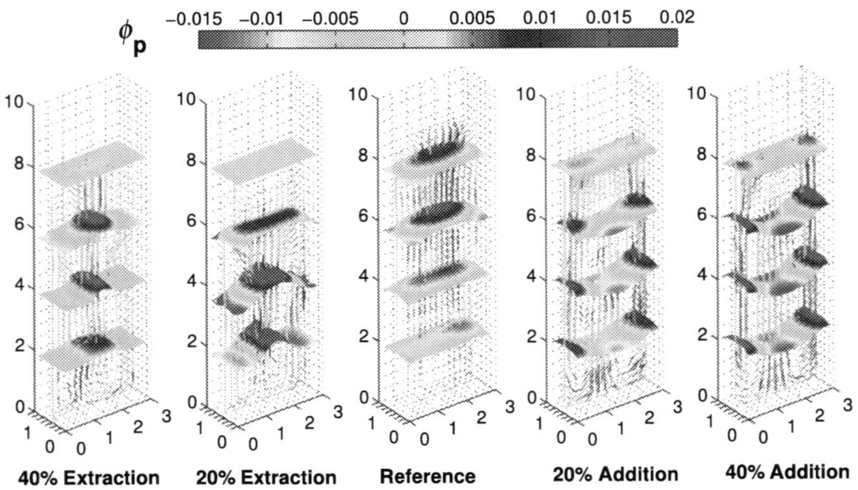

Figure 4.35 Solid circulation patterns in MAFBs with an outlet superficial gas velocity of 0.05 m/s and gas permeation via the left and right membrane walls (length in mm). *Reprinted from Tan et al. (2014) with permission from Elsevier.*

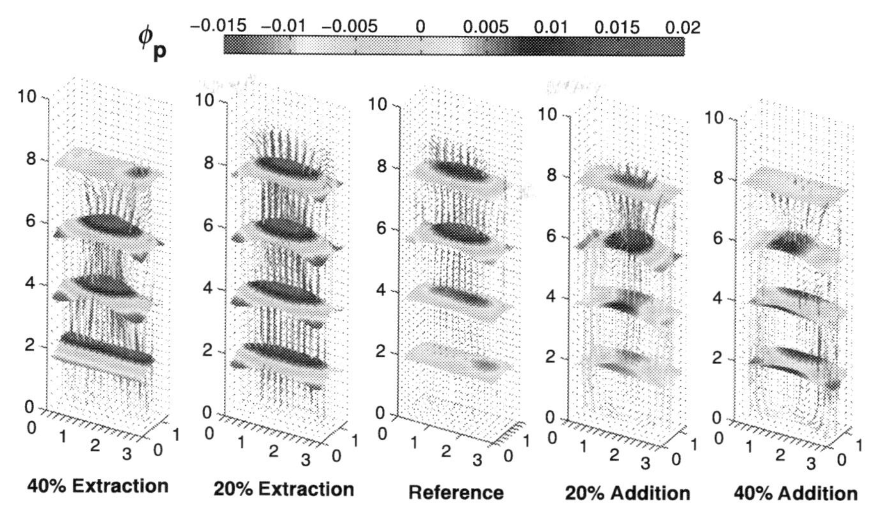

Figure 4.36 Solid circulation patterns in MAFBs with an outlet gas velocity of 0.05 m/s and gas permeation via the front and back membrane walls (length in mm). *Reprinted from Tan et al. (2014) with permission from Elsevier.*

distinguished in the vector plots of 80% + 20% and 60% + 40% cases, a careful observation reveals the emerging of a reversed solids circulation pattern close to the impermeable walls with increasing gas extraction ratio which has also experimentally been found in the previous Section 2.3.3.2 (Dang et al., 2014).

Figs. 4.35 and 4.36 clearly show that the effect of the gas permeation on the solids behavior increases with increasing gas permeation ratio and decreasing membrane area—both meaning the increase of gas permeation velocities. Moreover, it is also interesting to notice that there is hardly any difference between the solids circulation patterns of 120% − 20% (B) case and the reference case in Fig. 4.36 and an enhanced solids circulation in 120% − 20% (B) case probably indicates a better solids mixing.

Figs. 4.37 and 4.38 show the time-averaged solids holdup distribution with gas extracted or added via the membranes at different positions. Fig. 4.37 clearly shows the effect of gas extraction causing the formation of very densified zones close to membrane walls. Here, the solids holdup can reach values up to the value of a random packed bed with monodisperse particles of the same type. However, analogous to the experimental results presented in Section 2.3.3.2, the term "stagnant" is not accurate since some movement of the particles is still detected, and hence "densified zones" is a more appropriate term. Even though particles close to membrane walls are

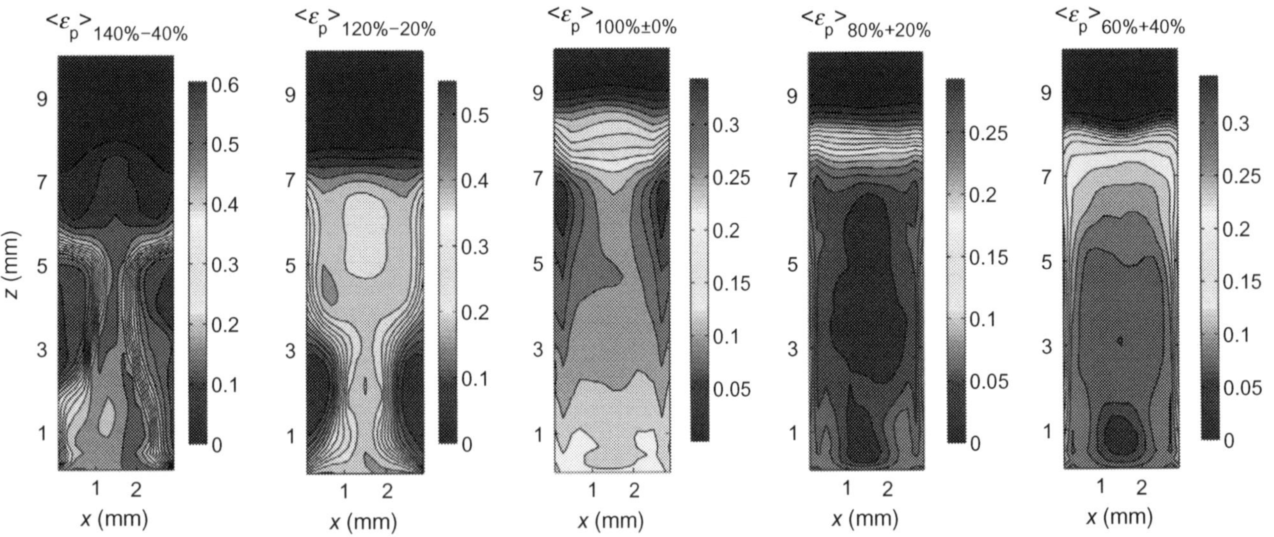

Figure 4.37 Time-averaged solids fraction in MAFBs with outlet gas velocity of 0.05 m/s and gas permeation via the left and right membrane walls (length in mm). *Reprinted from Tan et al. (2014) with permission from Elsevier.*

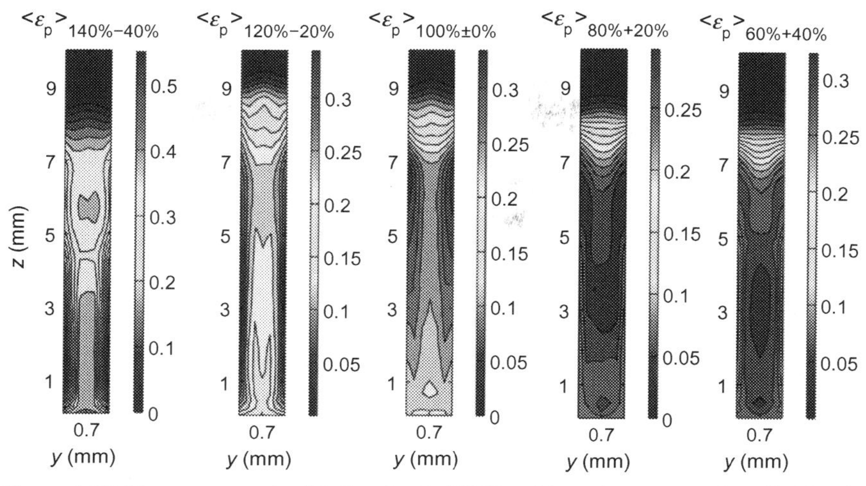

Figure 4.38 Time-averaged solids fraction in MAFBs with outlet gas velocity of 0.05 m/s and gas permeation via the front and back membrane walls (length in mm). *Reprinted from Tan et al. (2014) with permission from Elsevier.*

not stagnant, for the case with a high gas extraction ratio particles in the pronounced densified zones do have very low velocities as indicated in Fig. 4.35 for the 140% − 40% (A) case.

For the cases with gas permeation imposed via membranes built in the front and back walls, similar but much less pronounced phenomena can also be observed in Fig. 4.38 for the solids holdup distribution, i.e., relative densified zones for the cases with gas extraction and relatively lower solids holdup close to the membrane walls for the cases with gas addition, especially for the cases with a high gas permeation ratio of 40%. Note that the very densified zones close to membrane walls, where the values of solids holdup are close to the value of a random packed bed with monodisperse particles of the same type, can be seen in the 140% − 40% (A) case but not in the 120% − 20% (B) case in Fig. 4.38. Close comparison of the solids holdup distribution for the 140% − 40% (A) and 120% − 20% (B) cases tells that it is possible to avoid very high solids holdups close to membrane walls by increasing the membrane area.

The time-averaged solids holdup distribution in a fluidized bed operated in the bubbling fluidization regime results from the bubble dynamics, resulting in a relatively high solids holdup close to the walls due to bubble coalescence, causing migration of the bubbles to the bed center while rising through the bed. When gas extraction is imposed via membranes built in the vertically walls confining the bed, the solids holdup close to the membranes is even further increased causing enhanced bubble coalescence in the bed center, also observed in the experiments with larger beds and particle size discussed above.

In contrast to the cases of gas extraction, gas addition shows an opposite effect on the solids holdup distribution through the bed, resulting in relatively uniformly increased solids holdup throughout the bed center but decreased solids holdup close to the membrane walls. When increasing the membrane area, and thereby decreasing the permeation velocity, the effect of gas addition was not reduced (i.e., from cases indicated with (A) to cases indicated with (B)).

2.4. Hydrodynamics in FBMRs with Submerged Membranes

Extending the study of the previous sections which covered wall-mounted membranes, this section presents an investigation on the effect of a change in gas flow rate inside an FBMR containing an array of horizontal membrane tubes submerged in the gas–solid suspension.

2.4.1 Experimental Investigation

This section, due to the work by De Jong et al. (2013), focuses specifically on bubble formation/annihilation close to the membranes, bubble size distribution and particle mixing as a function of the gas permeation ratio, i.e., the ratio of gas added/extracted relative to the total gas feed, fed via the bottom distributor and the membranes. After a description of the experimental setup and the procedures used for data postprocessing, the PIV/DIA results for cases of gas extraction and gas addition for different membrane configurations with membranes of different tube diameters will be discussed and compared.

2.4.1.1 Experimental Setup

An experimental setup has been constructed to investigate the effects of gas permeation on the fluidized bed hydrodynamics. The bed width, depth, and height are 30 cm, 1.5 cm, and 1.5 m, respectively. Via 121 holes in the rear wall, membrane tubes consisting of a porous cylinder with a mean pore size of 10 μm can be inserted. These membranes are 1.4 cm in height and either 6.4 or 9.6 mm in diameter.

As base configuration, a fluidized bed containing sixty-one 9.6-mm membrane tubes in a staggered arrangement was chosen. For this case, two sets of experiments were carried out; in the first set, the background fluidization velocity was kept constant, while the amount of secondary gas added or extracted via the membranes was varied. In the second set of experiments, the amount of secondary gas added/extracted via the membranes was varied, while simultaneously the background fluidization velocity was adjusted so that the outlet flow rate remained constant. For the base case, permeation ratios of 10%, 20%, and 40% were investigated. Although the fluxes of gas through permselective dense membranes are typically much smaller than the cases used in this study, these values were chosen to emphasize the effect of permeation and to anticipate future improvements in membrane performance.

Next, the influence of the membrane diameter, the geometrical arrangement (i.e., staggered and inline), and the total number of membrane tubes (i.e., membrane pitch) has been investigated. For these cases, the permeation ratio was 20%. An overview of all experiments has been summarized in Table 4.13. Fig. 4.39 shows the four different membrane arrangements investigated.

As base case, a staggered membrane arrangement with sixty-one 9.6-mm membrane tubes was selected. The pseudo-2D bed was filled with glass beads to a height of 55 cm, referred to as the fixed bed height. Permeation ratios have been varied from 10% to 40% (relative to the background fluidization velocity). For the base case, all experiments have been performed both with constant background fluidization velocity (100%) as well as with constant outflow velocity (i.e., varying the inlet velocity from 60% to 140%, based on the amount of permeation via the membrane tubes, see also Table 4.13). Both time-averaged solid flux profiles (Section 2.4.1.2) and bubble properties (Section 2.4.1.3) will be discussed.

2.4.1.2 Solid Flux Profiles

Using the PIV/DIA technique to reconstruct the solids flow, the instantaneous solid flux profiles have been determined. An example of the time-averaged solid flux vector plot—in this case for the base case without permeation—is shown in Fig. 4.40. The figure reveals that the PIV reconstruction algorithm has some problems in determining the solid motion accurately in the immediate vicinity of the membranes. A mask was used to alleviate this problem, but nevertheless, some spurious vectors can be discerned in the resulting graph (the same holds for vectors immediately next to the walls).

Fig. 4.40A reveals that the overall solid circulation pattern does not deviate significantly from a fluidized bed without internals (see also De Jong et al., 2012a); solids rise in the center of the bed and are recirculated via both sides. In the bottom of the bed, as well as in between the upward and

Table 4.13 Overview of the Experimental Series (De Jong et al., 2013)

Measurement Series	Background Gas Velocity (m/s)	Membrane Gas Velocity (m/s)	Number of Images for		Arrangement	d_{membr} (mm)	Nr. of Tubes (–)
			DIA (–)	PIV (–)			
Reference	0.65	0	9900	8800	Stag.	9.6	61
100% – 10%	0.65	−0.106	5600	6000	Stag.	9.6	61
100% – 20%	0.65	−0.212	5600	6000	Stag.	9.6	61
100% – 40%	0.65	−0.424	5600	6000	Stag.	9.6	61
100% + 10%	0.65	+0.106	5600	6000	Stag.	9.6	61
100% + 20%	0.65	+0.212	5600	6000	Stag.	9.6	61
100% + 40%	0.65	+0.424	5600	6000	Stag.	9.6	61
110% – 10%	0.715	−0.106	5600	6000	Stag.	9.6	61
120% – 20%	0.78	−0.212	5600	6000	Stag.	9.6	61
140% – 40%	0.91	−0.424	5600	6000	Stag.	9.6	61
90% + 10%	0.585	+0.106	5600	6000	Stag.	9.6	61
80% + 20%	0.52	+0.212	5600	6000	Stag.	9.6	61
60% + 40%	0.39	+0.424	5600	6000	Stag.	9.6	61
Ref – 6.4 mm	0.65	0	4000	4000	Stag.	6.4	61

100% – 20% – 6.4	0.65	−0.318	4000	4000	Stag.	6.4	61
100% – 40% – 6.4	0.65	−0.636	4000	4000	Stag.	6.4	61
100% + 20% – 6.4	0.65	+0.318	4000	4000	Stag.	6.4	61
100% + 40% – 6.4	0.65	+0.636	4000	4000	Stag.	6.4	61
Ref – inline	0.65	0	4000	4000	Inline	9.6	60
100% – 20% – inl.	0.65	−0.215	4000	4000	Inline	9.6	60
100% – 40% – inl.	0.65	−0.431	4000	4000	Inline	9.6	60
100% + 20% – inl.	0.65	+0.215	4000	4000	Inline	9.6	60
100% + 40% – inl.	0.65	+0.431	4000	4000	Inline	9.6	60
Ref – few	0.65	0	4000	4000	Stag.	9.6	18
100% – 20% – few	0.65	−0.718	4000	4000	Stag.	9.6	18
100% – 40% – few	0.65	−1.437	4000	4000	Stag.	9.6	18
100% + 20% – few	0.65	+0.718	4000	4000	Stag.	9.6	18
100% + 40% – few	0.65	+1.437	4000	4000	Stag.	9.6	18

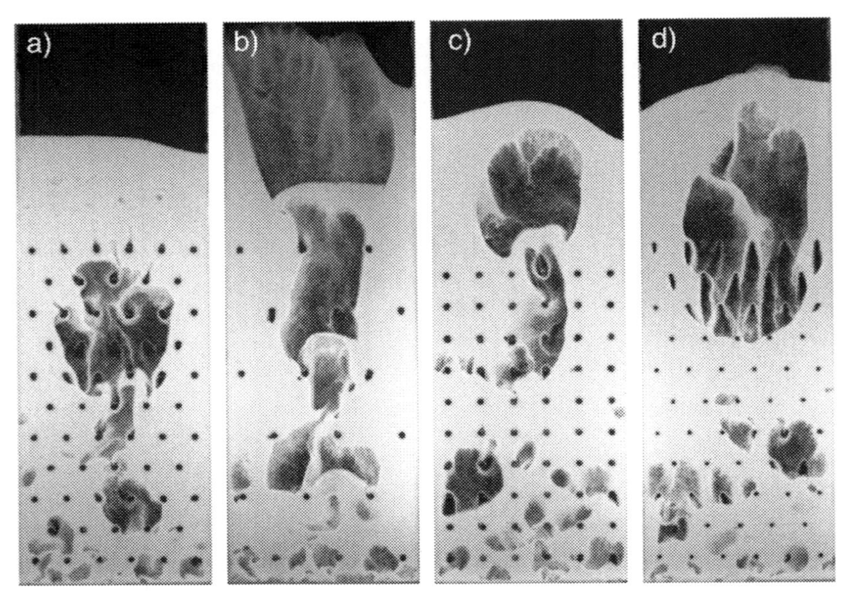

Figure 4.39 Snapshots of the four different membrane arrangements investigated in this study: (A) staggered arrangement with 9.6-mm membranes, (B) staggered arrangement with fewer 9.6-mm membranes (decreased membrane pitch), (C) inline with 9.6-mm membranes, and (D) staggered arrangement with 6.4-mm membranes. *Reprinted from De Jong et al. (2013) with permission from Elsevier.*

downward solid flow, regions exist with limited solid movement. However, the flux in the bed with submerged membrane tubes is approximately a factor three smaller than in the fluidized bed setup without membrane tubes. To assess the solid motion in a more quantitative way, in Fig. 4.40B the lateral profiles of the solid flux at several vertical positions in the bed have been shown. This figure not only contains the solid flux of the reference series but also includes the series with 40% gas extraction and addition for both cases where the total inlet or outlet superficial velocity was kept constant. The locations of the membrane tubes are also indicated in the graph.

The most conspicuous aspect in this graph is the difference in the magnitude of the solid fluxes between the part with immersed membrane tubes and the top part without membranes due to the expansion of the bed; in the part with immersed membranes, the solid flux is significantly smaller than in the upper part of the fluidized bed without membranes. The presence of the membranes results in a significant decrease in the solid velocity in the immediate vicinity of the membrane tubes irrespective whether the solids move upward or downward.

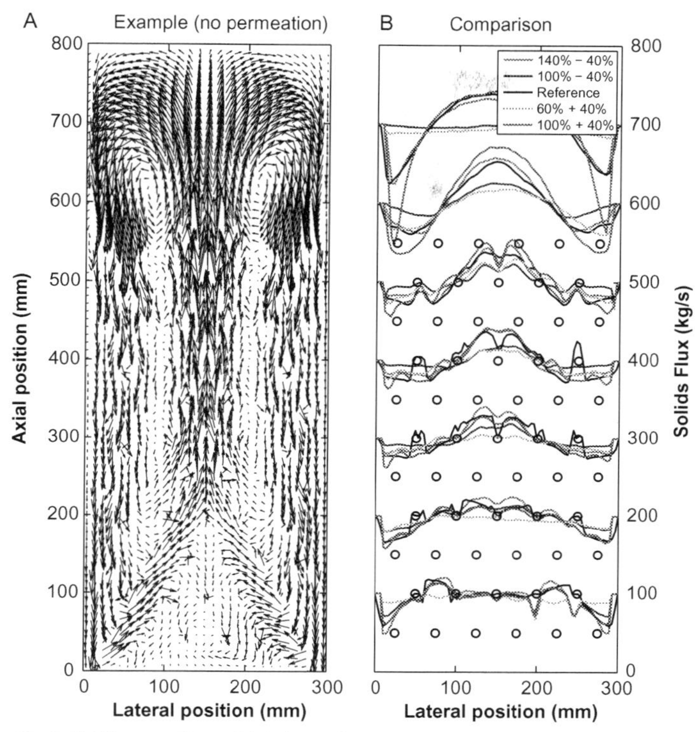

Figure 4.40 Solid flux profiles of (A) the reference case (without permeation) with staggered arrangement of 61 membranes of 9.6 mm presented as a vector field and (B) lateral profiles of the axial solid flux for the reference case without permeation and with a permeation ratio of 40% extraction and addition with constant inflow (100% − 40% and 100% + 40%, respectively) and constant outflow (140% − 40% and 60% + 40%, respectively) at seven different vertical positions (100, 200, 300, 400, 500, 600, and 700 mm). *Reprinted from De Jong et al. (2013) with permission from Elsevier.*

Second, we can conclude that the difference among the cases of gas extraction, no permeation, and gas addition is relatively small, especially in comparison to the case of the permeation of gas through the side walls. By comparing the results for the cases of a constant background fluidization velocity and a constant superficial outlet velocity, it becomes clear that not the extent of the permeation of gas through the membrane tubes, but the total flow of gas (i.e., the height-averaged superficial gas velocity) inside the fluidized bed determines the magnitude of the solid fluxes.

2.4.1.3 Bubble Properties

For the staggered arrangement and keeping the background fluidization velocity constant, Fig. 4.41 portrays the influence of the presence

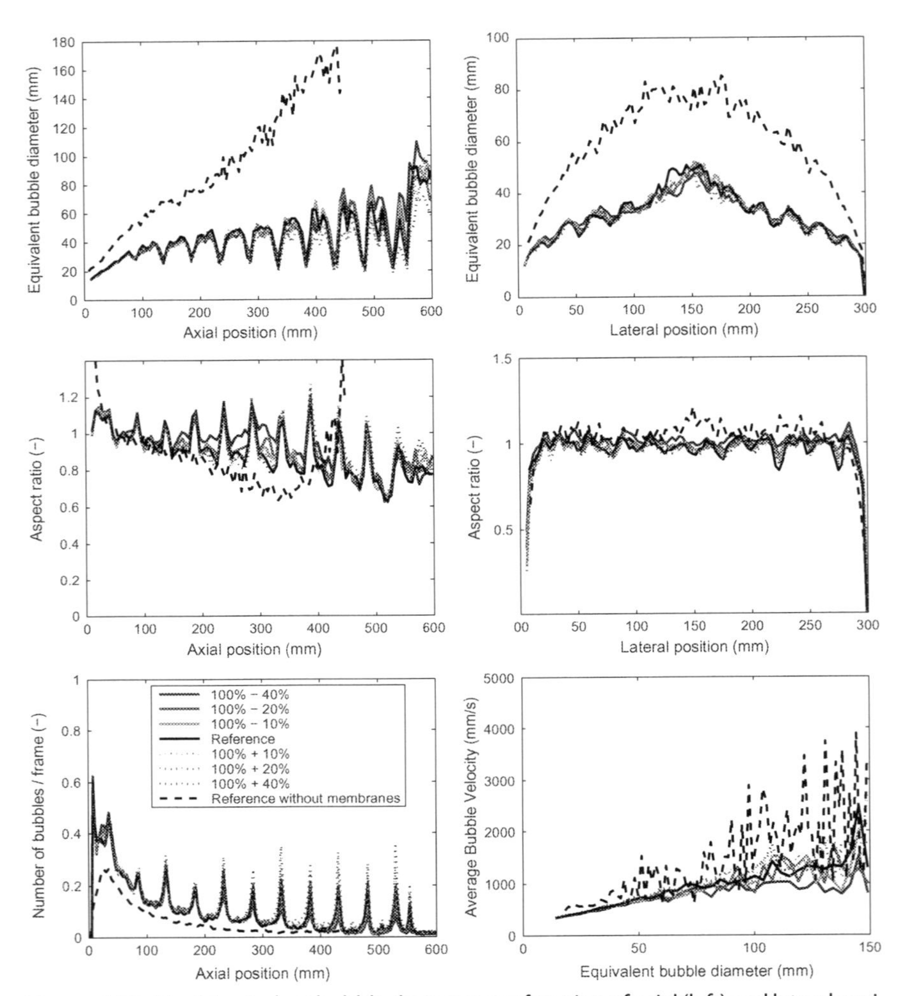

Figure 4.41 (Top) Equivalent bubble diameter as a function of axial (left) and lateral positions (right), (center) bubble aspect ratio as a function of axial (left) and lateral positions (right), and (bottom) number of bubbles per frame as a function of the axial position (left) and average bubble rise velocity as a function of the equivalent bubble diameter (right) for different gas permeation ratios of 0%, 10%, 20%, and 40% with respect to the background fluidization velocity. In all cases, the background fluidization velocity was kept constant (100%). *Reprinted from De Jong et al. (2013) with permission from Elsevier.*

of—and permeation of gas through—the membranes on the average equivalent bubble diameter (top) and bubble aspect ratio (middle) as a function of the axial and lateral positions, the average number of bubbles as a function of the axial position (bottom left) and the bubble velocity as a function of the equivalent bubble diameter (bottom right). In all graphs, the dashed black line represents the same fluidized bed without internals to highlight the difference between these two systems.

The most striking aspect is the fact that the average equivalent bubble diameter in a system with internals is much smaller compared to the system without internals, despite the fact that the other bubble properties (bubble aspect ratio and number of bubbles) are relatively similar. Compared to this difference in bubble size, the influence of the permeation of gas through the membrane tubes is of minor importance. The increased bubble breakup due to the presence of the membrane tubes can clearly be distinguished from the local decrease in bubble size.

Zooming in near the axial position of 40–50 cm, a small difference between the cases of gas addition and gas extraction can be discerned; gas extraction appears to be resulting in a slightly larger average bubble size compared to gas addition. This effect can be explained by inspecting the bottom left graph of Fig. 4.41, showing the number of bubbles as a function of the axial position.

In the experiments with gas addition, small bubbles are frequently formed right next to (often just underneath) the membrane tubes. Because these bubbles are relatively small, the average bubble size for the experiments with gas addition is slightly decreased.

It is worth taking a closer look at the effect of the presence of—and permeation of gas through—the membranes on a local level. Fig. 4.42 shows three snapshots; one during gas extraction and two during gas addition. In the latter cases, the frequent collapse of the bed results in significant gas pockets underneath the membranes (Fig. 4.42B), in particular close to large bubbles. Also, during bed expansion, the membrane tubes in the upper half of the bed regularly show gas voids above the membrane tubes (Fig. 4.42C). In all cases, voids quickly disappear again after they have been formed, indicating that the additional gas is taken up by the emulsion phase, if no larger bubble is present. Both phenomena were hardly observed during gas extraction (Fig. 4.42A).

2.4.1.4 Effect of Membrane Configurations

Several other membrane tube arrangements, i.e., inline versus staggered, number of membranes tubes and diameter of the membranes have been investigated, and the results are discussed below.

The following configurations have been investigated:

- membrane tube diameter of 6.4 mm (instead of 9.6 mm) in staggered arrangement

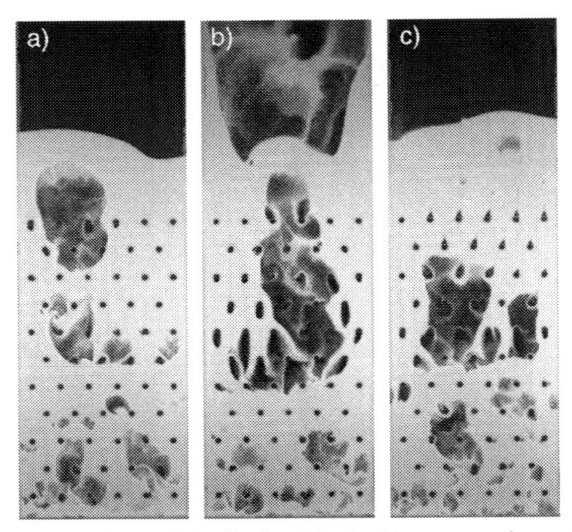

Figure 4.42 Three snapshots of the fluidized bed with staggered membrane arrangement of sixty-one 9.6-mm membrane tubes for (A) 40% gas extraction and (B and C) 40% gas addition, all with constant background fluidization velocity. *Reprinted from De Jong et al. (2013) with permission from Elsevier.*

- inline membrane tube arrangement (with 60 instead of 61 membrane tubes)
- 18 instead of 61 membrane tubes in staggered arrangement.

For all cases, the results for the 100% − 40% extraction case, the reference case without permeation and the 100% + 40% addition case are illustrated in Fig. 4.43. The membrane tubes are depicted in the graphs to highlight the differences between these cases. Please note that the arrangement with fewer membrane tubes (Fig. 4.43B) is not symmetric on a certain axial position, which is reflected in an asymmetric solid flux profiles. Furthermore, the inline arrangement has 10 rows of membrane tubes (compared to 11 rows of membrane tubes in the staggered arrangement) to keep the number of the membranes about the same (60 vs. 61).

The solid flux profiles of the arrangement with fewer membranes are somewhat smoother compared to the base case (Fig. 4.43A,B), resulting in a slightly higher maximum flux in the center and at both sides. The inline arrangement displays significantly larger solid flux profiles in between two columns of the membrane tubes compared to the base case.

Smaller membrane tubes display a smaller effect on the particle flux; while Fig. 4.43A already shows peaks in the profile at the axial position of 100 mm, the profile in Fig. 4.43D looks still relatively smooth at this axial position. At higher positions, the solid fluxes increase and the peaks in the lateral profiles increase as well.

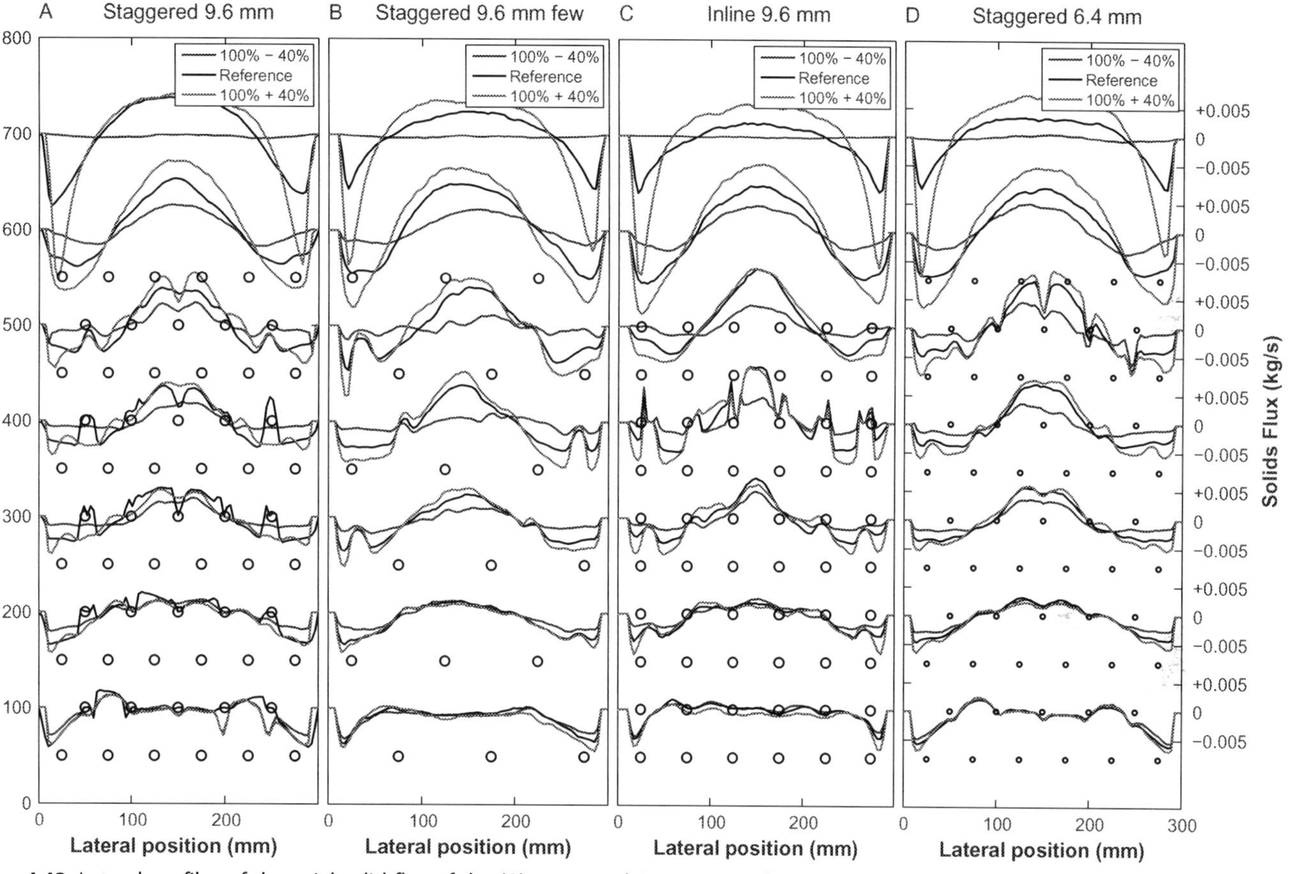

Figure 4.43 Lateral profiles of the axial solid flux of the (A) staggered 9.6-mm membrane arrangement with 61 membrane tubes, (b) staggered 9.6-mm arrangement with 18 membranes, (C) inline 9.6-mm arrangement, and (d) staggered 6.4-mm membrane arrangement, all for 40% gas extraction, no permeation, and 40% gas addition at seven different vertical positions (100, 200, 300, 400, 500, 600, and 700 mm). *Reprinted from De Jong et al. (2013) with permission from Elsevier.*

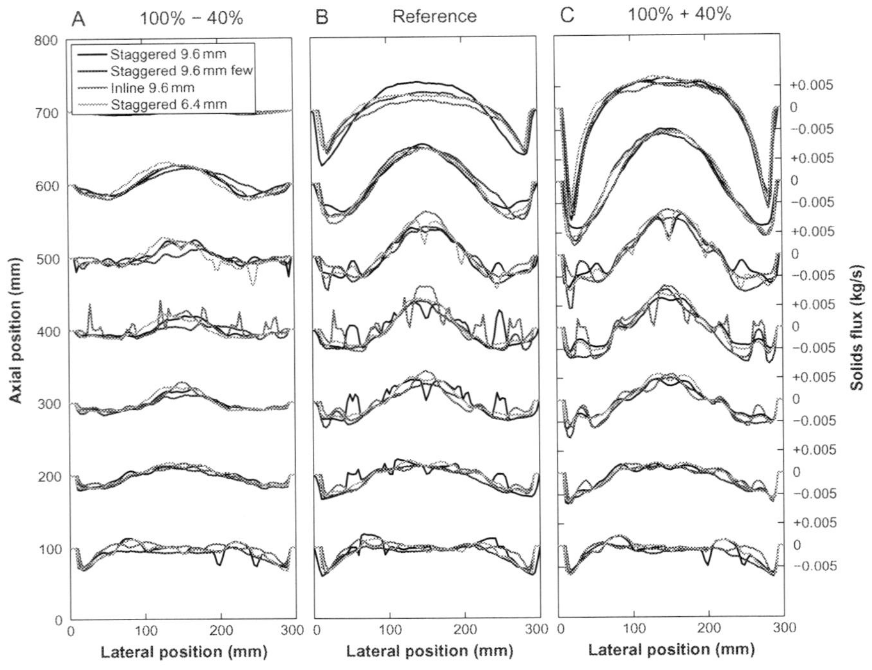

Figure 4.44 Lateral profiles of the axial solid flux of the four different membrane arrangements for (A) 40% gas extraction, (B) no permeation, and (C) 40% gas addition, all with constant background fluidization velocity. *Reprinted from De Jong et al. (2013) with permission from Elsevier.*

A more direct comparison between these cases is shown in Fig. 4.44, in which all four arrangements are combined in the same graphs. The most noticeable aspect is the fact that there is not much difference between the arrangements with respect to the solid fluxes. The exact locations of the peaks in solid fluxes differ slightly, depending on the location of the membrane tubes, but overall the difference is remarkably small.

Fig. 4.45 shows the equivalent bubble diameter as a function of lateral and axial positions, as well as the number of bubbles per frame for the different membrane configurations investigated.

For all membrane configurations, it can be concluded from the graphs that the average bubble size is directly related to the number of bubbles: if more bubbles prevail in the bed, these are the result of the presence of the membranes and therefore reflect either the formation of very small bubbles near the membranes or bubbles that have been split up by the membranes into several smaller bubbles. This results in a larger number of smaller bubbles and a decrease in average bubble size.

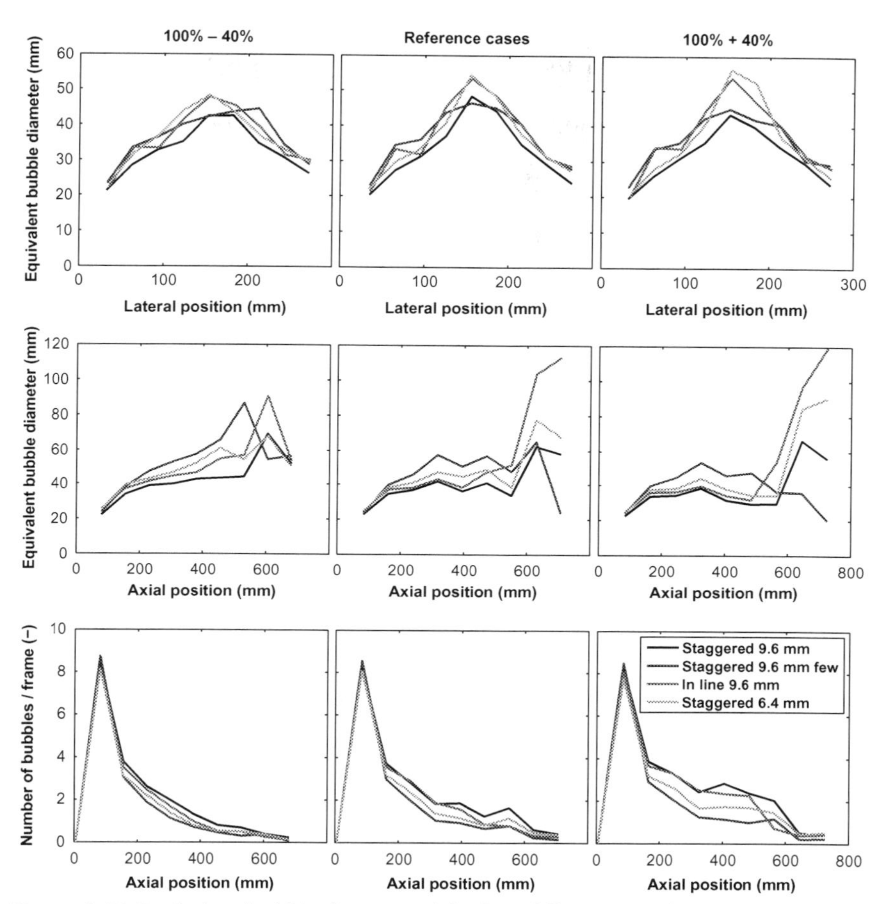

Figure 4.45 Equivalent bubble diameter of the four different membrane arrangements as a function of lateral position (top) and axial position (center) for 40% gas extraction (left), no permeation (center), and 40% gas addition (right), all with constant background fluidization velocity. The bottom row shows the number of bubbles per frame as a function of the axial position. *Reprinted from De Jong et al. (2013) with permission from Elsevier.*

Comparing the different membrane configurations, the base case shows the largest number of bubbles and the smallest average bubble diameter, closely followed by the inline tube arrangement. The configurations with smaller membrane tubes and fewer membrane tubes show a significant decrease in the number of bubbles, and thus an increase in average bubble size. The bubble diameter as a function of the lateral position confirms the finding for these results, but it seems to show a contradicting result for the case with fewer membrane tubes. This deviation is due to the fact that the bubble diameter of those experiments decreases in the upper half of the bed, while the other experimental series show an increase in bubble size in this region. Because the bubble size as a function of the lateral position is averaged over the entire height of the bed, the average bubble size of the series with fewer membranes is smaller than that of the inline and 6.4 mm series. Since small bubbles are preferred to decrease the mass transfer limitation for the gas from the bubble to the emulsion phase where the gases typically react, the base case is the most favorable arrangement of those investigated in this work.

2.4.2 Simulation Study

Simulations have been performed using a DPM extended with an IBM, as explained in Section 2.2.3. The IBM takes into account the presence of— and gas permeation through—cylindrical membranes inside the domain by means of a prescribed flux (the transport of gas through the membrane structure itself is not resolved here).

2.4.2.1 Simulation Setup

Three sets of five simulation cases have been performed: all simulations were carried out for a $9 \times 0.3 \times 27\text{-cm}^3$ domain containing approximately 200,000 glass particles (density 2500 kg/m^3) of 0.5 mm in diameter. For the first five cases, no membranes were inserted (wall); gas was added or extracted via the first 9 cm of the left and right side walls. These simulations allow to relate the results with immersed membranes to membranes that are set up in the wall which has been thoroughly discussed in Section 2.3. In the next five cases, an inline arrangement of nine membrane tubes was used (inline). Gas was added or extracted via these nine membranes. In the last five cases, gas was added or extracted via eight membranes in staggered arrangement (stag or staggered). For all simulations a time step of $\Delta t = 5 \cdot 10^{-5}$ s, a default fluidization velocity of 0.65 m/s and a grid of $90 \times 1 \times 270$ cells were used. Because only one grid cell is used in the depth direction, a free-slip boundary condition was applied for the front and rear wall. An overview of these simulation series is given in Table 4.14 and Fig. 4.46.

The 15 simulation cases were again divided into two sets: one keeping the background gas velocity constant (series 3 and 5), and one keeping the total gas outflow constant (series 2 and 4). Simulation series 1 is the reference series without additional in- or outflow (see Fig. 4.46). By means of a comparison between simulations of 10, 30, and 90 s of

Table 4.14 Overview of the Simulation Series Investigated in This Section

Measurement Number/Name		Background Gas Flow/Velocity		Total Membrane Flow/Velocity		Simulation Time (s)		
		(%)	(m/s)	(%)	(m/s)	Wall	Inline	Stag
1	Reference	100	0.65	0	0	44	45	47
2	60% + 40%	60	0.39	+40	+0.13	39	42	40
3	100% + 40%	100	0.65	+40	+0.13	44	30	30
4	140% − 40%	140	0.71	−40	−0.13	36	31	31
5	100% − 40%	100	0.65	−40	−0.13	31	34	33

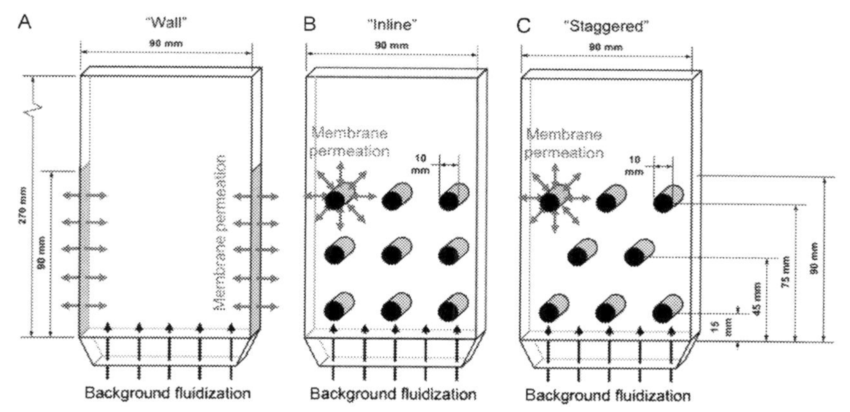

Figure 4.46 The different membrane geometries considered: gas permeation through (A) the left and right wall, (B) nine inline placed membranes, and (C) eight staggered membranes. *Reprinted from De Jong et al. (2012c) with permission from Elsevier.*

simulation time, it was established that 30 s was sufficient to obtain time-independent results (Fig. 4.47).

2.4.2.2 Solids Circulation Patterns

In order to provide a qualitative impression of the differences between the simulation series, Figs. 4.48 and 4.49 show the time-averaged solids fractions for the wall and staggered simulations (the inline cases are omitted here, because they are very similar to the staggered simulations). It becomes clear that the difference between the top row (constant background fluidization) and bottom row (constant outflow, thus varying background fluidization) in these figures is directly related to the total gas flow rate in the system.

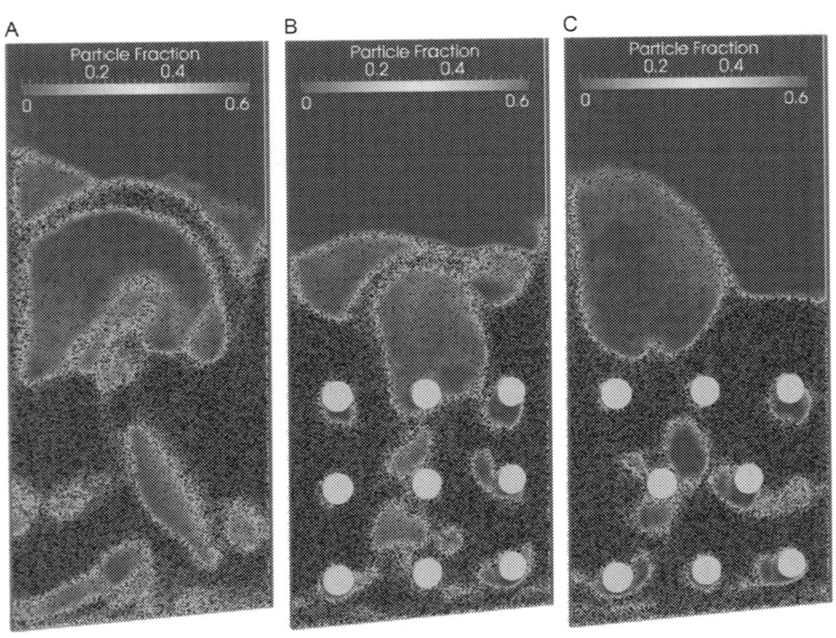

Figure 4.47 Snapshots of simulations of series 1 with (A) wall, (B) inline, and (C) staggered arrangements of the membranes. Only 50,000 particles (out of 200,000) are displayed in each image. *Reprinted from De Jong et al. (2012c) with permission from Elsevier.*

Although, for example, Fig. 4.48A and D shows very similar characteristics, the difference in fluidization velocity causes the shape of the stagnant zones to the left and right to be different, and with 140% background fluidization particles are ejected into the freeboard region to a larger extent.

Fig. 4.49 clearly shows the effect of the membrane tubes in the system. First, the regions with high solids concentration (cap) on top of each membrane tube, and the void underneath each tube can be easily distinguished. Despite the relative large amount of gas addition/extraction, no significant difference in those regions can be observed; the voids underneath the tubes and the dense regions on top of them hardly change when gas is added or extracted and the total gas flow rate in the system is equal (compare, for example, Fig. 2.46C and D, or Fig. 2.46A and F). It can be concluded that the total gas flow rate is more important to these regions than the gas permeation rate through the membranes.

Not only the average particle fraction gives a large amount of useful information regarding the solids behavior, but also the solids flux is of considerable importance in understanding the influence of the membranes. The

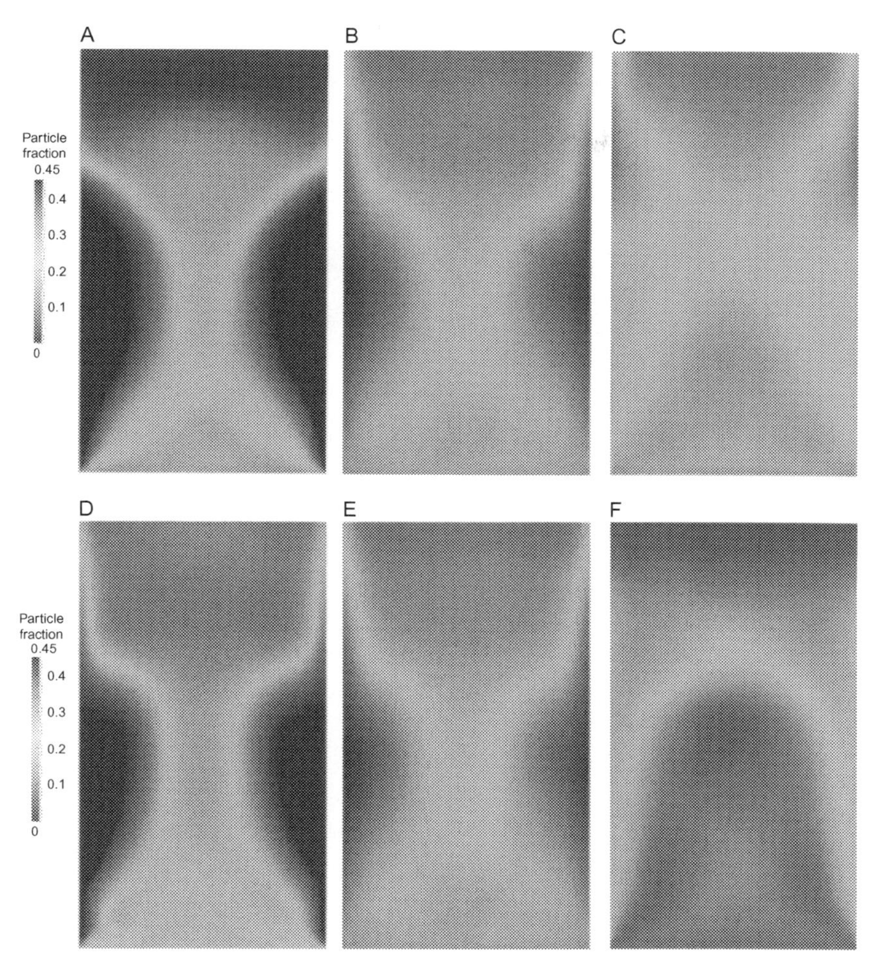

Figure 4.48 Time-averaged solids fraction of the wall simulations. Top row is for constant inflow, bottom row for constant outflow. (A) 100% − 40%, (B) reference, (C) 100% +40%, (D) 140% − 40%, (E) reference, and (F) 60%+40%. *Reprinted from De Jong et al. (2012c) with permission from Elsevier.*

solids flux profiles from the wall series is depicted in Fig. 4.50 (these effects have been covered earlier in this chapter), as a reference to the results for the simulations series with inline membrane arrangement (top) and staggered arrangement (bottom) which are shown in Fig. 4.51 (also for constant background fluidization velocity only). By examining the reference cases (B) and (F), an upward solids movement through the center and a downward movement near the wall can be discerned. This is comparable to the wall case without inserts, although it is obvious that the membrane tubes do

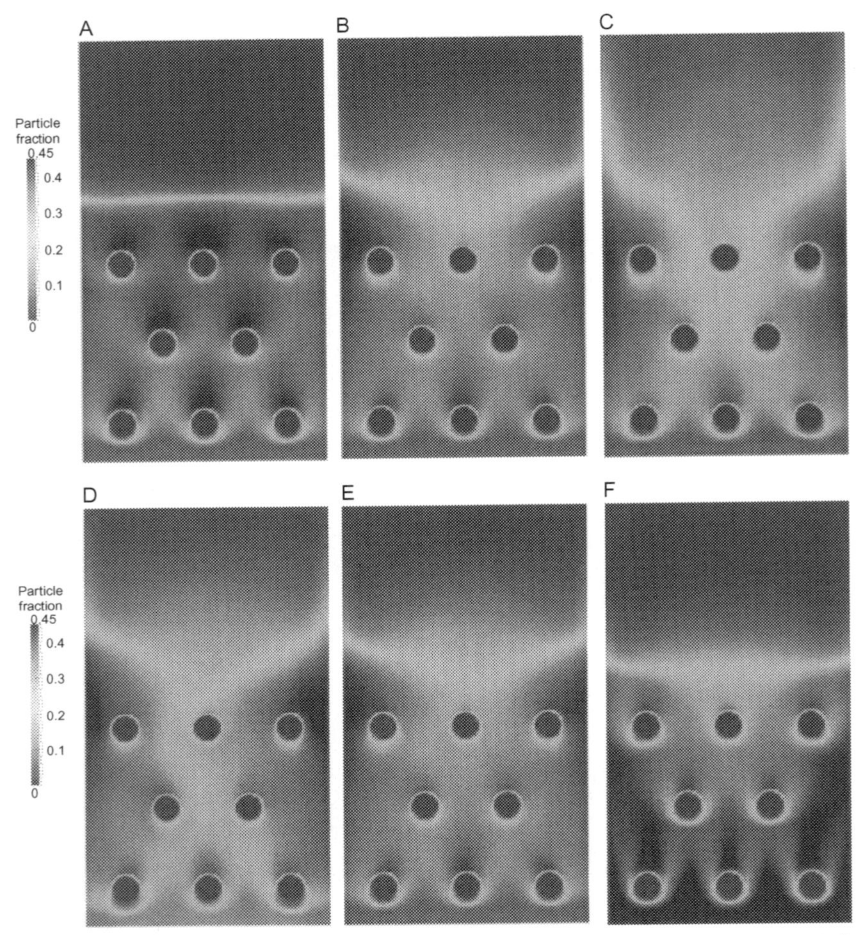

Figure 4.49 Time-averaged solids fraction of the Staggered simulations. Top row is for constant inflow, bottom row for constant outflow. (A) 100% − 40%, (B) reference, (C) 100% + 40%, (D) 140% − 40%, (E) reference, and (F) 60% + 40%. *Reprinted from De Jong et al. (2012c) with permission from Elsevier.*

significantly influence the pathway of the solids; for the inline simulations, solids move upward via the two possible pathways in between the membrane tubes. For the staggered case, the particles are forced through the single opening in between the central two membrane tubes, after which the solids split up again into two possible pathways.

With respect to the addition and extraction of gas via the membrane tubes, the permeation does not seem to have a pronounced effect on the solids circulation patterns; the only significant difference is the predicted

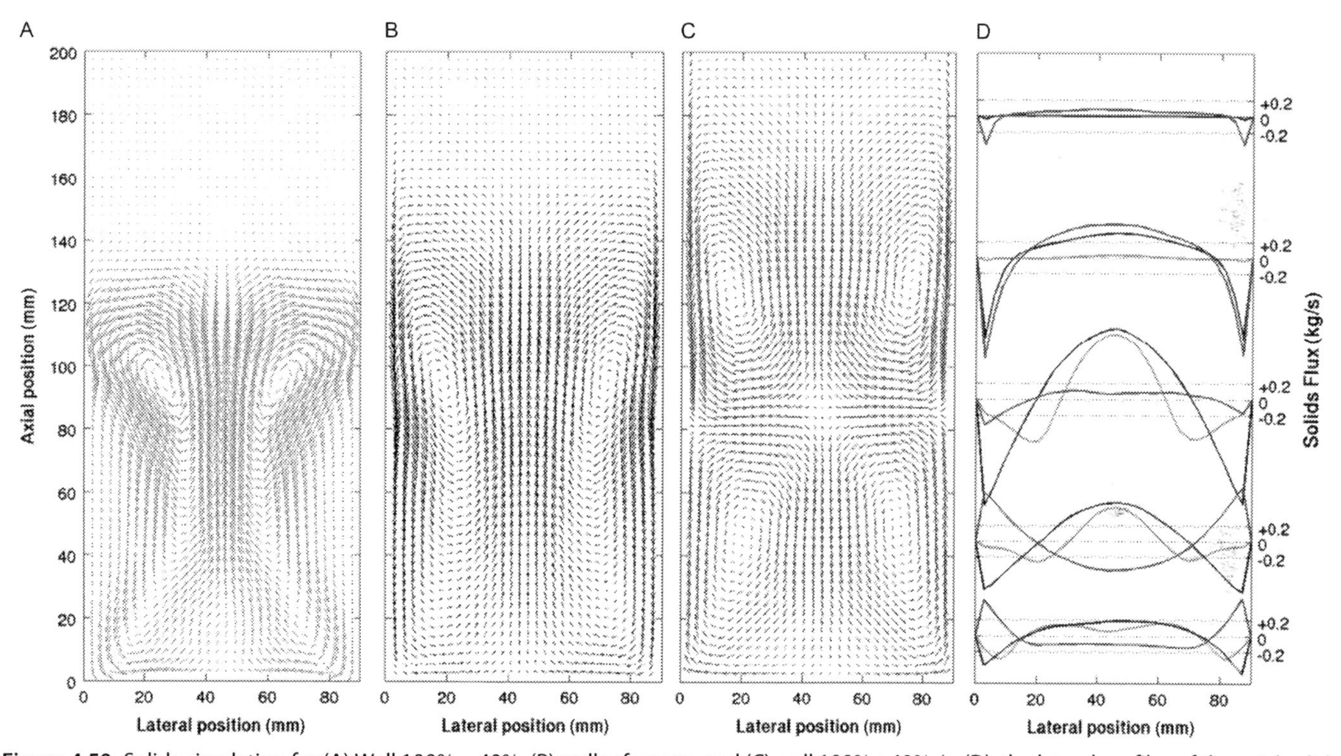

Figure 4.50 Solids circulation for (A) Wall 100% − 40%, (B) wall reference, and (C) wall 100% + 40%. In (D), the lateral profiles of the axial solids velocity for those three simulation cases is shown for five different axial positions in the bed (at a height of 15, 45, 90, 135, and 180 mm). *Reprinted from De Jong et al. (2012c) with permission from Elsevier.*

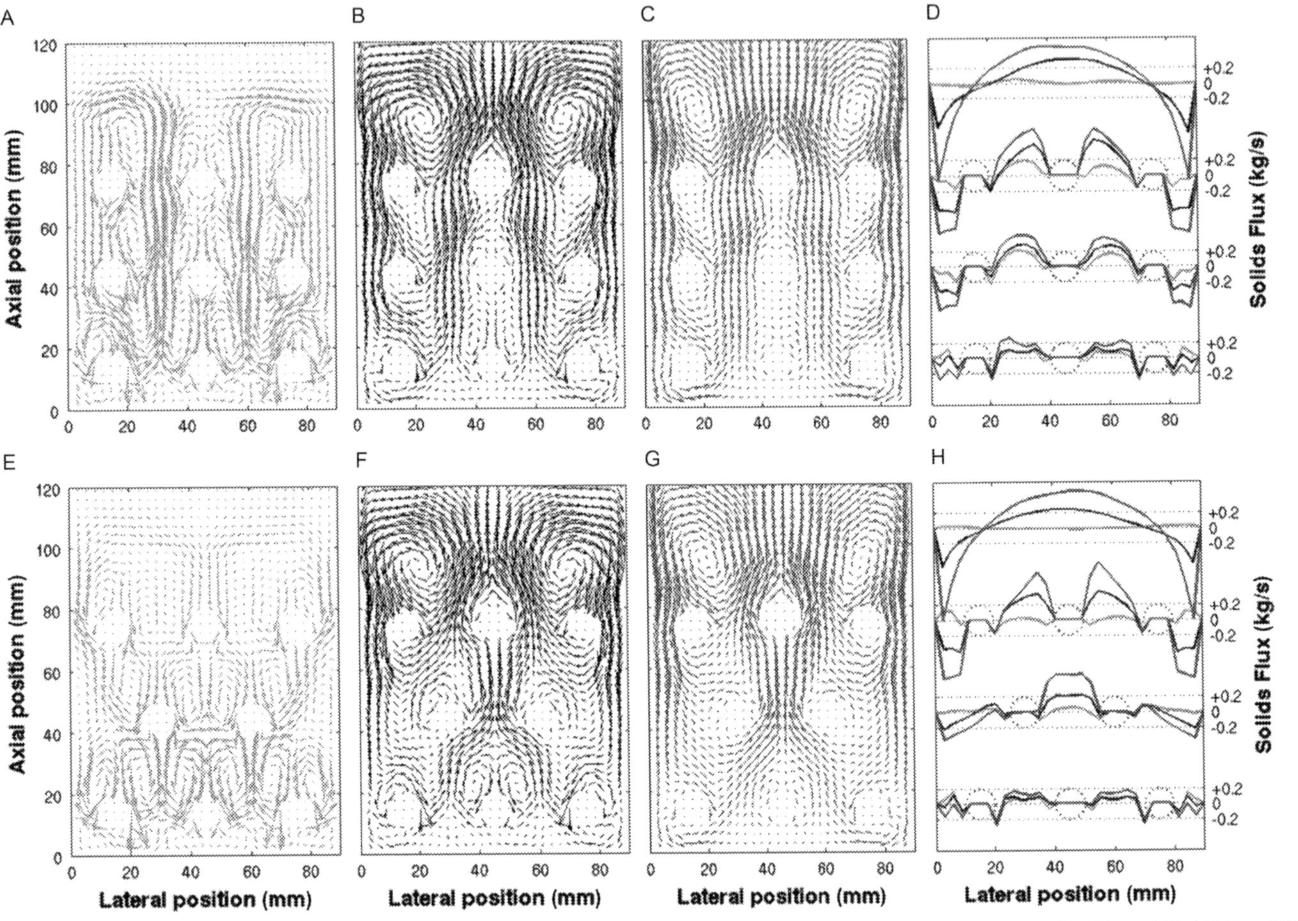

Figure 4.51 Solids circulation for inline simulations (top) and staggered simulations (bottom). For both series, the (A and E) 100% − 40%, (B and F) reference, and (C and G) 100%+40% cases are presented. In (D) and (H), the lateral profiles of the axial solids velocity for those three simulation cases are shown for four different axial positions in the bed (at a height of 15, 45, 75, and 105 mm). *Reprinted from De Jong et al. (2012c) with permission from Elsevier.*

bed height, which is higher in case of gas addition. However, this effect cannot be traced back to the permeation of gas through the membranes tubes, but is due to the overall increased gas flow rate in the system for the 100% + 40% cases (and to the smaller gas flow rate in the system for the 100% − 40% cases).

2.4.2.3 Bubble Size Distribution

The average bubble size has been examined and plotted as a function of the axial position in Fig. 4.52 and of the lateral position in Fig. 4.53. In both figures, central graphs B and E represent the reference case, i.e., without permeation. In the top graphs, the bottom flow rate is kept constant, while the bottom graphs show results for a varying bottom flow rate (and thus constant flow rate at the outlet).

The most obvious trend that can be inferred from Fig. 4.52 is the fact that the equivalent bubble diameter remains small in the presence of the membrane tubes compared to the wall simulation series; even without permeation (graphs B and E), the average bubble size is significantly reduced.

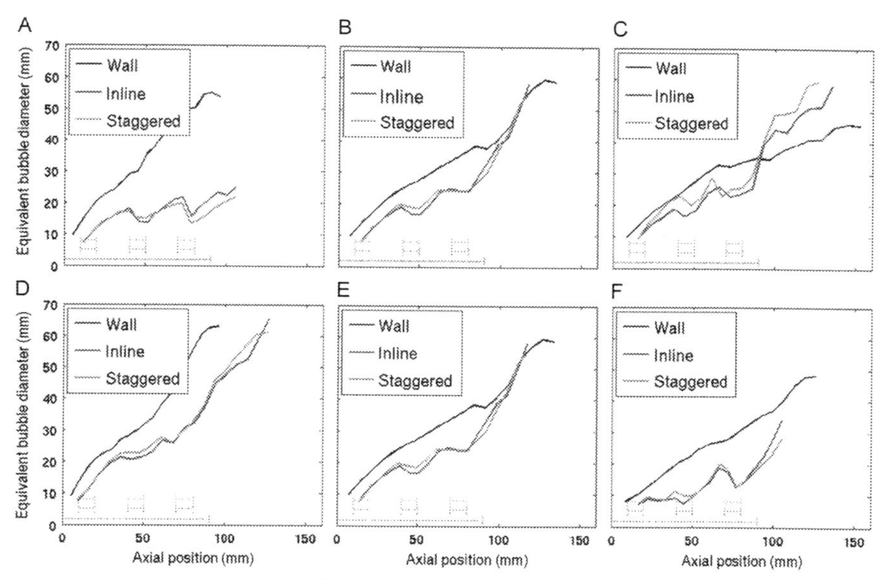

Figure 4.52 Comparison of bubble diameter as a function of axial position in the fluidized bed for wall, inline, and staggered configurations. Top graphs are for constant bottom inflow, bottom graphs are for varying bottom inflow (and thus constant outflow). Membrane locations are indicated by dotted lines. (A) 100% − 40%, (B) reference, (C) 100% + 40%, (D) 140% − 40%, (E) reference, and (F) 60% + 40%. *Reprinted from De Jong et al. (2012c) with permission from Elsevier.*

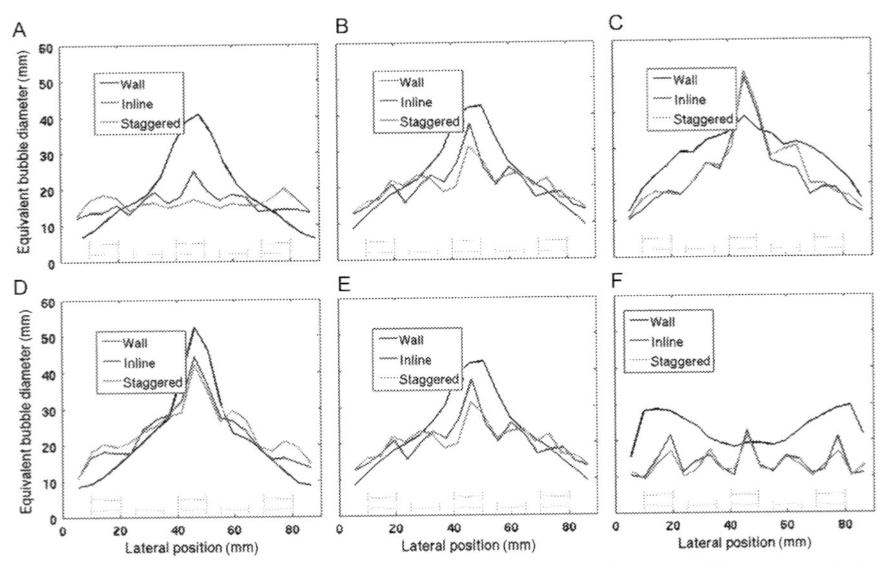

Figure 4.53 Comparison of bubble diameter as function of lateral position in the fluidized bed for wall, inline, and staggered configurations. Top graphs are for constant bottom inflow, bottom graphs are for varying bottom inflow (and thus constant outflow). Membrane locations are indicated by dotted lines. (A) 100% − 40%, (B) reference, (C) 100% + 40%, (D) 140% − 40%, (E) reference, and (F) 60% + 40%. *Reprinted from De Jong et al. (2012c) with permission from Elsevier.*

Note that in particular at the exact height of the membrane tubes, the bubble diameter decreases and increases until the next row of membrane tubes is encountered. Above the membrane area (above 80 mm), bubbles for the inline and staggered configurations quickly increase in size. Graph C shows that the bubble diameter surpasses even that of the wall series; this is due to the fact that for the wall series, bubbles are attracted toward the left and right membranes and split (see also Fig. 4.53F), resulting in a smaller average bubble diameter (De Jong et al., 2011).

A second conclusion that can be drawn from the results for the inline and staggered configurations shown in Fig. 4.52 is that the total gas flow rate in the system in the membrane zone has a strong effect on the bubble size; in Fig. 4.52A and F, the average gas flow rate in the zone containing the membrane tubes is only approximately 80% compared to the gas flow rate of the reference cases. On the other hand, for graphs C and D, the corresponding quantity amounts approximately 120% in that same region. One can clearly see that for the latter cases, the average bubble diameter in that region is larger than that of the reference, which in turn is larger than for the simulations of graphs A and F.

Fig. 4.53 shows the bubble diameter as function of the lateral position; in agreement to what has been discussed in Section 2.3, extracting gas through the wall membranes forces the gas through the center of the bed while feeding gas flattens the curve (Fig. 4.53C) and produces two smaller peaks near both membranes to form (graph F). Because the inline and staggered simulations permeate (extract or add) gas evenly over the width of the bed, these series do not show this trend. However, a trend found earlier in Fig. 4.52 can also be seen here: again, the total gas flow rate determines the shape of these curves. In Fig. 4.53A and F, the system contains a smaller amount of gas, while the system corresponding to graphs C and D contain relatively more gas than the reference cases. With a smaller gas flow rate, the systems display a more even bubble size distribution in the lateral direction. Contrary, with more gas present in the system (see C and D), the bubble diameter increases significantly in the center of the bed. It is worth mentioning that also from these graphs, it becomes obvious that the systems with membrane inserts yield smaller bubbles on average.

From both Figs. 4.52 and 4.53, it can be seen that the difference between the inline and staggered is not significant. Apparently, the geometrical arrangement of the membrane tubes is by far less important to the bubble diameter than their presence in the fluidized bed.

Fig. 4.54 shows the bubble rise velocity as a function of its size. It can be seen clearly that within the membrane region, in all cases there is no difference at all between the simulations with inserts and the simulations without inserts. Above the membrane area, the bubbles for the systems with inserts (reference system) and for the systems with inserts and gas addition seem to increase in size and exceed the bubble size for the corresponding system without inserts.

By examining the snapshots from the simulations, we noticed that the bubbles inside the membrane arrays exhibit a different shape compared to bubbles in a bed without inserts. The bubbles are often connected via membrane tubes or connected to the additional void underneath each membrane tube. Moreover, bubbles inside membrane arrays seem more irregularly shaped compared to the situation without inserts. To obtain a better insight in this phenomenon, the perimeter of each bubble was calculated by DIA, and the results are plotted as a function of the bubble size in Fig. 4.55. These graphs confirmed our visual observation; in almost all cases, the bubble perimeter of systems with membrane arrays (inline and staggered) exceeds the perimeter of the wall simulations above a bubble diameter of approximately 30 mm. Only for the 100% – 40% series (Fig. 4.55A), the bubbles did not reach a certain bubble size to allow a conclusion to be drawn

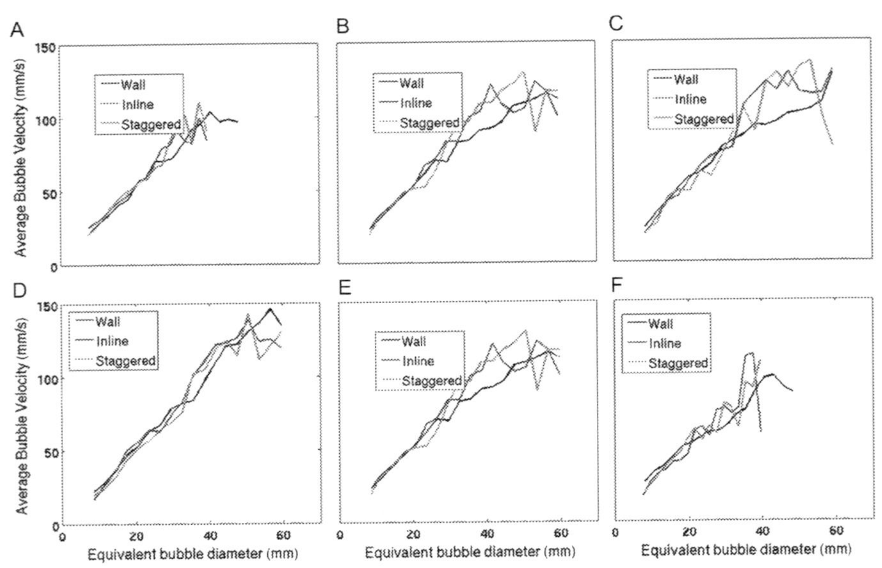

Figure 4.54 Comparison of bubble rise velocity as a function of the equivalent bubble diameter in the fluidized bed for wall, inline, and staggered simulation series. Top graphs are for constant bottom inflow, and bottom graphs are for varying bottom inflow (and thus constant outflow). (A) 100% − 40%, (B) reference, (C) 100 + 40%, (D) 140% − 40%, (E) reference, and (F) 60% + 40%. *Reprinted from De Jong et al. (2012c) with permission from Elsevier.*

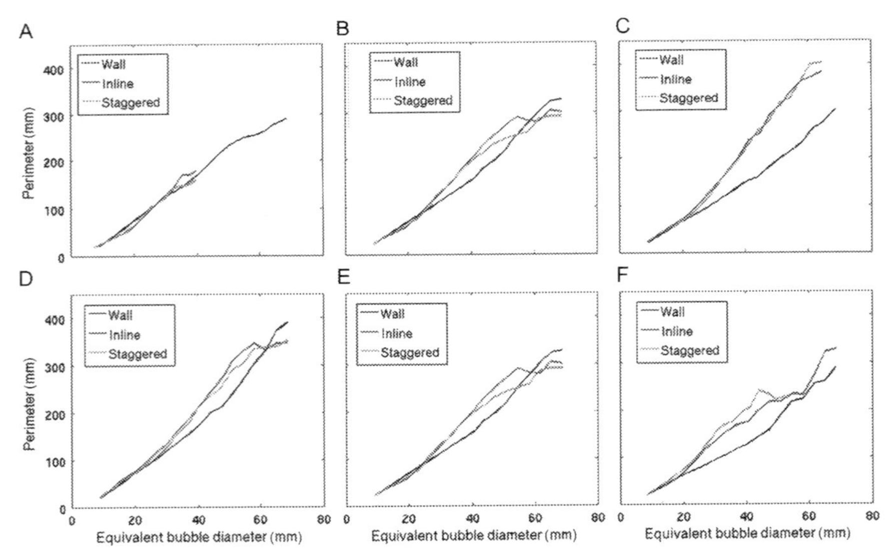

Figure 4.55 Comparison of bubble perimeter as function of axial position in the fluidized bed for wall, inline, and staggered simulation series. Top graphs are for constant bottom inflow, and bottom graphs are for varying bottom inflow (and thus constant outflow). (A) 100% − 40%, (B) reference, (C) 100% + 40%, (D) 140% − 40%, (E) reference, and (F) 60% + 40%. *Reprinted from De Jong et al. (2012c) with permission from Elsevier.*

from these graphs. Note that the largest bubbles should not be considered, because these are only present in the top section of the bed, well above the membrane tubes (recall Fig. 4.52). For small bubbles (smaller than 30 mm in diameter), no difference can be observed between the systems with and without internals.

3. MASS TRANSFER IN FBMRs

A thorough insight in the heat and mass transfer phenomena in FBMRs is essential to optimize their operation. In the previous section, it has been found that the extractive/additive fluxes through the membranes provide a significant influence on the hydrodynamics, including the formation of stagnant zones where mass transfer limitations can be anticipated. In order to confirm whether these stagnant zones are really detrimental to the process, noninvasive measurement and simulation techniques have to be devised that allow detection of concentration in the gas and emulsion phase.

This section discusses a novel technique that allows the measurement of concentration profiles in the gas phase in a fluidized bed.

3.1. Infrared-Assisted PIV/DIA

The gas exchange from bubble-to-emulsion phase in bubbling gas–solids fluidized beds may occur via the combined effects of gas diffusion, coherent gas flow, and solids motion carrying adsorbed gas atoms (Davidson and Harrison, 1963; Kunii and Levenspiel, 1991. Experiments of bubble-to-emulsion phase gas exchange generally use invasive techniques (such as gas injection/extraction), where bubbles containing a tracer gas are injected into an incipiently fluidized bed and the concentrations of the tracer gas are measured either in the emulsion phase or at the outlet of the bed (see, e.g., Deshmukh et al., 2003; Li and Weistein, 1989; Nguyen et al. 1981; Patil et al., 2003). The invasive probe obstructs the flow path of both gas and solids and inevitably disturbs the bed operation, while the injected bubbles differs significantly from naturally generated bubbles in terms of size, density, and dynamics. Recent developments are based on the imaging techniques like X-ray (Roels and Carmeliet, 2006) and MRI (Müller et al., 2006, 2007) which have allowed nonintrusive and fast measurements of the flow behavior. (Patil et al. (2003) applied an MRI technique to measure the mass transfer from bubble-to-emulsion phase in gas–solids fluidized beds by injection and tracing of laser-polarized xenon (^{129}Xe). However, the application of the MRI technique is limited by an extremely high investment cost and the required particles containing MR-sensitive nuclei. Moreover, the technique does not provide an instantaneous, whole-field measurement of the tracer gas, but only time-averaged information based on solid measurements. A qualitative description of gas flowing through a bubble was given by the early work by Rowe et al. (1964), who injected bubbles of nitrogen dioxide (visible brownish gas) into pseudo-2D beds. However, to the authors' knowledge, the bubble-to-emulsion mass exchange in fluidized bed has never been quantitatively investigated with optical techniques due to experimental difficulties associated with the opaqueness of the fluidized beds and transparency (like CO_2, He) or toxicity (NO_2) of tracer gases that hamper the detection of the gas streamlines. Related to these measurement difficulties, all the models in the literature (Davidson and Harrison, 1963) are based on the assumptions of a uniform concentration profile inside the bubble and a constant bubble volume and rise velocity. However, it has been demonstrated by CFD simulations (Patil et al., 2003) that the concentration inside the bubble is nonuniform; in particular, the gas concentration is lower in the center and at the bottom part of the bubbles. Thus, to more exactly quantify the mass exchange rate in bubbling fluidized bed systems, detailed information on the gas flow patterns as well as the concentration distribution inside the bubble are of great importance. This work aims at developing a new experimental technique that will allow the noninvasive, instantaneous whole-field concentration measurements to evaluate amongst others these assumptions used in the open literature, and to quantify the bubble-to-emulsion phase mass exchange.

With this aim, a novel optical technique based on thermal imaging with a high-speed infrared (IR) camera equipped with an optical filter centered at the IR absorbing wavelength of a tracer gas has been proposed (Dang et al., 2013), which will be described below.

3.1.1 Measurement Principle

Any object with a temperature above 0 K will emit radiation, both in the visual as well as in the IR wavelength range, where the radiation intensity depends on the temperature of the object. IR detectors correlate electrical signals to the incident infrared radiation intensity. Instruments using modern IR detectors and optics to gather and focus energy from the targets onto these detectors are capable of measuring target surface spot temperatures with sensitivities up to 0.1 °C and with a response time in the microsecond range (spot radiometers). An IR camera can combine this measurement capability and convert IR radiation measurements into digital level (DL) signals.

If a gas, such as CO_2, is present between the emitting object and the camera, this gas will absorb part of the radiation, where different components absorb radiation to a different extent at different wavelengths, and the signal of the camera will decrease (note that the gas itself also emits, but this intensity is significantly lower than that of the radiation absorbed). The extent of this decrease thus depends on the gas component and its concentration. The measurement technique described here is based on the principle of decreased transmission. First, a pseudo-2D fluidized bed column (i.e., with a relatively small depth of the column) is placed in between the emitting object and the IR camera. The column is made out of sapphire to reduce the absorption of IR radiation by the column itself as much as possible. Sapphire transmits IR in the same wavelength range in which the IR camera is sensitive, as well as in the visual range. Additionally, it is scratch resistant and available at reasonable size for a 2D fluidized bed column, contrary to other crystals on the market. A narrowband optical filter that filters the IR wavelength absorbed by CO_2 (4.26 μm) is then placed in front of the IR camera detector, which then measures an IR signal that relates to the CO_2 concentration.

3.1.1.1 Quantitative Analysis

The signal generated by the camera when a sapphire column flushed with nitrogen is placed in between the IR source and the camera, is indicated with DL_0. When CO_2 is fed into the column, it absorbs part of the radiation and the remaining signal transmitted to the camera is denoted with DL. From these two signals, the transmittance and absorbance can be defined as follows:

$$T = \frac{\mathrm{DL}}{\mathrm{DL}_0} \tag{4.39}$$

The absorbance is in theory linear to the CO_2 concentration according to Lambert Beer's law:

$$A = \varepsilon \ell C = aC \tag{4.40}$$

where C is theconcentration (mol/l), ℓ is the target length (cm), ε is the molar absorbance ($\mathrm{mol}^{-1}\ \mathrm{cm}^{-2}$), and $a = \varepsilon\ell$.

Not always can the linear relation between absorbance and concentration be obeyed, and the actual absorbance versus concentration correlation can be obtained from calibration experiments via a fitted polynomial (Buijs and Maurice, 1969), which is in detail discussed in Dang et al. (2013):

$$A = a_1 C + a_2 C^2 + a_3 C^3 + \ldots + a_n C^n \tag{4.41}$$

3.1.2 Experimental Setup

The experimental setup is schematically depicted in Fig. 4.56. An anodized aluminum slab ($0.15 \times 0.30 \times 0.02$ m) has been used as an IR source. The IR source is heated up via 15 electrical hot rods placed inside the slab and the temperature is controlled at 430 ± 5 °C. A good insulation (combined with the high thermal inertia of the plate and proper control) ensures a stable temperature during the experiments. A pseudo-2D sapphire column ($0.04 \times 0.2 \times 0.005$ m) was installed in front of the IR source. The 2D column allows IR measurements (sapphire being almost transparent to relevant IR signals) as well as PIV measurements with visual cameras. Wall absorbance and reflection were minimized by using a sapphire glass (purchased from the Technical Glass Company) with small thickness (3 mm) and polished surfaces. A porous metal plate with an average pore size of 10 μm and 3 mm thick was used as the gas distributor at the bottom. A nozzle (4 mm inner diameter) was used in the center of the distributor for CO_2 injection. A high-speed IR camera FLIR series SC7000 of 512×640 px allowing a maximum full frame rate of 100 frames/s was used to detect transmitted IR radiation. An optical CO_2 filter with a narrow band pass (4.26 ± 0.03 μm) has been used with the camera for CO_2 detection. The camera was positioned at a proper distance and orientation to be able to capture the interested area of the target column and to avoid camera self-reflection, causing the narcissus effect. The wavelength operation of the IR camera allows CO_2 to be detected at 4.26 ± 0.03 μm wavelength.

The experimental setup has been placed in a box with controlled ventilation to avoid (or decrease) the effects of the environment on the experimental results. In fact, for instance, also water is absorbing IR radiation and thus the humidity in the environment and in the feeding gases has to be controlled. The room is air conditioned, and the walls are uniformly emitting.

The new technique has first been calibrated and validated in single-phase (gas-phase) flow and subsequently extended to two-phase gas–solid flow applications. In gas–solid systems, the particles also absorb IR radiation; therefore, the particles need to be detected with the high-speed camera

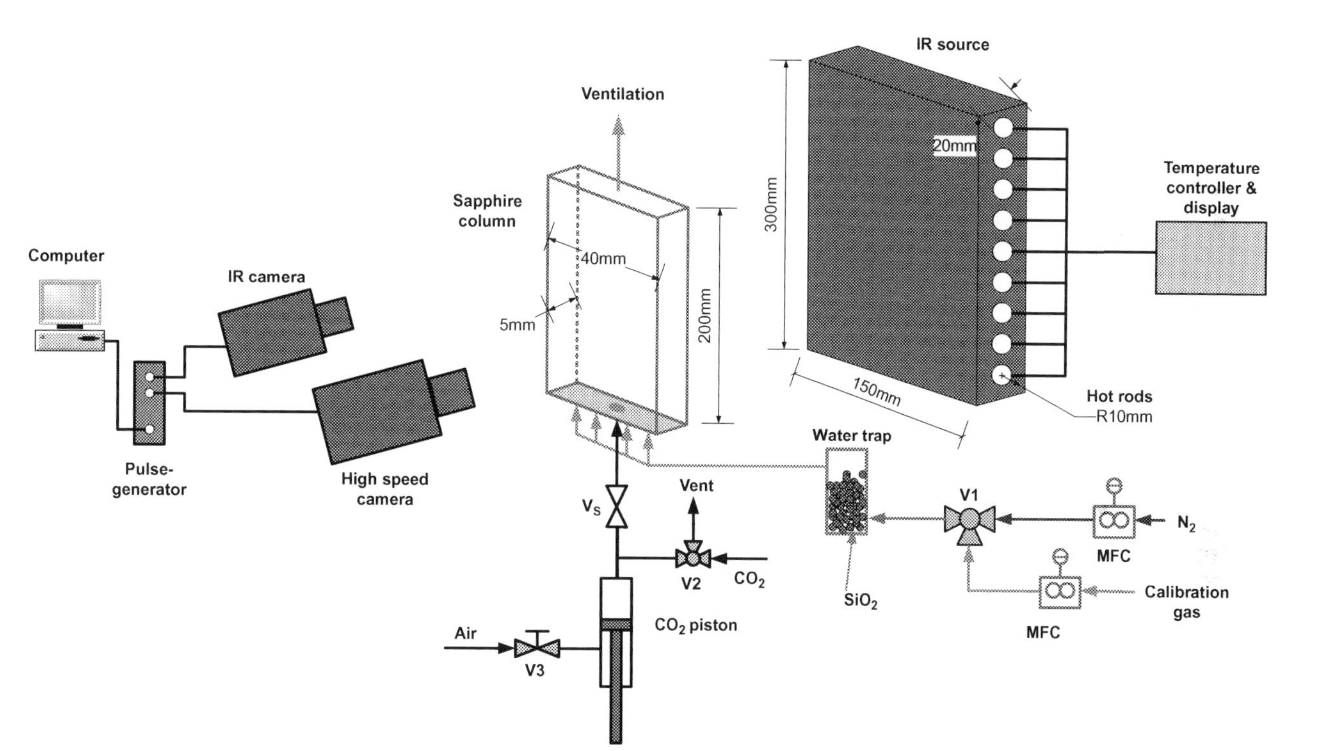

Figure 4.56 Scheme of the experimental setup (MFC, mass flow controllers; V, valves). *Reprinted from Dang et al. (2013) with permission from Elsevier.*

operated in the visual wavelength range and subsequently the IR image needs to be corrected for the presence of the particles (detailed below).

The gas flow rates have been controlled with Brooks MFCs. For CO_2 injection, a piston with adjustable volume has been used that allows fast injection of a measured amount of CO_2 via a nozzle ($d=0.004$ m) into the bed. Glass beads ($d_{particle}=400–600$ μm, density$=2525$ kg/m^3) are fluidized by dried N_2, ($u_{mf}=0.206$ m/s). CO_2 is filled in a piston at a certain prepressure via a three-way valve (V2 in Fig. 4.56). For fast bubble injection, CO_2 inside the piston is then compressed by air at high pressure (6 bar) for fast bubble injection. CO_2 at high pressure is subsequently released into the column through a solenoid valve (Vs) opened in 0.01 s. The amount of CO_2 injected is determined by different pressure before and after injection and the volume of CO_2 filled to the piston. The injection velocity depends on the amount of tracer gas and the injection time.

To avoid electrostatics buildup between particles and the column walls, the particles are first discharged by ionized air with an air gun, this avoids particles sticking to front and back walls of the column. The front and back walls have been pretreated with an antistatic agent.

3.1.3 Results and Discussion

3.1.3.1 Application of the IR Camera for Single-Phase Measurements

The biggest advantage of the novel optical technique is the possibility to measure, noninvasively, the whole-field concentration profile with high spatial and temporal resolution. This is evident from single gas-phase measurements reported in this section.

An experiment has been performed where pure N_2 was continuously fed into the sapphire column when at $t=t_i$ a certain volume of CO_2 is virtually instantaneously injected into the column. Fig. 4.57 (left) shows the obtained IR image in which the lower signals indicate the presence of CO_2. The apparent absorbance is computed (center image) from that IR image and subsequently translated into CO_2 concentration by applying the calibration curve. The whole-field concentration map is shown in the right image, where the CO_2 jet is apparent.

3.1.3.2 Extension of the Novel Technique To Gas–Solid Flows

The extension of the IR transmission technique to gas–solid flows would make it possible to measure noninvasively concentration profiles inside bubbles rising in gas–solid fluidized beds and thus gather important information on the bubble-to-emulsion gas exchange rate to be compared with literature findings.

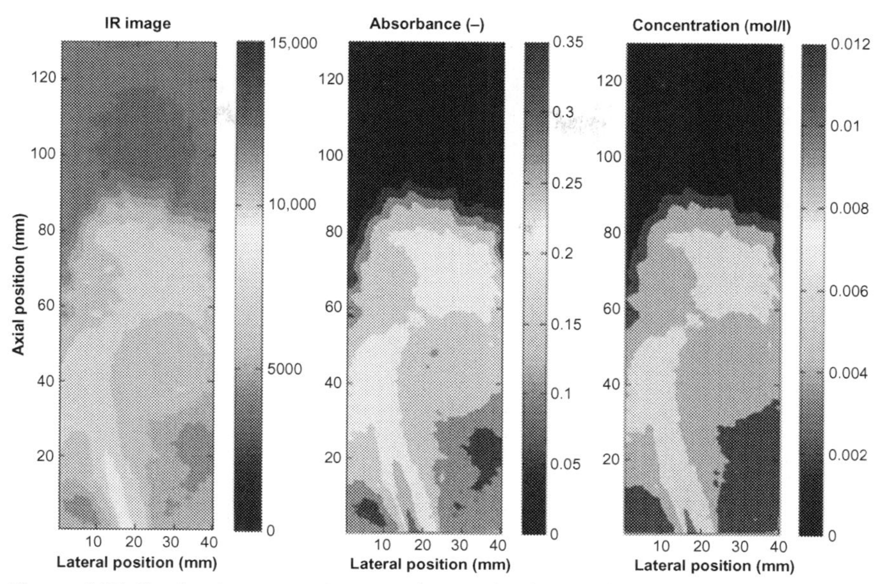

Figure 4.57 Single-phase experiment with N_2 as background gas and CO_2 injection via a nozzle installed in the middle of the porous distributor at the bottom: IR image (left), apparent absorbance (middle), and CO_2 concentration field (right). *Reprinted from Dang et al. (2013) with permission from Elsevier.*

The main difficulty in extending the technique to bubbling fluidized beds is represented by the presence of particles that absorb part of the IR radiation. Moreover, as bubbles rise in the bed, particles are raining from the roof of the bubbles into the bubble itself. As the IR technique is only able to detect changes in IR signal, the presence of particles in the bubbles would lead to erroneous concentration profiles, when not properly corrected for.

The solution of this problem is to couple the IR technique with the high-resolution high-speed visual measurements which allows detecting individual particles in the fluidized bed.

A visual high-speed camera (VIS-2016 × 2016 px@1600 Hz from Lavision) is coupled with the IR camera via a pulse generator that sends a trigger to both IR and VIS cameras, for taking coupled IR–VIS images simultaneously. Subsequently, DIA is applied to identify and remove particles inside bubbles before correlating the IR signal to the CO_2 concentration. Fig. 4.58 shows all steps of the coupling VIS–IR–DIA with the VIS image in Fig. 4.58A. The positions of particles inside the bubble are easily detected from the visual images based on the particle pixel intensity. Subsequently, these particle positions are identified in the IR images and the pixel intensity

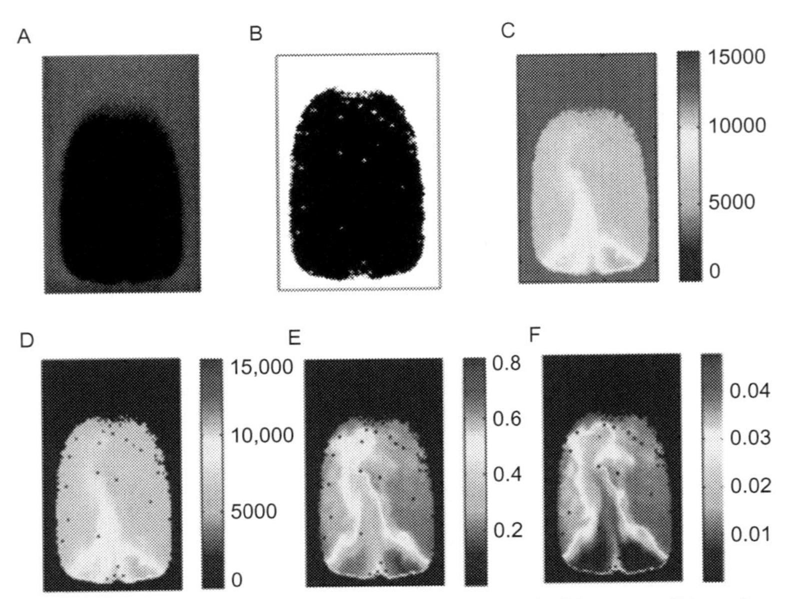

Figure 4.58 Steps for gas concentration measurements inside gas–solids systems via combination of VIS and IR images: VIS camera image (A), particle detection from VIS image (B), IR camera image (C), particle removal by stitching VIS–IR image (D), apparent absorbance (E), and concentration profile (F). *Reprinted from Dang et al. (2013) with permission from Elsevier.*

in the IR image, at the positions where particles are present, are replaced by zero intensity. The IR image after removal of the particles contains only gas phase which is used to compute the concentration profile, by applying the calibration curve. The results of the individual IR–VIS/DIA steps are shown in Fig. 4.58, for a case where CO_2 is injected into a gas–solid fluidized bed at incipient fluidization conditions using N_2 as a fluidizing agent.

The next step in the DIA technique is to average the concentration inside the whole bubble as well as to determine the equivalent bubble diameter. The snapshots of concentration profiles (Fig. 4.58E) show some dark "spots" caused by the presence of some particles inside the bubble, which will not be taken into account for when determining the averaged concentration.

3.1.3.3 Mass Exchange from Bubble-to-Emulsion Phase in Gas–Solids Fluidized Beds

In this section, the new analytical technique is applied to study the mass exchange coefficient from bubble-to-emulsion phase in bubbling fluidized beds for a case of a single bubble injection from the bottom distributor into a bed at incipient fluidization conditions. It should be noted that the actual limitations of the experimental setup do not allow the detection of the gas concentration in the emulsion phase. For this reason, a simplified approach will be used here also assuming negligible concentration in the emulsion phase. The technique will be further extended in the future by adopting IR transparent particles to be able to detect the CO_2 also in the emulsion phase. Furthermore, the influence of particle size and the bubble diameter on the mass exchange rate is evaluated and compared with different semiempirical correlations:

$$K_{be} = \frac{4}{d_b}\left(0.6D_{CO_2}^{0.5}\left(\frac{g}{d_b}\right)^{0.25} + \frac{2U_{mf}}{\pi}\right) \tag{4.42}$$

$$K_{be} = \frac{4}{d_b}\sqrt{\frac{4D_{CO_2}\varepsilon_{mf}^2 U_b(\alpha - 1)}{\pi D_c \alpha}} \tag{4.43}$$

Fig. 4.59 shows snapshots of the tracer gas concentration profiles inside the bubble at different moments in time after injection of a CO_2 bubble into the fluidized bed at minimum fluidization conditions with N_2 as a background fluidizing agent. At $t = 0.01$ s, a certain amount of tracer gas is released into the fluidized bed with a nozzle velocity of 17 m/s, creating a single bubble with high CO_2 concentration. Once released from the Vs valve (opening in 0.01 s), the injected CO_2 creates a single bubble with flat shape at the bottom, due to the effect of the gas distributor.

During the injection, N_2 background fluidization gas enters the bubble and dilutes the CO_2 inside the bubble. The bubble is subsequently growing and rising through the bed, while exchanging gas with the emulsion phase. The concentration field of tracer gas inside the bubble shown in Fig. 4.59 indicates that the mass transport of gas is dominated by convection of fluidization gas rather than diffusion, as expected for the used particle type and size. It is also clear that the tracer gas concentration inside the bubble is not uniform and in particular, it is lower in the lower part and in the center of the bubble. The results also clearly indicate that the CO_2 in the center part of the bubble is relatively quickly exchanged with the emulsion phase, but significant amount of CO_2 remain in the center of the vortices at the left and right sides of the bubble. This finding is in good agreement with CFD simulations performed by Patil et al. (2003). This first result shows the great potential of the new technique and allows the measurement of concentration inside the whole bubble without disturbing the flow with intrusive probes.

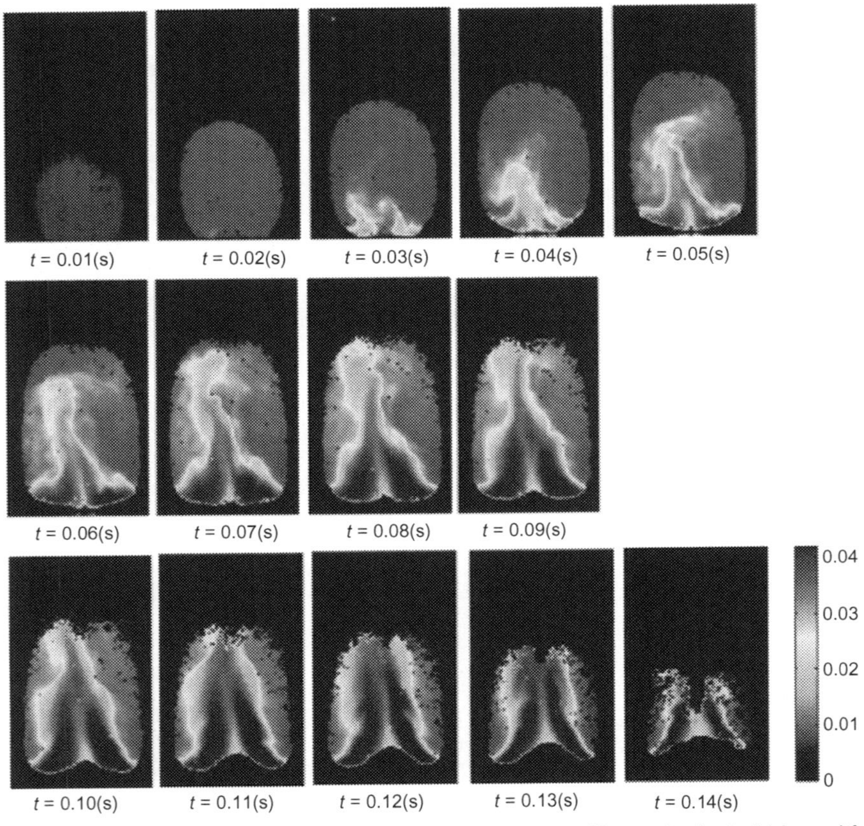

Figure 4.59 Snapshots of the tracer gas concentration profile inside the bubble at different times after bubble injection (snapshots are taken every 0.01 s with an injection velocity of 17 m/s). CO_2 concentration is presented in mol/l. *Reprinted from Dang et al. (2013) with permission from Elsevier.*

The evolution of the averaged CO_2 concentration inside the bubble and the equivalent bubble diameter as function of time are shown in Fig. 4.60A and B, respectively. As explained above, the concentration inside the bubble decreases in time, due to the mass exchange with gas in the emulsion phase. At the same time, the bubble is rising and growing, reaching its maximum size of about 42 mm after about 0.07 s after injection. The decrease in the bubble diameter after the maximum is reached is due to the increase in the amount of particle raining inside the bubble and the increase of bubble wake. Eventually, the wake and the particles raining merge together and the bubble is split into two smaller bubbles. As can be seen, the concentration curve against time reaches a plateau within 0.13 s, indicating that the mass

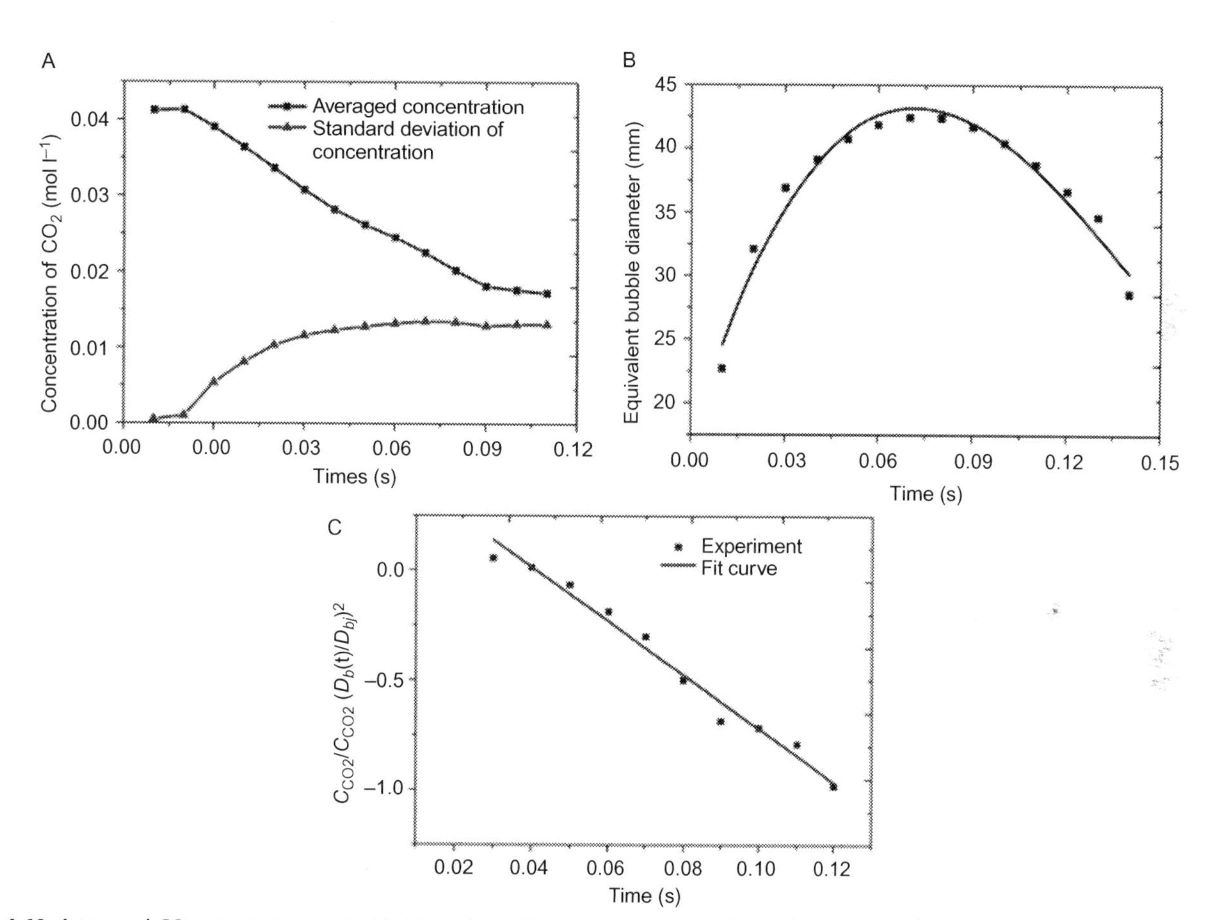

Figure 4.60 Averaged CO_2 concentration inside the bubble (A), the equivalent bubble diameter as a function of time (B), and the linearity correlation for mass transfer coefficient (C). *Reprinted from Dang et al. (2013) with permission from Elsevier.*

exchange rates decreases after this time. The results indicate that the gas exchange is due to a combination of effects where first convection takes place and afterward the diffusion from the vortices to the emulsion.

Fig. 4.60A plots the standard deviation of concentration in time which indicate that the very low standard deviation of concentration is obtained for the initial period, where CO_2 is suddenly injected into the column, and the bubble is detached from the distributor (without gas exchange with the emulsion phase). Afterward, N_2 from the emulsion phase dilutes the bubble, leading to an increased standard deviation, until the end of the evolution. The standard deviation approaches the plateau at $t=0.11$ s, when about 50% of CO_2 inside the bubble is replenished by N_2. For time longer than $t=0.11$ s, the concentration decreases slowly, due to the diffusion contribution. Fig. 4.60 shows that the concentration inside the bubble during the phase exchange is nonuniform. This means that, once the concentration in the emulsion phase can be detected, a more detailed model should be used to elucidate the gas exchange between bubble and emulsion. For the scope of this chapter, a simplified method is used.

3.1.3.4 Bubble-to-Emulsion Phase Mass Transfer Coefficient

The bubble-to-emulsion phase mass transfer can be described with a convection term and a diffusion term. The convection term describes the flow pattern of the fluidization gas from the emulsion phase inside the bubble which is dominant for the first part of our experiments. The molecular diffusion of CO_2 from the bubble phase (especially from the vortices) to the dense phase also influences the mass transfer process as discussed by Kunii and Levenspiel.

From the experimental results, the gas exchange rate from bubble-to-emulsion phase can be estimated by analyzing the tracer concentration inside the bubble and solving the mass balance equation for the tracer gas, which eventually yields for a bubble with initial volume $V_{\mathrm{b,i}}$ and with concentration $C_{\mathrm{CO_2,i}}$ at minimum fluidization conditions (Dang et al (2013)):

$$\frac{C_{\mathrm{CO_2,b}}(t)}{C_{\mathrm{CO_2,i}}}\left(\frac{d_{\mathrm{b}}(t)}{d_{\mathrm{b,i}}}\right)^2=\exp[-K_{\mathrm{be}}t] \tag{4.44}$$

From the experiments, the averaged CO_2 concentration in the bubble $C_{\mathrm{CO_2,b}}$ and the equivalent bubble diameter d_{b} have been determined and fitted as a function of time. It should be noted that this is a simplification used to be able to compute a gas exchange coefficient and compare it with literature data. When extending the technique to freely bubbling beds and to turbulent regime, a more accurate model should be developed also taking into account that the actual bubble concentration is not uniform. By subsequently plotting $\ln\left(\frac{C_{CO_2,b}(t)}{C_{CO_2,i}}\left(\frac{d_b(t)}{d_{b,i}}\right)^2\right)$ as a function of time, the mass transfer coefficient K_{be} can be calculated from the slope of the curve which yields to the K_{be} of 12.29 s^{-1}.

3.1.3.5 Effect of Bubble Diameter On The Bubble-to-Emulsion Mass Transfer Rate

Fig. 4.61 shows the snapshots of the CO_2 concentration profiles inside the bubble after injection with different volume of CO_2 injected in the bed resulting in different initial bubble sizes. For small injected bubble sizes of 31 mm (Fig. 4.61, top row), the gas exchange takes place right after the bubble has been injected. Smaller bubbles have a much lower bubble rise velocity and thus the background N_2 introduced into the column at minimum fluidization velocity enters into and travels through the bubble more easily resulting in a faster dilution of the CO_2 in the bubble. Moreover, a small

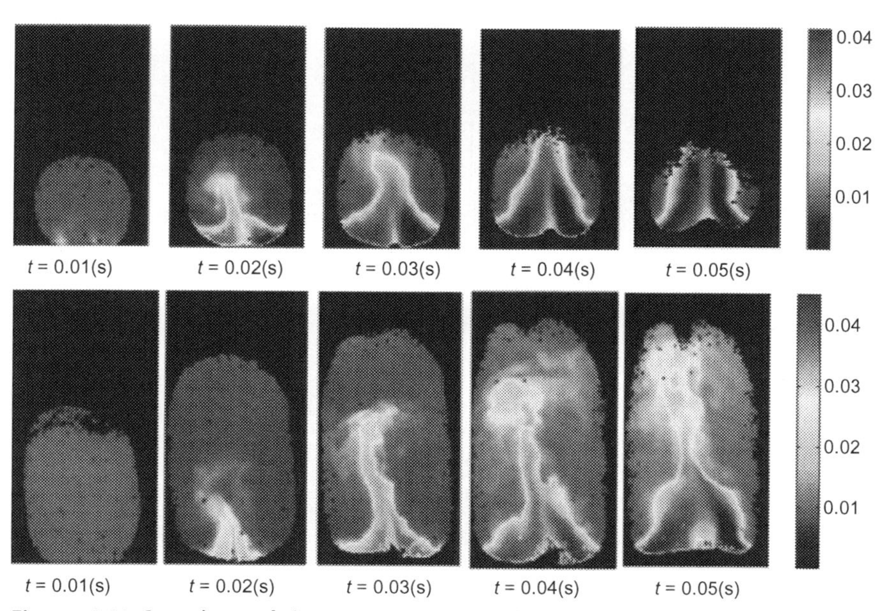

Figure 4.61 Snapshots of the concentration profile inside the bubbles at different moments in time for cases with different injected bubble sizes. The averaged bubble sizes 31 mm (injection velocity 9.75 m/s, top row) and 47 mm (20.5 m/s, bottom row). *Reprinted from Dang et al. (2013) with permission from Elsevier.*

bubble is less influenced by wall effects and thus maintains the spherical shape and bubble aspect ratio. As can be seen, a circular bubble is formed initially; however, the fluidization gas carries some particles into the lower section (wake) of the bubble transforming the circular shape into a kidney shape, in accordance to Rowe et al. (1964) and Murray (1965).

For the case of injection of a bubble with a larger equivalent bubble size (47 mm, Fig. 4.61, bottom row) the bubble rises with higher velocity and the ratio u_b/u_{mf} increases; in this case, a lower amount of N_2 enters and travels through the bubble resulting in a much slower mass exchange rate. As bubbles are bigger and the reactor is narrow, the wall effects increase, resulting in a strongly elongated bubble with a very high bubble aspect ratio (dy/dx). For both small and large bubbles however, the exchange of gas is dominated at the beginning of the experiment by the convection of N_2 from the bottom of the bubble, while afterward diffusion of gas from the vortices at the right and left side of the bubble to the emulsion phase rules the mass exchange.

The dependency of the gas exchange coefficient on the bubble diameter is shown in Fig. 4.62 in which a faster gas interchange rate is obtained from a smaller bubble. The experimental findings have been compared with the correlation developed by Davidson and Harrison and show a good agreement. The differences between experiment and model have a maximum

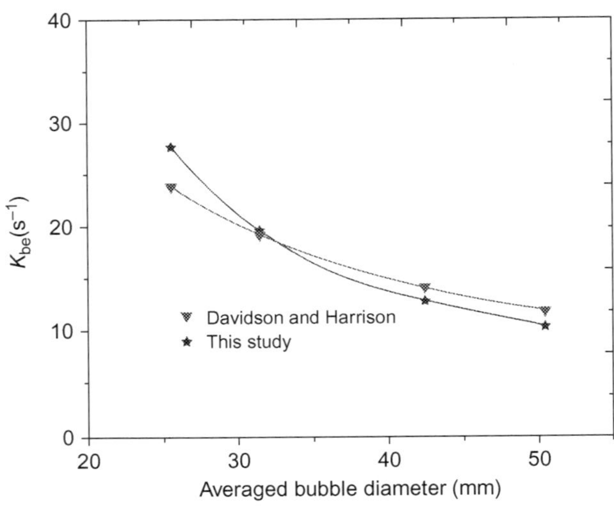

Figure 4.62 Mass exchange coefficient K_{be} as a function of the equivalent bubble diameter. *Reprinted from Dang et al. (2013) with permission from Elsevier.*

error of about 16%. This can be attributed to wall effects in the experiments and the assumptions from the theoretical model which was developed for a 3D fluidized bed, constant volume of bubble, constant bubble rise velocity, and uniform concentration inside bubble. In particular, the last three assumptions are in contrast with the findings of this experimental investigation as detailed above.

3.1.4 Conclusions

A novel optical technique, using an IR and VIS high-speed cameras, has been developed to measure CO_2 concentrations in gas and gas–solid flows. The new technique is noninvasive and allows a whole-field virtually instantaneous measurement of the CO_2 concentration with high temporal and spatial resolution. The technique has been calibrated in single-phase systems and the calibration has been verified with different calibration mixtures, resulting in relative errors lower than 0.5%. The application of the technique has been subsequently extended to gas–solid bubbling fluidized beds to determine the CO_2 concentration inside the bubbles. A single bubble of CO_2 has been injected into a fluidized bed at minimum fluidization conditions with N_2 as a fluidizing agent. In contrast to the assumptions often used in theory and semiempirical correlations, experiments have shown that the concentration inside the bubble is not uniform and the mass exchange is firstly controlled by convection and afterward controlled by diffusion between the vortices at the right and left side of the bubble and the emulsion phase. A new model for bubble-to-emulsion phase mass transfer is thus required and the developed technique can be used to perform more detailed experiments. Moreover, the new technique allows fast and precise measurements that can be applied for studying back-mixing and mass exchange in fluidized beds in both bubbling and turbulent regimes. It can be very easily extended with PIV.

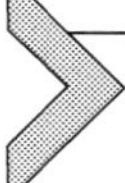

4. FUTURE WORK

4.1. High-Temperature PIV

Since many reactions that may benefit from the use of MAFBRs are performed at elevated temperatures, one of the important steps is to develop a nonintrusive measurement technique that allows insight in the fluidization behavior under reactive conditions. It can be anticipated that the fluidization behavior changes not only due to a different temperature but also due to the addition or removal of gas due to chemical reactions.

The main difficulty is that reactive conditions are typically not within the recommended operating conditions such as temperature, and also the presence of reactive gasses cannot safely be combined with generic high-speed cameras. For this reason, Rottier et al. (2010) coupled a high-temperature endoscope with a PIV CCD camera. They applied endoscopic PIV (ePIV) on a laboratory-scale furnace operating in the flameless mild combustion regime. The velocity field was obtained by seeding methane and air with zirconium dioxide particles.

In the work due to Campos Velarde et al. (2013), a novel ePIV/DIA using both optical and laser high-temperature endoscopes is validated at room temperature with comparison with standard PIV/DIA measurements. This is done by comparing the hydrodynamic parameters obtained with the endoscopic-laser PIV/DIA with the standard PIV/DIA illuminated with LED lights and using standard camera lenses, which is set as benchmark. Then, the design of the endoscopic-laser PIV/DIA high-temperature setup is described.

4.1.1 Experimental Setup

A typical setup for PIV/DIA measurements consists of a high-speed camera placed in front a pseudo-2D fluidized bed. The required illumination is obtained from a set of LED lights (Fig. 4.2).

The first step to extend this technique to high temperature is to find the right heating mechanism. Due to the high heat losses of the transparent pseudo-2D bed, the only way to supply homogeneous heating is to place the column inside a furnace. Because the high-speed camera is placed outside of the furnace, visual access to the furnace internal is required. However, a glass window makes it difficult to capture the whole bed, while heat losses through the window are also large, creating a risk for the camera. Therefore, a high-temperature endoscope was used. High-temperature endoscopes typically are used for inspection of glass furnaces and can thus work at very high temperatures ($>1000\ ^{\circ}$C).

The first attempt to run ePIV/DIA was done by coupling the high-speed camera with the high-temperature endoscope to the furnace at room temperature. The illumination was provided by LED lights, but it has quickly turned out that the illumination is poor due to light intensity losses in the endoscope. In order to fill the lack of illumination, a double-pulse Nd: Yag laser (SOLO IV 50 mJ/pulse) has been used to illuminate the surface of the pseudo-2D fluidized bed; therefore, a laser-cone is required for illumination.

Additionally, an inherent property of the pumped Nd:YAG lasers is the presence of pulse to pulse differences in coherency. When the beam subsequently passes through the lenses, this difference is enhanced. In order to tackle the issues caused by the use of a double-pulse Nd:Yag laser, a homogenizer was developed by the Bayerisches Laserzentrum to remove the pulse-to-pulse differences and to diverge the laser light into a cone. The laser beam is divided into beamlets by two microlenses arrays. The beamlets are overlapped with the help of a plan-convex lens in a plane behind this lens. To supply illumination inside the furnace, the double-pulse laser was coupled also to a customized high-temperature endoscope.

The high-temperature endoscope used by Campos Velarde et al. (2013) is the 38-mm double jacket manufactured by Cesyco Kinoptic Endoscopy. This endoscope has a water-cooled jacket and an air-cooled end tip-lens, making possible to use it up to 1000 °C. A sketch of these cooling features is displayed in Fig. 4.63. In addition, OptoPrecision developed a customized high-temperature endoscope to use the homogenizer and to couple it with the laser (Fig. 4.64).

4.1.2 Results and Discussion

At this time, the ePIV/DIA technique is still under development and only a few verification cases have been obtained, which claim good correspondence between the ePIV results versus the conventional PIV results.

The results of the validation of the endoscopic-laser PIV/DIA are presented in Fig. 4.65. The solids profiles of the ePIV/DIA follow the same trend as the benchmark. The deviation of the solids fluxes estimated with the ePIV/DIA is within the 10% of deviation of the standard PIV/DIA. The major

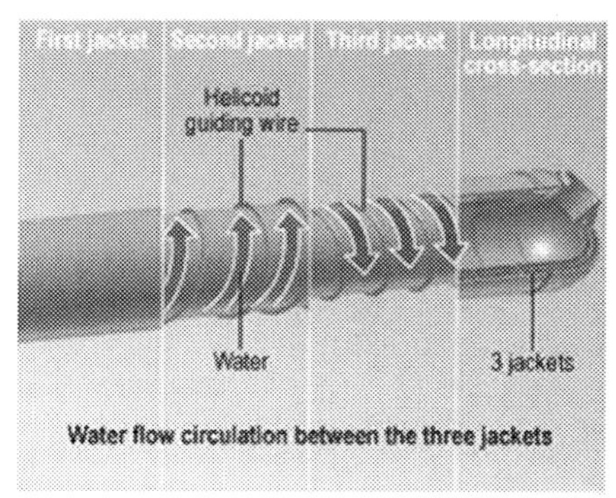

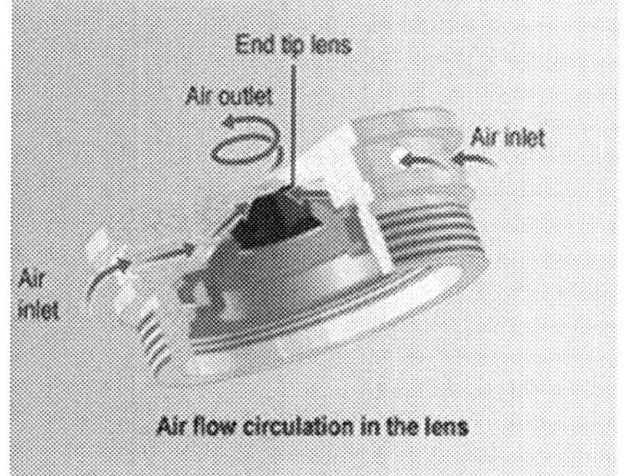

Figure 4.63 High-temperature endoscope. *Reprinted from Campos Velarde et al. (2013). Courtesy of Cesyco Kinoptic Endoscopy, France.*

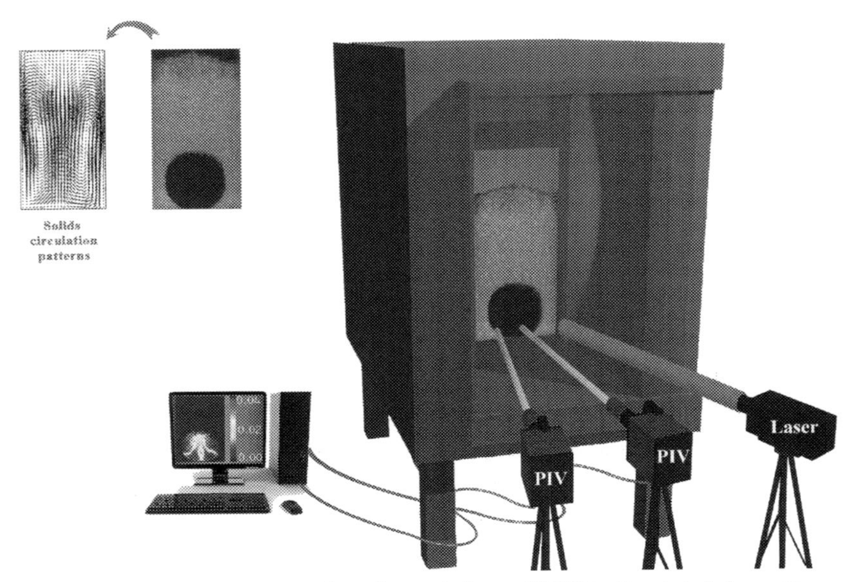

Figure 4.64 Schematic view of endoscopic-laser PIV/DIA. *Reprinted from Campos Velarde et al. (2013).*

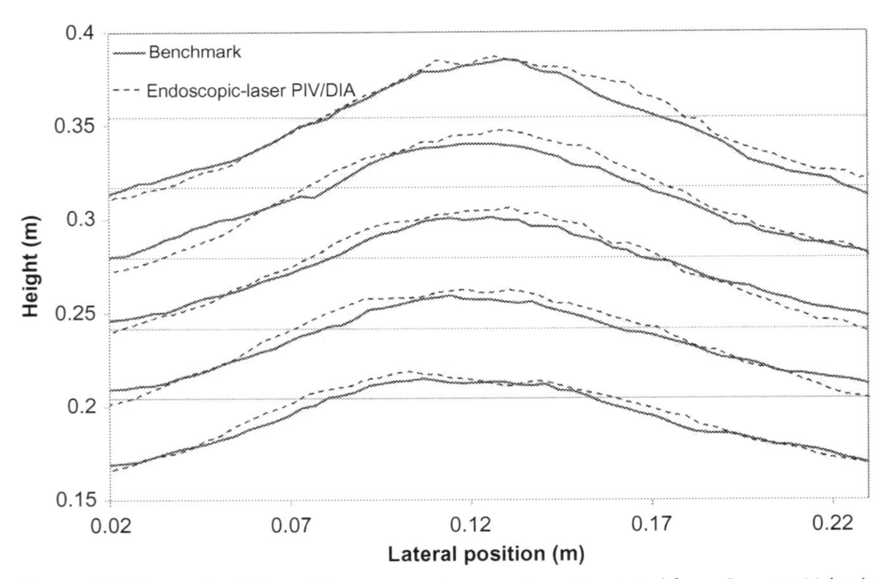

Figure 4.65 Lateral solids profiles at room temperature. *Reprinted from Campos Velarde et al. (2013).*

variation is due to the difference in illumination. The endoscopic-laser PIV/DIA has a single source of illumination, while four LED lights provide the illumination in the standard PIV/DIA. There is good agreement between the two cases, proving that it is possible to run PIV/DIA measurements using the optical endoscope and the illumination provided by the Nd: Yag laser.

4.2. Simulations of Mass Transfer in FBMRs

While the experimental IR-PIV/DIA technique allows to measure concentration profiles in the gas phase in fluidized beds, simulations would be able to capture the concentration profiles even within the interstitial gas flow. For this reason, the DPM code (discussed in Section 2.2.1) has been extended with species transport equations (based on convection-diffusion), which are solved combined with the rest of the DPM model.

At this time, a number of simulations are being performed that allow an even more detailed study into the bubble-to-emulsion phase mass transfer, because also the emulsion phase concentration can be taken into account (Fig. 4.66).

With a validated description for mass transfer in DPM/TFM type models, the effect of densified zones on the mass transfer toward the membranes can finally be understood.

5. CONCLUSIONS

This chapter has discussed our recent investigations on gas-solid FBMRs. Several novel integrated reactor concepts that use an FBMR, for instance for the production of ultrapure hydrogen, have been discussed. The use of FBMRs is advantageous for various reasons, among which a direct separation of the product, shifting the chemical reaction equilibrium to the product side and a possibility of separating waste gases such as CO_2 *in situ*. However, the influence of the presence and permeation of gas through submerged membranes in FBMRs must be better understood.

For this reason, several experimental and simulation techniques have been outlined. Optical techniques, such as PIV or DIA, can acquire whole-field data on the emulsion phase and bubble phase behavior simultaneously. DPMs and TFMs can be used to accurately simulate fluidized beds at small scales. These techniques have been used to investigate the effect of extraction or addition of gas via membranes into an FBMR. When the membranes were mounted in the vertical walls, it was observed and quantified that flow field

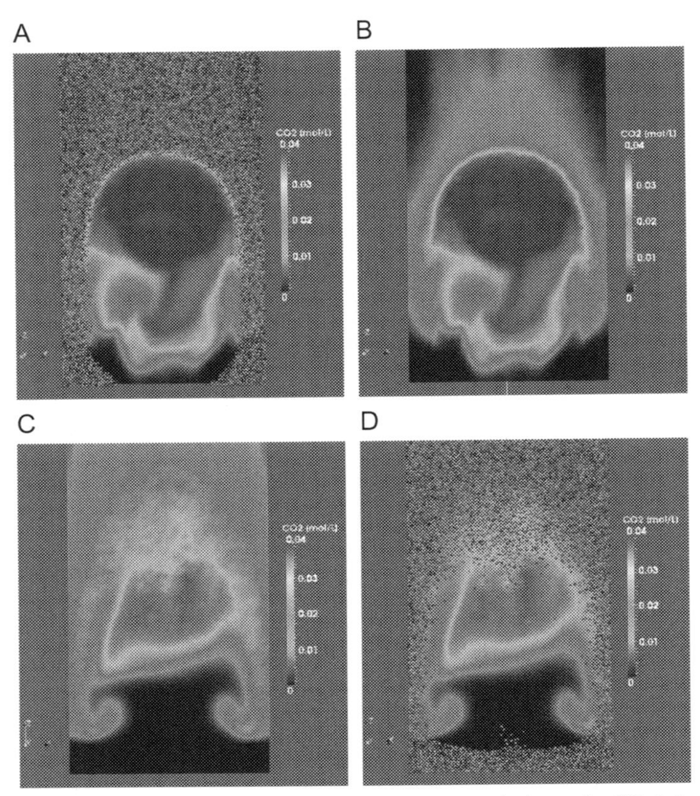

Figure 4.66 Snapshots of the DPM-species transport simulations of a CO_2 injection in a N_2 fluidized bed. (A and B) 0.05 s after injection and (C and D) 0.11 s after injection. Panels (A) and (D) show the particles, and panels (B) and (C) show only the concentration profile.

reversal, gas bypassing, or densified zones may result, which may strongly deteriorate the reactor performance. These effects have been outlined for different permeation velocities, and several methods to prevent this, such as the use of microscale FBRs and/or operation in the turbulent fluidization regime, have been indicated. Additionally, the effect of horizontally submerged membranes has been investigated numerically and experimentally.

The latest achievements include the development of a noninvasive measurement technique that allows to measure the gas-phase concentration profile of a CO_2 jet injected in a N_2 fluidized bed, via the extinction of the IR signal through the bed. This requires the combination of a high speed for the visual and IR spectrum. Ultimately, it has been shown that the bubble-to-emulsion phase mass transfer can be obtained experimentally via bubble injection.

Future work includes the development of noninvasive experimental techniques that are capable of measuring in realistic reactive conditions, such as ePIV/DIA. On the simulation point, extension of DPM and TFM models with species transport equations will aid the understanding and provide accurate closure equations for bubble-to-emulsion phase mass transfer.

The research efforts discussed in this chapter allow one to design FBMRs taking into account all hydrodynamic implications of the membranes, so that processes that incorporate an FBMR can optimally benefit from the integrated separation step.

REFERENCES

Abashar MEE, Alhabdan FM, Elnashaie SSEH: Discrete injection of oxygen enhances hydrogen production in circulating fast fluidized bed membrane reactors, *Int J Hydrog Energy* 33:2477–2488, 2008.

Adris AEM, Elnashaie SSEH, Hughes R: A fluidized-bed membrane reactor for the steam reforming of methane, *Can J Chem Eng* 69:1061–1070, 1991.

Adris AEM, Lim CJ, Grace JR: The fluidized bed membrane reactor system: a pilot scale experimental study, *Chem Eng Sci* 49:5833–5843, 1994.

Adris AEM, Lim CJ, Grace JR: The fluidized-bed membrane reactor for steam methane reforming: model verification and parametric study, *Chem Eng Sci* 52:1609–1622, 1997.

Ahchieva D, Peglow M, Heinrich S, Morl L, Wolff T, Klose F: Oxidative dehydrogenation of ethane in a fluidized bed membrane reactor, *Appl Catal A Gen* 296:176–185, 2005.

Al-Sherehy F, Grace JR, Adris AEM: The influence of distributed reactant injection along the height of a fluidized bed reactor, *Chem Eng Sci* 60:7121–7130, 2005.

Andrés M-B, Boyd T, Grace JR, et al: In-situ CO_2 capture in a pilot-scale fluidized-bed membrane reformer for ultra-pure hydrogen production, *Int J Hydrog Energy* 36(6):4038–4055, 2011.

Beetstra R, van der Hoef MA, Kuipers JAM: Drag force of intermediate Reynolds number flow past mono- and bidisperse arrays of spheres, *AIChE J* 53(2):489–501, 2007.

Bokkers GA, van Sint Annaland M, Kuipers JAM: Mixing and segregation in a bidisperse gas-solid fluidised bed: a numerical and experimental study, *Powder Technol* 140:176, 2004.

Bokkers GA, Laverman JA, van Sint Annaland M, Kuipers JAM: Modelling of large-scale dense gas–solid bubbling fluidised beds using a novel discrete bubble model, *Chem Eng Sci* 61(17):5590–5602, 2006.

Boon J, Pieterse JAZ, Dijkstra JW, Van Sint Annaland M: Modelling and systematic experimental investigation of mass transfer in supported palladium-based membrane separators, *Int J Greenhouse Gas Control* 115:S122–S129, 2012.

Brunetti A, Caravella A, Drioli E, Barbieri G: Process intensification by membrane reactors: high-temperature water gas shift reaction as single stage for syngas upgrading, *Eng Technol* 35(7):1238–1248, 2012.

Buijs K, Maurice MJ: Some consideration on apparent deviation of Lambert-Beer's Law, *Anal Chim Acta* 47(3):469–474, 1969.

Campos Velarde I, van Heck LJ, Gallucci F, van Sint Annaland M: Development of a high-temperature endoscopic laser PIV/DIA technique for the study of hydrodynamics of gas-solid fluidized beds. In *Proceedings of the 13th International Conference on Fluidization Engineering*, 2013.

Capecelatro J, Desjardins O, Fox RO: Numerical study of collisional particle dynamics in cluster-induced turbulence, *J Fluid Mech* 747:R2, 2014.

Chen JAK, Al-Dahhan Muthanna H, Dudukovic Milorad P, Lee DJ, Fan LS: Comparative hydrodynamics study in a bubble column using computer-automated radioactive particle tracking (CARPT)/computed tomography (CT) and particle image velocimetry (PIV), *Chem Eng Sci* 54:2199–2207, 1999.

Chen ZX, Po F, Grace JR, et al: Sorbent-enhanced/membrane-assisted steam-methane reforming, *Chem Eng Sci* 63:170–182, 2008.

Christensen D, Nijenhuis J, van Ommen JR, Coppens MO: Influence of distributed secondary gas injection on the performance of a bubbling fluidized-bed reactor, *Ind Eng Chem Res* 47:3601–3618, 2008a.

Christensen D, Nijenhuis J, van Ommen JR, Coppens MO: Residence times in fluidized beds with secondary gas injection, *Powder Technol* 180:321–331, 2008b.

Christensen D, Vervloet D, Nijenhuis J, van Wachem BGM, van Ommen JR, Coppens MO: Insights in distributed secondary gas injection in a bubbling fluidized bed via discrete particle simulations, *Powder Technol* 183:454–466, 2008c.

Coroneo M, Montante G, Paglianti A: Numerical and experimental fluid-dynamic analysis to improve the mass transfer performances of Pd−Ag membrane modules for hydrogen purification, *Ind Eng Chem Res* 49:9300–9309, 2010.

Dang TYN, Kolkman T, Gallucci F, van Sint Annaland M: Development of a novel infrared technique for instantaneous, whole-field, non invasive gas concentration measurements in gas–solid fluidized beds, *Chem Eng J* 219(1):545–557, 2013.

Dang Nhi TY, Gallucci Fausto, van Sint Annaland Martin: Micro-structured fluidized bed membrane reactors: solids circulation and densified zones distribution, *Chem Eng J* 239(1):42–52, 2014.

Darton RC, Lanauze RD, Davidson JF, Harrison D: Bubble-growth due to coalescence in fluidized-beds, *Trans Inst Chem Eng* 55:274–280, 1977.

Davidson JF, Harrison D: *Fluidized particles,* New York, 1963, Cambridge University Press.

De Jong F, van Sint Annaland M, Kuipers JAM: Experimental study on the effects of gas permeation through flat membranes on the hydrodynamics in membrane-assisted fluidized beds, *Chem Eng Sci* 66:2398–2408, 2011.

De Jong JF, Odu SO, van Buijtenen MS, Deen NG, van Sint Annaland M, Kuipers JAM: Development and validation of a novel digital image analysis method for fluidized bed particle image velocimetry, *Powder Technol* 230:193–202, 2012a.

De Jong JF, van Sint Annaland M, Kuipers JAM: Membrane-assisted fluidized beds—part 1: development of an immersed boundary discrete particle model, *Chem Eng Sci* 84(24):814–821, 2012b.

De Jong JF, van Sint Annaland M, Kuipers JAM: Membrane-assisted fluidized beds—part 2: numerical study on the hydrodynamics around immersed gas-permeating membrane tubes, *Chem Eng Sci* 84(24):822–833, 2012c.

De Jong JF, Dang TYN, van Sint Annaland M, Kuipers JAM: Comparison of a discrete particle model and a two-fluid model to experiments of a fluidized bed with flat membranes, *Powder Technol* 230:93–105, 2012d.

De Jong JF, van Sint Annaland M, Kuipers JAM: Experimental study on the hydrodynamic effects of gas permeation through horizontal membrane tubes in fluidized beds, *Powder Technol* 241:74–84, 2013.

Deen NG, van Sint Annaland M, Kuipers JAM: Multi-scale modeling of dispersed gas–liquid two-phase flow, *Chem Eng Sci* 59(8–9):1853–1861, 2004.

Deen NG, van Sint Annaland M, Kuipers JAM: Detailed computational and experimental fluid dynamics of fluidized beds, *Appl Math Model* 300(11):1459–1471, 2006.

Deshmukh SARK, van Sint Annaland M, Kuipers JAM: Effect of fluidization conditions on the membrane permeation rate in a membrane assisted fluidized bed, *Chem Eng J* 96:125–131, 2003.

Deshmukh SARK, Laverman JA, Cents AHG, van Sint Annaland M, Kuipers JAM: Development of a membrane-assisted fluidized bed reactor. 1. Gas phase back-mixing and bubble-to-emulsion phase mass transfer using tracer injection and ultrasound experiments, *Ind Eng Chem Res* 44:5955–5965, 2005a.

Deshmukh SARK, Laverman JA, van Sint Annaland M, Kuipers JAM: Development of a membrane-assisted fluidized bed reactor. 2. Experimental demonstration and modeling for the partial oxidation of methanol, *Ind Eng Chem Res* 44:5966–5976, 2005b.

Deshmukh SARK, van Sint Annaland M, Kuipers JAM: Gas back-mixing studies in membrane assisted bubbling fluidized beds, *Chem Eng Sci* 62:4095–4111, 2007a.

Deshmukh SARK, Heinrich S, Mörl L, van Sint Annaland M, Kuipers JAM: Membrane assisted fluidized bed reactors: potentials and hurdles, *Chem Eng Sci* 62:416–436, 2007b.

Dijkhuizen W, Bokkers GA, Deen NG, van Sint Annaland M, Kuipers JAM: Extension of PIV for measuring granular temperature field in dense fluidized beds, *AIChE J* 53:108, 2007.

Drioli E, Curcio E: Membrane engineering for process intensification: a perspective, *J Chem Technol Biotechnol* 82(3):223–227, 2007.

Drioli E, Criscuoli A, Curcio E: Membrane contactors and catalytic membrane reactors in process intensification, *Chem Eng Technol* 26(9):975–981, 2003.

Drioli E, Brunetti A, Di Profio G, Barbieri G: Process intensification strategies and membrane engineering, *Green Chem* 14(6):1561–1572, 2012.

Fox RO: Large-Eddy-simulation tools for multiphase flows, *Annu Rev Fluid Mech* 44:47–76, 2012.

Fox RO: On multiphase turbulence models for collisional fluid-particle flows, *J Fluid Mech* 742:368–424, 2014.

Gallucci F, van Sint Annaland M, Kuipers JAM: Autothermal reforming of methane with integrated CO_2 capture in a novel fluidized bed membrane reactor. Part 1: experimental demonstration, *Top Catal* 51:133–145, 2008.

Gallucci F, van Sint Annaland M, Kuipers JAM: Chapter 10 modeling of membrane reactors for hydrogen production and purification. In *Membrane engineering for the treatment of gases: volume 2: gas-separation problems combined with membrane reactors*, The Royal Society of Chemistry, 2011. pp. 1–39, ISBN:978-1-84973-239-0, http://dx.doi.org/10.1039/9781849733489-00001.

Gallucci F, Fernandez E, Corengia P, van Sint Annaland M: Recent advances on membranes and membrane reactors for hydrogen production, *Chem Eng Sci* 92:40–66, 2013.

Gidaspow D: *Multiphase flow and fluidization: continuum and kinetic theory descriptions,* Boston, MA, 1994, Academic Press.

Gimeno MP, Wu ZT, Soler J, Herguido J, Li K, Menendez M: Combination of a two-zone fluidized bed reactor with a Pd hollow fiber membrane for catalytic alkane dehydrogenation, *Chem Eng J* 155:298–303, 2009.

Godini HR, Trivedi H, de Villasante AG, et al: Design demonstration of an experimental membrane reactor set-up for oxidative coupling of methane, *Chem Eng Res Des* 91(12):2671–2681, 2013.

Grace JR, Clift R: 2-Phase theory of fluidization, *Chem Eng Sci* 29:327–334, 1974.

Grace JR, Harrison D: Behavior of freely bubbling fluidised beds, *Chem Eng Sci* 24:497–508, 1969.

Grace JR, Elnashaie SSEH, Lim CJ: Hydrogen production in fluidized beds with in-situ membranes, *Int J Chem React Eng* 3(1), 2005.

Hogue C, Newland D: Efficient computer-simulation of moving granular particles, *Powder Technol* 78(1):51–66, 1994.

Hoomans BPB, Kuipers JAM, Briels WJ, van Swaaij WPM: Discrete particle simulation of bubble and slug formation in a two-dimensional gas-fluidised bed: a hard-sphere approach, *Chem Eng Sci* 51(1):99–118, 1996.

Johnson PC, Jackson R: Frictional collisional constitutive relations for antigranulocytes-materials, with application to plane shearing, *J Fluid Mech* 176:67–93, 1987.

Kuipers JAM, van Duin KJ, van Beckum FHP, van Swaaij WPM: A numerical model of gas-fluidized beds, *Chem Eng Sci* 47:1913, 1992.

Kunii D, Levenspiel O: *Fluidization engineering,* Newton, MA, 1991, Butterworth-Heinemann series in Chemical Engineering.
Ladd AJC: Lattice-Boltzmann simulations of particle fluid suspensions. Part 1. Theoretical foundation, *J Fluid Mech* 271:285, 1994.
Ladd AJC, Verberg R: Lattice-Boltzmann simulations of particle fluid suspensions, *J Stat Phys* 104:1191, 2001.
Laverman JA: *On the hydrodynamics in gas phase polymerization reactors* (Ph.D. thesis) 2010, Eindhoven University of Technology.
Laverman JA, Roghair I, van Sint Annaland M, Kuipers JAM: Investigation into the hydrodynamics of gas–solid fluidized beds using particle image velocimetry coupled with digital image analysis, *Can J Chem Eng* 86:523–535, 2008.
Li J, Weistein H: An experimental comparison of the gas back mixing in fluidized beds across the regime spectrum, *Chem Eng Sci* 44:1697–1705, 1989.
Link J, Zeilstra C, Deen NG, Kuipers JAM: Validation of a discrete particle model in a 2-D spout-fluid bed using non-intrusive optical measuring techniques, *Can J Chem Eng* 82:30, 2004.
Mahecha-Botero A, Grace J, Elnashaie SSEH, Lim CJ: Comprehensive modeling of gas fluidized-bed reactors allowing for transients, multiple flow regimes and selective removal of species, *Int J Chem React Eng* 4:1–21, 2006.
Mahecha-Botero A, Boyd T, Gulamhusein A, et al: Pure hydrogen generation in a fluidized-bed membrane reactor: experimental findings, *Chem Eng Sci* 63:2752–2762, 2008.
Marin P, Hamel C, Ordonez S, Diez FV, Tsotsas E, Seidel-Morgenstern A: Analysis of a fluidized bed membrane reactor for butane partial oxidation to maleic anhydride: 2D modelling, *Chem Eng Sci* 65:3538–3548, 2010.
Mleczko L, Ostrowski T, Wurzel T: A fluidised-bed membrane reactor for the catalytic partial oxidation of methane to synthesis gas, *Chem Eng Sci* 51:3187–3192, 1996.
Müller CR, Davidson JF, Dennis JS, Fennell PS, Gladden LF: Real time measurement of bubbling phenomena in a 3-D gas-fluidized bed using ultra-fast magnetic resonance imaging, *Phys Rev Lett* 96:154–504, 2006.
Müller CR, Davidson JF, Dennis JS, Fennell PS, Gladden LF: Oscillations in gas-fluidized beds: ultra-fast magnetic resonance imaging and pressure sensor measurements, *Powder Technol* 177:87–98, 2007.
Murray JD: On the mathematics of fluidization: steady motion of fully developed bubble, *J Fluid Mech* 22:57–80, 1965.
Nguyen HV, Potter OE, Dent DC, Whitehead AB: Gas back-mixing in large fluidized beds containing tube assemblies, *AIChE J* 27:509–514, 1981.
Patil DJ, van Sint Annaland M, Kuipers JAM: Gas dispersion and bubble-to-emulsion phase mass exchange in gas–solid bubbling fluidized bed: a computational and experimental study, *Int J Chem React Eng* 1:1–22, 2003.
Patil CS, Van Sint Annaland M, Kuipers JAM: Design of a novel auto-thermal membrane assisted fluidized bed reactor for the production of ultra-pure hydrogen from methane, *Ind Eng Chem Res* 44:9502–9512, 2005.
Patil CS, van Sint Annaland M, Kuipers JAM: Fluidised bed membrane reactor for ultrapure hydrogen production via methane steam reforming: experimental demonstration and model validation, *Chem Eng Sci* 62:2989–3007, 2007.
Patterson Emily E, Halow Jack, Daw Stuart: Innovative method using magnetic particle tracking to measure solids circulation in a spouted fluidized bed, *Ind Eng Chem Res* 49:5037–5043, 2010.
Peskin CS: The immersed boundary method, *Acta Numer* 11:479–517, 2002.
Rahimpour MR, Elekaei H: A comparative study of combination of Fischer–Tropsch synthesis reactors with hydrogen-permselective membrane in GTL technology, *Fuel Process Technol* 90:747–761, 2009.

Rahimpour MR, Bayat M, Rahmani F: Dynamic simulation of a cascade fluidized-bed membrane reactor in the presence of long-term catalyst deactivation for methanol synthesis, *Chem Eng Sci* 65:4239–4249, 2010.

Rakib MA, Grace JR, Lim CJ, Elnashaie SSEH, Ghiasi B: Steam reforming of propane in a fluidized bed membrane reactor for hydrogen production, *Int J Hydrog Energy* 35:6276–6290, 2010.

Rakib MA, Grace JR, Lim CJ, Elnashaie SSEH: Modeling of a fluidized bed membrane reactor for hydrogen production by steam reforming of hydrocarbons, *Ind Eng Chem Res* 50(6):3110–3129, 2011.

Roels S, Carmeliet J: Analysis of moisture flow in porous materials using microfocus X-ray radiography, *Int J Heat Mass Transf* 49(25–26):4762–4772, 2006.

Rottier C, Godard G, Corbin F, Boukhalfa AM, Honore D: An endoscopic particle image velocimetry system for high-temperature furnaces, *Meas Sci Technol* 21:115404, 2010.

Rowe PN, Partridge BA, Lyall E: Cloud formation around bubbles in gas fluidized beds, *Chem Eng Sci* 19:973–985, 1964.

Seville J: "A single particle view of fluidization". In "*The 13th International Conference on Fluidization - New Paradigm in Fluidization Engineering*", 2010, Sang Done Kim, Korea Advanced Institute of Science and Technology, Korea; Yong Kang, Chungnam National University, Korea; Jea Keun Lee, Pukyong National University, Korea; Yong Chil Seo, Yonsei University, Korea Eds, ECI Symposium Series, Volume RP6. http://dc.engconfintl.org/fluidization_xiii/7

Shen LH, Johnsson F, Leckner B: Digital image analysis of hydrodynamics two-dimensional bubbling fluidized beds, *Chem Eng Sci* 59:2607–2617, 2004.

Srivastava A, Sundaresan S: Analysis of a fractional-kinetic model for gas–particle flow, *Powder Technol* 129(1–3):72–85, 2003.

Tan L, Roghair I, Van Sint Annaland M: Simulation study on the effect of gas permeation on the hydrodynamic characteristics of membrane-assisted micro fluidized beds, *Appl Math Model*, 2014. http://dx.doi.org/10.1016/j.apm.2014.04.044 (in press).

Tiemersma TP, Patil CS, Sint Annaland MV, Kuipers JAM: Modelling of packed bed membrane reactors for autothermal production of ultrapure hydrogen, *Chem Eng Sci* 61(5):1602–1616, 2006.

Tsuji Y, Tanaka T, Ishida T: Lagrangian numerical-simulation of plug flow of cohesionless particles in a horizontal pipe, *Powder Technol* 71(3):239–250, 1992.

Uhlmann M: An immersed boundary method with direct forcing for the simulation of particulate flows, *J Comput Phys* 209:448–476, 2005.

van der Hoef MA, Ye M, van Sint Annaland M, Andrews AT, Sundaresan S, Kuipers JAM: Multi-scale modeling of gas-fluidized beds, *Adv Chem Eng* 31:65–149, 2006.

Van Sint Annaland M, Nijmeijer A, Assink J, Kramer GJ: Process and reactor for the production of hydrogen and carbon dioxide and a fuel cell system. Pending European patent application nr. 06116419, 2006.

Wang J, Tan L, van der Hoef MA, van Sint Annaland M, Kuipers JAM: From bubbling to turbulent fluidization: advanced onset of regime transition in micro-fluidized beds, *Chem Eng Sci* 66:2001–2007, 2011.

Westerweel J: Theoretical analysis of the measurement precision in particle image velocimetry, *Exp Fluids* 290:S3–S12, 2000.

INDEX

Note: Page numbers followed by "*f*" indicate figures and "*t*" indicate tables.

B

C

D

E

F

G

H

I

K

L

M

N

O

P

CONTENTS OF VOLUMES IN THIS SERIAL

Volume 1 (1956)

J. W. Westwater, *Boiling of Liquids*
A. B. Metzner, *Non-Newtonian Technology: Fluid Mechanics, Mixing, and Heat Transfer*
R. Byron Bird, *Theory of Diffusion*
J. B. Opfell and B. H. Sage, *Turbulence in Thermal and Material Transport*
Robert E. Treybal, *Mechanically Aided Liquid Extraction*
Robert W. Schrage, *The Automatic Computer in the Control and Planning of Manufacturing Operations*
Ernest J. Henley and Nathaniel F. Barr, *Ionizing Radiation Applied to Chemical Processes and to Food and Drug Processing*

Volume 2 (1958)

J. W. Westwater, *Boiling of Liquids*
Ernest F. Johnson, *Automatic Process Control*
Bernard Manowitz, *Treatment and Disposal of Wastes in Nuclear Chemical Technology*
George A. Sofer and Harold C. Weingartner, *High Vacuum Technology*
Theodore Vermeulen, *Separation by Adsorption Methods*
Sherman S. Weidenbaum, *Mixing of Solids*

Volume 3 (1962)

C. S. Grove, Jr., Robert V. Jelinek, and Herbert M. Schoen, *Crystallization from Solution*
F. Alan Ferguson and Russell C. Phillips, *High Temperature Technology*
Daniel Hyman, *Mixing and Agitation*
John Beck, *Design of Packed Catalytic Reactors*
Douglass J. Wilde, *Optimization Methods*

Volume 4 (1964)

J. T. Davies, *Mass-Transfer and Inierfacial Phenomena*
R. C. Kintner, *Drop Phenomena Affecting Liquid Extraction*
Octave Levenspiel and Kenneth B. Bischoff, *Patterns of Flow in Chemical Process Vessels*
Donald S. Scott, *Properties of Concurrent Gas–Liquid Flow*
D. N. Hanson and G. F. Somerville, *A General Program for Computing Multistage Vapor–Liquid Processes*

Volume 5 (1964)

J. F. Wehner, *Flame Processes—Theoretical and Experimental*
J. H. Sinfelt, *Bifunctional Catalysts*
S. G. Bankoff, *Heat Conduction or Diffusion with Change of Phase*
George D. Fulford, *The Flow of Lktuids in Thin Films*
K. Rietema, *Segregation in Liquid–Liquid Dispersions and its Effects on Chemical Reactions*

Volume 6 (1966)

S. G. Bankoff, *Diffusion-Controlled Bubble Growth*
John C. Berg, Andreas Acrivos, and Michel Boudart, *Evaporation Convection*
H. M. Tsuchiya, A. G. Fredrickson, and R. Aris, *Dynamics of Microbial Cell Populations*
Samuel Sideman, *Direct Contact Heat Transfer between Immiscible Liquids*
Howard Brenner, *Hydrodynamic Resistance of Particles at Small Reynolds Numbers*